"MESMERISM IS THE KEYSTONE OF ALL THE OCCULT SCIENCES."

PART I.]   [COMPLETE IN SIX MONTHLY PARTS.

"ANIMAL MAGNETISM;"

OR,

MESMERISM & ITS PHENOMENA.

BY THE LATE

WILLIAM GREGORY, M.D., F.R.S.E.

(Professor of Chemistry at Edinburgh University).

DEDICATED BY THE AUTHOR BY PERMISSION TO
HIS GRACE THE DUKE OF ARGYLL.

TOGETHER WITH

AN INTRODUCTION BY "M.A. (OXON.)"

LONDON:
THE PSYCHOLOGICAL PRESS ASSOCIATION,
AND
E. W. ALLEN, 4, AVE MARIA LANE.

PRICE SIXPENCE.

ANIMAL MAGNETISM;

OR,

MESMERISM AND ITS PHENOMENA.

BY THE LATE
WILLIAM GREGORY, M.D, F.R.S.E.

Professor of Chemistry in the University of Edinburgh.

DEDICATED BY THE AUTHOR, BY PERMISSION, TO HIS GRACE
GEORGE-DOUGLAS CAMPBELL, DUKE OF ARGYLL.
K.T., F.R.S.E.

Third, and slightly Revised and Abridged, Edition.

LONDON:
THE PSYCHOLOGICAL PRESS ASSOCIATION,
AND
E. W. ALLEN, 4, AVE MARIA LANE.
1884.

PREFACE TO THE THIRD EDITION.

My friend, Mrs. Makdougall Gregory, has honoured me with a request that I should write a short introduction to the third edition of her husband Professor Gregory's, well-known "Animal Magnetism." By this flattering request I understand her to desire that I should commend the book to the attention of Spiritualists. For though I may address those who are of my way of thinking in these matters, with a hope of acceptance born of much experience of kindly attention in the past, I have no title to intrude on that wider world of science which is just beginning to bestir itself with a little affectation of attention to the subject of Mesmerism under the guidance of the Society for Psychical Research.

A wish from one who has laboured so long and unweariedly, who has spent herself without stint in doing that which she found for her hand to do in furthering the cause of Spiritualism, comes to me with the force of a command. For she was working with her own tireless energy in the interests of Spiritualism when I was dimly groping my way to the light. She had then laid Spiritualism under heavy obligations, and, since I have been labouring in the same cause, her zeal has never once abated. When, therefore, she offers us a new edition of Professor Gregory's work on what is well described as "the keystone of all the occult sciences," she has a right to use any means that she thinks serviceable for securing from intelligent Spiritualists a study of this solution of some of our mysteries.

The time, I trust, is past when Spiritualists will be content to gaze open-mouthed at some recurrent marvel, without making any reasonable attempt at studying its nature, or accounting for its presence under certain observed conditions. The time is come, I hope, when the intelligent observer of the phenomena of Spiritualism will not consider that he has satisfactorily accounted for everything abnormal when he has referred it to the action of "spirits." He will remember that we are all "spirits" incarnate, and will deem the inherent powers of his own human spirit worthy of careful study, as well as the abnormal powers of certain members of the race who are to us, we believe, vehicles for communication with the great world of spirit that lies about and around us. His distinctive belief as a Spiritualist will run no risk of being shattered, if he will clear the ground carefully by such a course of study as the Society for Psychical Research, for instance, is now pursuing. Far otherwise: for, by such means alone, he will be able to give an intelligent reason for the faith that he professes, and will find himself able to meet attacks upon its essential principles as he certainly would not succeed in doing without such a course of preparation.

Among the aids to such study as I have indicated is an acquaintance with those states and conditions of the human spirit which Mesmerism reveals. Among the standard works on that subject, this volume of the late Dr. William Gregory, F.R.S.E., Professor of Chemistry in the University of Edinburgh, has always held its place.

The present edition is identical with the second, which differed from the first only in the omission of some matter which has now, happily, become irrelevant and unnecessary. The first three chapters of the original edition have been omitted, because it is now no longer needful to defend the reality of the mesmeric phenomena with which Professor Gregory deals under the name of *Animal Magnetism*. That title, which most investigators have now agreed to abandon, is retained in deference to the author's choice, and because the work is well known under its present name.

I do not conceive it to be any part of my present duty to point out in detail the various branches of this study, which may profitably engage the attention of the student. He will find most, if not all of them, treated in more or less detail in this volume; and among them he will very soon find clues to the interpretation of some of those occult phenomena which present such a perplexing field of research, and respecting which it is usually safe to say that the explanation that lies most obviously on the surface is least likely to be the true and complete solution of the difficulty.

I may, however, say that such students as I conceive myself to be primarily addressing, will find in the remarks made on Clairvoyance, Lucid Prevision, Prediction of changes in the patient's state of health at certain fixed times, and in the prescription of medical remedies to meet those new conditions, matter both intrinsically interesting, and having its own plain bearing on some of the familiar phenomena of Spiritualism.

The Mesmeric Trance, again, accidental or induced ; the Ecstatic state, in which the mesmerised subject seems to enjoy communion with the world of spirit, and to live in a state sometimes entirely detached from the world of sense ; will readily be seen to have their bearing on such experiences as those of Andrew Jackson Davis, and on the familiar state of Trance into which almost all well-developed psychics are accustomed to pass while utterances purporting to come from an alien spirit are made through their lips ; or while their vital forces are being utilised for the production of such phenomena as, for instance, those of Materialisation, or Form-manifestation.

But to the student of these obscure phenomena on whom the question is often forced "What is the use, the practical benefit, of these phenomena, assuming them to be what they are alleged to be ?" the culminating interest of this study will rest in the therapeutic use of Mesmerism. The employment of clairvoyance in the diagnosis of disease ; of mesmeric treatment in alleviating pain, and in curing obstinate and longstanding cases of actual disease ; and even the efficacy of the mesmeric pass in rendering the patient insensible to pain that must temporarily be endured, or in preventing it in such cases as surgical operations,—these will come home to the reader as undeniable instances of the beneficial action of Mesmerism; and as furnishing a clear answer to the *Cui bono* question, so often, and, in many cases, so foolishly forced into prominence.

The volumes of *The Zoist*, to which I may be permitted to make special reference, are a storehouse of fact, containing cases recorded, usually by medical men, with every possible exactness and attention to scientific detail. The works of Dr. Esdaile and Mr. Capern also bear abundant testimony to the curative effect of Mesmerism. And since the now distant time when these books were published vast numbers of records have accumulated, most of which will be found in the excellent library of the Society for Psychical Research, which is especially rich in works on this subject.

It remains that I add, if it be permitted me, a word of caution to any readers of this volume who may propose to follow out an experimental study of mesmeric phenomena on his own responsibility. The subject is not one to be played or trifled with. Dealing as it does with obscure conditions of the whole being, it presents various difficulties and not a few dangers to the inexperienced or rash experimenter. Harm may easily be done to the unconscious patient, and unmerited obloquy may be cast on the study of the subject by experiments ignorantly or heedlessly conducted. I would venture, therefore, to urge care in experimenting, and the acquisition of a certain amount of elementary knowledge before an attempt is made at eliciting any of the rarer and more delicate phenomena of the mesmeric state. With simple knowledge and ordinary care the course is clear, and danger need not be apprehended; but delicate states of the sensitive patient, respecting which we are as yet comparatively ignorant, do not fitly lend themselves to experiments on the part of the merely curious and uninstructed investigator.

M.A. (Oxon.)

London, November, 1884.

CONTENTS.

CHAPTER 1.
PAGE

First Effects Produced by Mesmerism—Sensations—Process for Causing Mesmeric Sleep—The Sleep or Mesmeric State—It Occurs Spontaneously in Sleep-walkers—Phenomena of the Sleep—Divided Consciousness—Senses affected—Insensibility to Pain 1

CHAPTER II.

Control Exercised by the Operator over the Subject in Various Ways—Striking Expression of the Feelings in Look and Gesture—Effects of Music—Truthfulness of the Sleeper—Various Degrees of Susceptibility—Sleep Caused by Silent Will ; and at a Distance—Attraction Towards the Operator—Effect in the Waking State of Commands Given in the Sleep 11

CHAPTER III.

Sympathy—Community of Sensations: of Emotions—Danger of Rash Experiments—Public Exhibitions of Doubtful Advantage—Sympathy with the Bystanders—Thought-reading—Sources of Error—Medical Intuition—Sympathetic Warnings—Sympathies and Antipathies—Existence of a Peculiar Force or Influence 22

CHAPTER IV.

Direct Clairvoyance or Lucid Vision, without the Eyes—Vision of Near Objects : through Opaque Bodies : at a Distance—Sympathy and Clairvoyance in Regard to Absent Persons—Retrovision—Introvision . 34

CHAPTER V.

Lucid Prevision—Duration of Sleep, etc., Predicted—Prediction of Changes in the Health or State of the Seer—Prediction of Accidents, and of Events Affecting Others—Spontaneous Clairvoyance—Striking Case of it—Spontaneous Retrovision and Prevision—Peculiarities of Speech and of Consciousness in Mesmerised Persons—Transference of Senses and of Pain 47

CHAPTER VI.

Mesmerism, Electro-Biology, Electro-Psychology, and Hypnotism, essentially the same—Phenomena of Suggestion in the Conscious or Waking State—Dr. Darling's Method and its Effects—Mr. Lewis's Method and its Results—The Impressible State—Control Exercised by the Operator—Gazing—Mr. Braid's Hypnotism—The Author's Experience —Importance of Perseverance—The Subject must be Studied . . 63

CHAPTER VII.

Trance, Natural and Accidental; Mesmeric—Trance Produced at Will by the Subjects—Col. Townsend—Fakeers—Extasis—Extatics not all Impostors — Luminous Emanations — Extasis often Predicted—M. Cahagnet's Extatics—Visions of the Spiritual World . . . 77

CHAPTER VIII.

Phreno-Mesmerism—Progress of Phrenology—Effects of Touching the Head in the Sleep—Variety in the Phenomena—Suggestions—Sympathy—There are Cases in which these Act, and others in which they do not Act—Phenomena Described—The Lower Animals Susceptible of Mesmerism—Fascination Among Animals—Instinct—Sympathy of Animals—Snail Telegraph Founded on It 87

CHAPTER IX.

Action of Magnets, Crystals, etc., on the Human Frame—Researches of Reichenbach—His Odyle is Identical with the Mesmeric Fluid of Mesmer, or with the Influence which Causes the Mesmeric Phenomena—Odylic or Mesmeric Light—Aurora Borealis Artificially Produced—Mesmerised Water—Useful Applications of Mesmerism—Physiological, Therapeutical, etc.—Treatment of Insanity, Magic, Divination, Witchcraft, etc., explained by Mesmerism, and Traced to Natural Causes—Apparitions—Second Sight is Waking Clairvoyance—Predictions of Various Kinds 98

CHAPTER X.

An Explanation of the Phenomena Attempted or Suggested—A Force (Odyle) Universally Diffused, Certainly Exists, and is Probably the Medium of Sympathy and Lucid Vision—Its Characters—Difficulties of the Subject—Effects of Odyle—Somnambulism—Suggestions, Sympathy—Thought-Reading—Lucid Vision—Odylic Emanations—Odylic Traces followed up by Lucid Subjects—Magic and Witchcraft—The Magic Crystal, and Mirror, etc., Induce Waking Clairvoyance—Universal Sympathy—Lucid Perception of the Future 119

CHAPTER XI.

Interest Felt in Mesmerism by Men of Science—Due Limits of Scientific Caution—Practical Hints—Conditions of Success in Experiments—Cause of Failure—Mesmerism a Serious Thing—Cautions to the Student—Opposition to be Expected 138

CHAPTER XII.

Phenomena Observed in the Conscious or Waking State—Effects of Suggestion on Persons in an Impressible State—Mr. Lewis's Experiments With and Without Suggestion—Cases—Dr. Darling's Experiments—Cases—Conscious or Waking Clairvoyance, Produced by Passes, or by Concentration—Major Buckley's Method—Cases—The Magic Crystal Induces Waking Lucidity, when Gazed at—Cases—Magic Mirror—Mesmerised Water—Egyptian Magic. . . . 149

CHAPTER XIII.

Production of the Mesmeric Sleep—Cases—Eight out of Nine Persons Recently Tried by the Author Thrown into Mesmeric Sleep—Sleep Produced Without the Knowledge of the Subject—Suggestion in the Sleep—Phreno-Mesmerism in the Sleep—Sympathetic Clairvoyance in the Sleep—Cases—Perception of Time—Cases ; Sir J. Franklin ; Major Buckley's Case of Retrovision 169

CHAPTER XIV.

Direct Clairvoyance—Cases—Travelling Clairvoyance—Cases—Singular Visions of Mr. D.—Letters of Two Clergymen, with Cases—Clairvoyance of Alexis—Other Cases 195

CHAPTER XV.

Trance—Extasis—Cases—Spontaneous Mesmeric Phenomena—Apparitions—Predictions 227

CHAPTER XVI.

Curative Agency of Mesmerism—Concluding Remarks, and Summary . 244

ANIMAL MAGNETISM.

Chapter I.

FIRST EFFECTS PRODUCED BY MESMERISM—SENSATIONS—PROCESS FOR CAUSING MESMERIC SLEEP—THE SLEEP OR MESMERIC STATE—IT OCCURS SPONTANEOUSLY IN SLEEP-WALKERS—PHENOMENA OF THE SLEEP—DIVIDED CONSCIOUSNESS—SENSES AFFECTED—INSENSIBILITY TO PAIN.

IF you will try the experiment of drawing the points of the fingers of your right hand, without contact, but very near, over the hands of several persons, downwards from the wrist, the hands being held with the palms upwards, and your fingers either all abreast, or one following the other, and repeat this, slowly, several times, you will most probably find one or more who distinctly perceive a peculiar sensation, which is not always the same in different persons. Some will feel a slight warmth, others a slight coolness, others a pricking; some a tingling; others a numbness. Such as perceive these sensations most distinctly may then be tested, and will be found, probably, very clear and consistent with themselves, even if blindfolded. But sometimes, blindfolding produces at once a state of nervous disturbance, most unfavourable to clear perception. All this I have often tried and seen, and Reichenbach, as well as many others, has minutely described it.

You may now, having found a person susceptible to a certain extent, proceed to try the effect of passes, made slowly with both your hands, downwards from the crown of the patient's head, over the face, to the pit of the stomach, or even down to the feet, always avoiding contact, but keeping as near as possible without contact. Or you may make the passes laterally, and so downwards over the arms. It is necessary to act with a cool, collected mind, and a firm will, while the patient is perfectly passive and undisturbed by noise or otherwise. He ought to look steadily at the eyes of the operator, who, in his turn, ought to gaze firmly on his subject. The passes should be continued, patiently, for some time, and will generally excite the sensations above mentioned, warmth, coolness, pricking, tingling, creeping of the skin, or numbness, according to the individual operated on. When these sensations are very marked, the subject will, in all probability, turn out a good one. It is probable that, with patience and perseverance, a vigorous, healthy operator, would finally succeed in affecting all persons; but in some cases, which have afterwards become very susceptible, the subjects have been only affected with great difficulty, and only after much perseverance, or even have not been at all affected on the first trial, nay, even for many successive trials. The operator must not be discouraged. If he perseveres, the chances of success are much increased, while he will often meet with cases in which a few minutes suffice to produce strong effects.

Another, and in some cases a more successful method, is to sit down, close before the patient, to take hold of his thumbs in your thumbs and fingers, and, gently pressing them, to gaze fixedly in his eyes, concentrating your mind upon him, while he does the same. This is, at least in the beginning, less fatiguing than making the unaccustomed motions of passes, although, with a little practice, it is easy to make several hundreds of passes uninterruptedly. I cannot give decided preference to either method. Both will occasionally fail, and both are often successful. They may be combined, that is, alternated, and often with advantage.

Two things are desirable. First, a passive and willing state of mind in the patient, although faith in mesmerism is not at all indispensable; but a *bonâ fide* passivity, or willingness to be acted on. This, however, signifies little in susceptible cases. Secondly, intense concentration on the part of the operator. It is self-evident that, to attain this, perfect silence is essential. Even the noises in the street will often distract both parties from the necessary attention, and still more, whispering among the company, moving about, the rustling of a lady's dress, &c., &c. The time required varies from a minute or two to an hour or more, but usually diminishes on repetition.

Intent gazing alone, especially if practised by both parties, will often produce the sensations above described, without close proximity. I have often seen Mr. Lewis, who likes this mode of operating, namely, gazing at a certain distance, with intensity and a firm volition, produce these sensations, and even stronger effects, in the space of five minutes, on a considerable proportion of the company, varying perhaps from 5 to 20 or 25 per cent., according to circumstances. But his power of concentration is truly astonishing, and is strongly indicated in his whole gesture, and in the expression of his countenance, while operating.

Lastly, these sensations may be produced by gazing, on the part of the patient alone, either at a small object in his hand, as practised by Dr. Darling with great success, or at an object placed above and before the eyes, as is done with equal success by Mr. Braid in producing hypnotism. Indeed, one difficulty in these cases is, to prevent the subject from going further, and becoming unconscious.

The same processes, when continued longer, give rise to phenomena still more striking; and I shall now proceed to these, while it will be unnecessary to repeat the detail of the processes, which, as already described, suffice to produce the whole train of mesmeric phenomena.

The first is, a twitching of the eyelids, which begin to droop, while, even when the eyelids remain open, there is in many cases, a veil, as it were, drawn before the eyes, concealing the operator's face and other objects. Now also comes on a drowsiness, and, after a time, consciousness is suddenly lost, and on awaking the patient has no idea whatever how long it is since he fell asleep, nor what has occurred during his sleep. The whole is a blank, but he generally wakes, with a deep sigh, rather suddenly, and says he has had a very pleasant sleep, without the least idea whether for five minutes or for five hours. He has been, more or less deeply, in the mesmeric sleep, which I shall now describe more particularly. I do so because in many cases of ordinary mesmerism, by passes or by gazing, it is the first marked result obtained, and in most of them it occurs immediately after the sensations formerly described.

I am aware that many very beautiful phenomena occur in the conscious state, but to produce them in that state, we must operate in a peculiar way; whereas, by operating as above described, we generally produce the sleep, in which all the same phenomena may be observed, and indeed we may produce them all in the conscious state, in this way also, by stopping short of the sleep. I shall consider their production in the conscious state, after I have described the sleep and its phenomena.

I have just said, that the sleeper wakes, without a recollection of what may have passed in his sleep. But we are not to suppose, because it now appears a blank to him, that it has really been a mere torpid, insensible, unconscious slumber. It is only an unconscious state, in reference to the ordinary waking condition; for the sleeper may have been actively engaged in thinking, observing, and speaking, during the whole period of sleep. This it is which renders the sleep so interesting a phenomenon. Let us now consider its characteristics a little more fully.

1. It is a state of somnambulism, sleep-walking, or more correctly sleep-waking. It is a sound, calm, undisturbed sleep; that is, it is not broken by gleams of ordinary consciousness. But the sleeper answers when spoken to by the operator, and answers rationally and sensibly. He frequently doubts, and therefore frequently uses the words "I don't know," and appears most anxious not to affirm or deny anything of which he is not quite sure. If desired, he will rise and walk, and, according to the particular stage in which he may be, he walks with more or less confidence and security, his eyes being always closed, or if found open, either turned up, or insensible to light. In short, he is a somnambulist, and possesses some means, not possessed in the ordinary state, of becoming aware of the presence of objects. Whether this depend on a preternatural acuteness of the senses of

touch, hearing, and smell, or on a more occult perceptive power, or on both, is a question which shall be discussed hereafter. I shall here only remark, that the variety in the phenomena, in different cases, and in different phases of the same case, is so great, that I am inclined to believe that both causes may be in operation; but that, sometimes, we have positive evidence that the external senses are entirely closed; while the numerous accounts given of spontaneous somnambulism would lead us to conclude that such is generally the case in that state. No one who has ever seen a case of natural sleep-walking, and who subsequently examines one of artificially excited somnambulism ever hesitates a moment in recognising the essential and complete identity of the two phenomena. I have not, myself, as yet had the good fortune to see a natural sleep-walker; but I have heard such cases often described by those who had seen and studied them, and who invariably, when allowed to see a case of mesmeric somnambulism, acknowledged that the phenomena were the same.

2. The sleeper sometimes, but not always, nor in all stages of the sleep, hears with increased acuteness, and that to an extent apparently marvellous. It is possible that this may depend, as in the blind, at all events in part, on the fact that, the eye being no longer active, nor indeed sensible to light, while the senses of touch, taste, and smell, are probably quiescent till objects are presented to them, the whole attention of the sleeper is concentrated on the sense of hearing. I have no knowledge, at present, of whether this sense is thus affected in natural sleep-walking, but I should expect that it will sometimes be found more acute, and at other times closed, as certainly occurs in the mesmeric sleep. Many cases of sleep-walking are recorded, in which no sound, however loud, was heard by the somnambulist, and some, in which very loud noises suddenly and dangerously awoke him, whereas less loud sounds had not been noticed. The state of utter deafness to all sounds, however loud, such as shouting, or firing a pistol, or ringing a large bell, close to the ear, is very common in the mesmeric sleep, and may, I believe, be produced in every case at some stage of it, or, by the will of the operator, at almost any stage.

3. When the sleeper has become fully asleep, so as to answer questions readily without waking, there is almost always observed a remarkable change in the countenance, the manner, and the voice. On falling asleep at first, he looks, perhaps, drowsy and heavy, like a person dozing in church, or at table, when overcome by fatigue, or stupefied by excess in wine, or by the foul air of a crowded apartment. But when spoken to, he usually brightens up, and although the eyes be closed, yet the expression becomes highly intelligent, quite as much so as if he saw. His whole manner seems to undergo a refinement, which, in the higher stages, reaches a most striking point, insomuch that we see, as it

were, before us, a person of a much more elevated character than the same sleeper seems to be when awake. It would seem as if the lower or animal propensities were laid to rest, while the intellect and higher sentiments shone forth with a lustre that is undiminished by aught that is mean or common. This is particularly seen in women of natural refinement and high sentiments, but it is also observed in men of the same stamp, and more or less in all. In the highest stages of the mesmeric sleep, the countenance often acquires the most lovely expressions, surpassing all that the greatest artists have given to the Virgin Mary, or to Angels, and which may fitly be called heavenly, for it involuntarily suggests to our minds the moral and intellectual beauty which alone seems consistent with our views of heaven. As to the voice, I have never seen one person in the true mesmeric sleep, who did not speak in a tone quite distinct from the ordinary voice of the sleeper. It is invariably, so far as I have observed, softer and more gentle, well corresponding to the elevated and mild expression of the face. It has often a plaintive and touching character, especially when the speaker speaks of departed friends or relations. In the highest stages, it has a character quite new, and in perfect accordance with the pure and lovely smile of the countenance, which beams on the observer, in spite of the closed eyes, like a ray of heaven's own light and beauty. I speak here of that which I have often seen, and I would say that, as a general rule, the sleeper, when in his ordinary state, and when in the deep mesmeric sleep, appears not like the same, but like two different individuals.

4. And it is not wonderful that it should be so. For the sleeper in the mesmeric state, has a consciousness quite separate and distinct from his ordinary consciousness. He is, in fact, if not a different individual, yet the same individual in a different and distinct phase of his being; and that phase, a higher one.

As a general rule, but not a rule without some exceptions, the sleeper does not remember, after waking, what he may have seen, felt, tasted, smelled, heard, spoken, or done during his sleep; but when next put to sleep, he recollects perfectly all that has occurred, not only in the last sleep, but in all former sleeps, and, as in the ordinary state, with greater or less accuracy, although usually very accurately indeed. He lives, in fact, a distinct life in the sleep, and has what is called a double or divided consciousness: of course, sleepers differ in their powers of memory in the mesmeric state, as they do in their ordinary state, if not to the same extent.

But, when in the mesmeric state, the sleeper is not always entirely cut off from his usual state, even in those cases in which he has no trace, on waking, left in his mind, of the actions or sensations of the sleep. On the contrary, he often speaks in the mesmeric sleep, with accuracy, of things known to him in his usual state. It is remarkable, that he finds, in general, a great difficulty, or even an impossibility, in *naming* persons or things in this way. He will define and describe them, but very often either cannot, or will not name them. If you name them, he will assent, but would rather not do it himself.

He often loses, in the mesmeric sleep, his sense of identity, so that he cannot tell his own name, or gives himself another, frequently that of the operator; while yet he will speak sensibly and accurately on all other points. He very often gives to his operator, and to other persons, wrong names, but always, so far as I have seen, the same name to the same person.

The phenomenon of double or divided consciousness, has frequently been described as a spontaneous one, and persons have lived, for years, in an alternation of two consciousnesses, in the one of which they forget all they had ever learned in the other, and have had, therefore, to be educated, like a child, in the former.

The same thing occasionally happens in mesmerism. The sleeper has often to learn, as a child, things with which, in his usual state, he is quite familiar, such as reading or writing; but this is by no means always observed; possibly, it is seldom looked for.

The phenomenon of which we have spoken, divided or double consciousness, more or less perfect, is one of the most surprising and beautiful in the whole series of mesmeric phenomena. As it is very easily observed, that is, if we have confidence in the sleeper (and, without confidence in his veracity, nothing can be ascertained even in regard to his ordinary consciousness), it ought to be among the first to be verified by the sceptical but truth-loving inquirer, who desires to ascertain the reality of the mesmeric sleep, as well as its peculiar characters.

5. The sleeper, with closed eyes, yet often speaks as if he saw certain objects, when his attention is directed to them. He even makes an apparent effort to see, or to look at them, while his eyes are only more firmly closed. But he very often feels them in his hand, and whether by the acuteness of his touch or by some other means, describes them as if he saw them. Or he places them on his forehead, on the summit of his head, or on the occiput, or on the epigastrium, and then describes them, which perhaps he could not do when they were held by the operator before his closed eyes. He talks of seeing them, and evidently makes an exertion to apply his internal or cerebral vision to their examination. In this he often succeeds, but often also finds great difficulty, especially in the earlier stages of the sleep. In fact, we have here the dawning of clairvoyance, which only reaches its noon-day brightness in the highest stage of the sleep. In the stage to which, at present, our remarks are confined, the object must be, in some way in contact with, or at least very close to the sleeper; he is incapable, other-

wise, of describing it. The subject of clairvoyance, in its multitudinous forms, is one of so great interest, and of so great extent, that it must be considered separately; in fact, as belonging to a different, or higher stage of the sleep. I find it quite impossible to draw any definite line between the various degrees or stages of the sleep, save only between those in which clairvoyance, or else a very high degree of sympathy, is present, which we may call the higher or later stages of the sleep, and those which we may call the lower or earlier stages, in which these powers are absent. The state or stage of perfect trance or extasis, may be regarded as the third and highest; but of this I am not yet qualified to speak from personal observation. In many cases, the sleeper passes at once into the higher, the lucid, or clairvoyant stage, without arresting himself in the lower or non-lucid state, which yet he may formerly, at an earlier period of the investigation, have exhibited distinctly. Most of the facts observed in that earlier stage, continue to appear in the later, but naturally attract less attention when we are occupied with the astounding facts of sympathy or of clairvoyance. As I have already stated, these will be afterwards separately considered. In the meantime, having noticed the first glimmerings of an unusual mode of vision, we shall go on to describe other phenomena, which appear without clairvoyance, while they often continue also in the lucid state.

6. The sleeper is very often deaf to every sound, save the voice of the operator. This is not, however, always the case. I have seen subjects, who readily heard and answered every question addressed to them by any of the persons present, without being in contact with them, or being purposely placed *en rapport* with them. In some of these cases, the subjects, either spontaneously, or at the will of the operator, or by passes, &c., made by him, pass into a higher state, and then instantly become deaf to all sounds, except his voice. Nay, I have seen and examined one very remarkable case, in which the sleeper, when she had passed spontaneously into a higher state of lucidity, became deaf even to the operator's voice, unless he spoke to her through the tips of her fingers, holding his mouth, while speaking, so as to touch them. When this was done, she started, and, after a moment, answered questions thus put, as readily as before. You might bellow in her ear, or fire off a pistol, without her countenance indicating the slightest change, or without her ceasing for an instant to dwell on and describe what she was engaged in looking at, which she readily did without questions being asked at all. Any one else could converse with this subject in the same way, and I did so for an hour or two. In some similar cases, any person, besides the operator, must be placed *en rapport* with the subject, physically or mentally, by the operator, before they are heard or answered. In others, again, we must address our words to the epigastrium, or to the head. There is here, as in all these phenomena, an endless variety in the details.

In many cases where the sleeper hears and answers any one, he may be instantly and completely deprived of this power by the will, whether silent or expressed, of the operator. Hence, when we are endeavouring to produce the sleep in a new subject, who happens to be much disturbed by the noises in the room or in the street, we may often, by commanding him not to hear those noises, greatly accelerate the arrival of the true sleep. This can only be done when the operator, as often happens, acquires, in the earlier stages, the power of controlling the sensations of the subject. This control, as we shall see, may be acquired over subjects in the conscious state.

7. The sleeper often becomes entirely insensible to pain; that is, he is rendered insensible to impressions of touch and other forms of feeling, as he was before to sounds. In many cases, where this does not spontaneously happen, but not in all, it may be effected by the will, expressed or silent, of the operator. Many persons who produce the sleep are not aware of this, and hence imagine that their subjects cannot be rendered insensible to pain.

It must, I think, be admitted, that of all the methods now known and used to produce insensibility to pain, this is the safest, and, *cæteris paribus*, the best. The mesmeric insensibility is never, so far as I have seen, followed by any unpleasant symptoms. On the contrary, every sleeper whom I have seen, feels better after the sleep than before. If, in a few cases, the production of the mesmeric sleep has either been followed by any discomfort, or if it has been found difficult to awake a person from the sleep, this has arisen solely from the inexperience of the operator, who has rashly produced a state which he knows not how to control. It happens only when inexperienced persons, out of curiosity, or for amusement, cause the mesmeric sleep. They are at first astonished and a little alarmed at their success. But when, on trying to rouse the sleeper they find him deaf, and obstinate in sleeping, they become agitated and nervous. Their state of mind is communicated, by sympathy, to the patient, who appears to suffer, and may even be seized with spasms or convulsions. This terrifies the operators still more. Matters become worse and worse, and at last a doctor is sent for, who is equally inexperienced, and only does harm by his attempts to rouse the poor sleeper.

There are two rules which should be borne in mind when any such case occurs, although the best rule of all is not to attempt mesmerism without at least the presence of an experienced mesmerist. The first is, that the sleep in itself is salutary, and that when the proper mode of ending it, namely, by upward or reversed passes, or by wafting, is not known, hurried and nervous proceedings will almost infallibly do harm. The operator ought to become cool, and then employ reversed passes. No one else should interfere with the patient, for cross mesmerism is generally hurtful. Secondly, if the operator cannot become collected and cool, so as to make the upward passes calmly, *let the patient sleep it out*. This is always safe, if he be not interfered with. The sleep may last an hour, or two, three, four, or twelve hours, or twenty-four, or even forty-eight hours, but it rarely lasts, if left entirely to take its own course, more than an hour or two. In the cases where it has lasted longest, there has always been improper interference and cross mesmerism. In all cases, if no interference have taken place, the state of the pulse and of the respiration may satisfy us that nothing is wrong, and we shall find the mesmeric sleep no more dangerous, and no more likely to be of indefinite duration, than our usual nightly sleep.

But to return to the use of mesmerism to produce Anæsthesia, or insensibility to pain. I have said that I regard it, *cæteris paribus*, as the best known method for doing this. It is the safest, and the sleep may last as long as the operator requires, without the necessity of renewing the operation. I do not hesitate to say, that, in proper and experienced hands, it is free from all danger.

There is, however, one objection, or rather difficulty, which applies to it. We cannot, in all cases, be sure of producing the sleep, and when an accident happens, we have no time to try long experiments. Now this is true to a certain extent. But if we had practised and powerful mesmerists, and if mesmerism were generally tried, it would be found, even among ourselves, to succeed far more frequently than is supposed, and in persons under the effects of disease or accident, often at the first attempt. Still it appears to be certain, that the natives of this country are not so easily and certainly mesmerised as those of others, for example, of Bengal. At Calcutta, Dr. Esdaile, who has now performed hundreds of painless operations, never fails to mesmerise the natives, while he has sometimes failed with Europeans. Nay more, Dr. Esdaile is not only himself successful, but has numbers of native assistants who mesmerise for him with perfect ease and success. We cannot at present expect the same measure of success in England, but we can at all events use mesmerism where it is efficacious; we can try it in all cases; we can, in chronic diseases, and in the period preceding accouchement, endeavour to acquire the necessary influence over our patients, so as to be prepared for the hour of the operation or the delivery; we can persuade healthy persons to have themselves brought under the influence of mesmerism, that accident or disease may not find them unprepared; and finally, we can, by investigating the subject scientifically and experimentally, endeavour to discover some means of increasing mesmeric power, some mesmeric battery, which shall enable us to mesmerise any one at pleasure. The researches of Reichenbach tend to show that such an expectation is far from chimerical.

With the same view, it would perhaps be advisable to begin early, and to mesmerise young persons, who are in general more susceptible than adults, just as we teach them to swim, that they may be able, if necessary, to save their own lives or those of others. If once mesmerised, the effect would be easily kept up. Were this done generally, not only should we gain our object in regard to the persons mesmerised, but we should acquire so vast a mass of interesting observations, that the progress of mesmeric science would be greatly promoted and accelerated. I shall continue the description of the phenomena in my next chapter.

Chapter II.

CONTROL EXERCISED BY THE OPERATOR OVER THE SUBJECT IN VARIOUS WAYS—STRIKING EXPRESSION OF THE FEELINGS IN LOOK AND GESTURE—EFFECT OF MUSIC—TRUTHFULNESS OF THE SLEEPER—VARIOUS DEGREES OF SUSCEPTIBILITY—SLEEP CAUSED BY SILENT WILL ; AND AT A DISTANCE—ATTRACTION TOWARDS THE OPERATOR—EFFECT IN THE WAKING STATE OF COMMANDS GIVEN IN THE SLEEP.

8. THE sleeper is usually very much under the control of the operator, in reference to the duration of the sleep. The operator may fix any time, long or short, and if the sleeper promise to sleep for that period, he will do so to a second. He then wakes up, and is instantly quite free from all effect, without any further process. The utility of this power is very obvious, especially in cases of pain or surgical operations.

But if no time be fixed by the operator, the sleeper awakes spontaneously, after a longer or shorter interval, generally from half an hour to two hours, at least in the cases I have seen. Sometimes, and especially if urged with many questions, requiring exertion to answer, the sleeper declares that he is fatigued, and begs to be awakened. It is always best to yield to this wish, and to avoid fatiguing the subject, since over-exertion has a most unfavourable influence on his powers.

9. Whether the time of sleeping be fixed by the operator, or left to nature, the sleeper, in a large number of cases, can tell when asked, and generally very readily, precisely how long he has to sleep ; and if he be repeatedly asked at different times, he will always be found correct as to the time still remaining. This is a truly remarkable phenomenon ; for, in the power of telling how long he is to sleep, we may see, and especially where no time has been fixed, the first glimmering of the power of prevision, and it is sometimes the only indication of this power. Different subjects give different accounts of how they become aware of the point of time at which their sleep is to terminate. But many of them declare, that they see the figures indicating the number of minutes, or divisions which they can count, by which means they can give the desired information. I shall mention, under the head of clairvoyance, some remarkable details which I noted down as they occurred, in a case of much interest, under my own management.

10. The sleeper, often when he is first put to sleep, and still oftener after several times, will answer a variety of questions as to the best and most effectual method of mesmerising him, whether

by passes or otherwise ; as to the powers which he will hereafter possess ; and as to the time when he shall acquire those powers, or exhibit certain phenomena. He will often fix with precision, and as it afterwards appears, with exactness, the number of times that he must be mesmerised or put to sleep, in order to produce certain effects ; and whether this should be done once a day, or twice a day, or less frequently. Here again we have a dawning of prevision, which, in a higher stage, as we shall see, enables him to predict certain occurrences in reference to his own state of health, for example. But this also must be reserved for the section on clairvoyance.

11. Although the sleeper, in general, has no recollection when awake of what has passed in the sleep, this is far from being an uniform occurrence. Some remember a part, others the whole, of what has taken place. But even in many of those cases, in which there is, naturally, no remembrance of it, the operator, if he choose, may command his subject, during his sleep, to remember a part or the whole of what has occurred, which will then be remembered accordingly. I have already alluded to this under divided consciousness. I do so again here, as a proof of the influence of the mesmerist on the sleeper. In these experiments, it is often desirable to enable the subject to remember, when awake, certain things ; and it is probable that the mesmerist will find that he has equally the power, when he tries it, of causing the sleeper to forget all, or part, of what has occurred in the sleep, in those cases in which he naturally remembers it. This, also, is often very desirable. It has already been mentioned, that when asleep, the subject is in connexion with the previous sleeps he may have had, and remembers them more or less perfectly, according to his natural powers of memory.

It is exceedingly probable, although it has not, so far as I know, been ascertained, that in ordinary spontaneous somnambulism, the sleep-walker remembers his previous acts of somnambulism.

We must not confound this forgetting what occurs in the sleep, after waking, or remembering in the sleep what has occurred in previous sleeps, which are the results of divided or double consciousness, with the loss of memory which may be produced, for the time, in either state, by the will of the operator. We shall see, when we come to the effects producible in the conscious state, that the memory may become entirely subject to the will of the mesmerist. Here we proceed to notice ;

12. That the subject, while asleep, may be made to forget anything that he would otherwise remember, by the will of the operator. He may be made to forget, not only what has happened in the former sleeps, but even that he has ever slept, or been mesmerised before. He often forgets spontaneously his own name, and if not, can be made to do so. This is another proof of the control exercised by the mesmerist on his subject.

13. This control is further shown by the power which the operator has of producing in the sleeper, inability to move the arm or leg, to speak, to rise up or to sit down, by his will. It is shown in the production of partial or general cataleptic rigidity and its removal. It is shown, in short, in the complete command of all the voluntary muscles of the subject acquired by the operator.

14. It further appears in the power of causing the sleeper instantaneously to imitate, with the most perfect and admirable mimicry, every gesture of the operator, and every tone of his voice. If the mesmerist speaks German or Italian, languages perhaps quite unknown to the subject, and with the greatest rapidity, the sleeper will speak after him so exactly, that it is often impossible, when his ear is acute in catching the minute shades of the sound, to perceive the slightest difference. If the mesmerist laughs, he instantly laughs; if the former make any gesture, however ridiculous, the latter imitates it exactly, and all this with closed eyes, and when the operator is behind him, so that he cannot be seen. The same subject, when awake, will often, indeed generally, be found to fail miserably in his attempts at this instantaneous mimicry, and indeed to fail even when he takes more time to it.

15. The sleeper, if naturally insensible to the voice or to the actions of all but his mesmerist, may be put *en rapport* with any other person. This may be done by simply giving him the person's hand, in many cases. In others, the sleeper requires to be told to communicate with that person, and this having been done, he becomes as completely and exclusively *en rapport* with him as he before was with the mesmerist. It often happens, that the stranger thus placed *en rapport* with the subject, must again retransfer him to the mesmerist, before the latter can communicate with him. The transference from one to another, in such cases, is usually attended with a start on the part of the sleeper, but he does not awake.

16. All the feelings, propensities, and talents of the sleeper, may be excited to action by the mesmerist, and that in various ways, either by merely touching the corresponding parts of the head, as in what is called Phreno-mesmerism, to be hereafterward considered, or, as comes naturally to be considered in this place, by the expressed will of the operator.

The subject may be rendered happy and gay, or sad and dejected; angry, or pleased; liberal, or stingy; proud, or vain; pugnacious, or pacific; bold, or timid; hopeful, or despondent; insolent, or respectful; &c., &c. He may be made to sing, to shout, to laugh, to weep, to act, to dance, to shoot, to fish, to preach, to pray, to deliver an eloquent oration, or to excogitate a profound argument. All this the mesmerist, in many cases, can cause him to do, and indeed a great deal more, by commanding him to do it, as I have often seen, nay, as I have myself done. I have heard a sleeper give a lecture on temperance, or on mesmerism; I have heard the most beautiful prayers, the most poetic imagery, from the mouths of persons who, in their ordinary state, were quite unequal to such things. And as we shall see hereafter, all this can also be done when the subject is in the conscious state.

17. In all such experiments I have observed, and it has been observed and recorded by others, that the gestures and voice, the manner and expression, in short, the whole physiognomical and natural language, is extremely perfect. The attitudes of pride, humility, anger, fear, kindness, pugnacity, devotion, or meditation, and all others, are, with peculiarities in each case, depending on the idiosyncrasy of the individual, beautiful studies for the artist. The most accomplished actor or mimic, a Garrick, or a Mathews, falls short of the wonderful truth and nature of these attitudes and gestures, as I have seen them in numerous cases, and most frequently in persons of limited intellectual cultivation, who, in their waking state, showed no peculiar talent for pantomime.

I have already stated, and may here repeat, that subjects of a superior refinement of character, exhibit, as all do more or less, an exaltation of refinement when in the mesmeric sleep; I now add, that they further, when the higher sentiments are intentionally excited, exhibit a purity, beauty, and sublimity of gesture, attitude, and expression of countenance, equalling, nay, far surpassing, all that the greatest artist has ever conceived or executed. Did all artists know, as some do, how precious a fountain of inspiration exists in these mesmeric phenomena, they would spend hours in studying them. It is not improbable that some of the great masters did so, and, at all events, the appearance of mesmerised persons constantly recalls to us, as an imperfect imitation of what we see, the saints, angels, and virgins of Raphael, Guido, Corregio, Murillo, &c. I am convinced, that ere long, artists will have recourse to mesmerism for expression, as they now have to the nude subject for forms.

A most beautiful case of this kind lately occurred to a mesmerist of my acquaintance, when a gentleman, highly distinguished as an artist and a man of taste, was so enchanted with what he saw, that it was arranged that the mesmerist should produce the effects at a subsequent time, to be studied and copied by an artist of great talent. This has not, I believe, yet been possible, owing to the absence of some of the parties from the somewhat remote scene of the experiments.

I have myself seen one case, of a young and pretty girl, thirteen or fourteen years of age, belonging to a family in a humble station, whose countenance became, in the mesmeric sleep, and especially when devout feelings were excited, and when music was performed, lovely and heavenly in expression, to a degree beyond my power to describe. Her face beamed with a spiritual ethereal beauty, such as I had previously never even conceived. In that case, the organisation of the brain was, in the coronal region, the seat of the organs of the higher sentiments, particularly fine. The organs of the intellectual faculties were well developed, while those of the lower propensities were much below the average proportion.

In short, the characteristic of the phenomena thus obtained is their entire truthfulness; and this, strange to say, is often the cause of doubts as to their genuineness, in the minds of those who see them for the first time. If the subject be uneducated, there will always remain, even while he becomes improved and refined in manner to a considerable extent in the sleep, a certain something, which marks the uncultivated mind. Hence his performance, although true to nature, is not perfect, and looks very like acting, precisely because the best acting is that which approaches nearest to nature, and yet can never reach it. When experiments are made with a person of fine natural disposition and highly cultivated mind, the results are so beautiful as to delight all spectators.

18. I must here mention a circumstance, which I have remarked in every case in which I have tried the experiment, or seen it tried. It is, that the sleeper is invariably much more strongly affected by music than when in his ordinary state. All the subjects on whom I have seen it tried, have been agreeably influenced by it. Their faces brighten, and they usually assume attitudes and gestures corresponding to the character of the music. Thus, a reel or a quadrille will set them dancing, and those of fine temperament do so with a singular grace, while the clownish stump about with much vigour, but little elegance. I have seen this occur in persons of both kinds, who had never learned to dance, except from nature. A solemn strain, again, will readily cause them to kneel and pray, or to join in the devotional music. A warlike march or quick step will cause them to march and strut about, and often to exhibit a very pugnacious pantomime. All this will take place, more or less, in persons who have in their ordinary state, no love for music, or care, at all events, little for it. It would appear also, from the observations of Mr. H. E. Lewis, that a strain of soft music often assists in inducing the sleep in new subjects. This agrees with the recorded fact, that music has always formed a part of the magician's arrangements. When a sorcerer wished to cause those who consulted him to see visions, that is, to become somnambulists, he always used soft music and fumigations.

19. Not only are the attitudes and gesture, the tone of voice and the expression of the face, true to nature, in the expression of every feeling that is excited, but this truthfulness extends to all that is said by the sleeper. As a general, perhaps invariable rule, he refuses, whatever questions may be asked, or suggestions made, to go beyond what he feels sure of, in describing his own sensations, or his visions, if we call them so. The spectator often

unconsciously does his best to mislead him by leading questions, and also, by such as arise from a misconception of his meaning. Yet of all things observed in the sleep, that which most constantly recurs, and most forcibly strikes us, is the frequent repetition of the words, "I don't know exactly;" "I cannot say for certain;" "I cannot see whether it is so or not;" "I must not say what I do not see, or feel, or know," and the like; while, when the sleeper once sees, feels, or knows a thing, he adheres firmly to it. This truthfulness gives great value to experiments properly made.

I have always admitted the possibility of deceit in mesmerism, when practised for the sake of money. And I believe that cases have occurred of genuine somnambulists, who under certain circumstances, have been guilty of imposture. Let us suppose a person who is really possessed of certain powers, in the mesmeric sleep, but who is greedy of gain, and vain of his powers. If, as I have shown to be probable, he should, on some public occasion, find his power much less than usual, or should be deprived of it by over exertion in previous experiments, which have succeeded, not only is his vanity hurt, but his prospects of gain are diminished, and if, as may happen, he does not possess the highest sense of truth and honour, he may try to make up for deficient power by deceit. I have some reason to believe, that individuals, of whose power at times no doubt can reasonably be entertained, have, when over-fatigued, or by some chance, less lucid than usual, endeavoured to cover failure by deceit. Of course, although I might believe such persons to possess great lucidity at certain times, and to have acted thus dishonestly, simply from the desire to escape the confession of failure and the loss of expected gain, yet I should not use the evidence derived from such cases. It is best to reject all evidence to which any suspicion can attach. There is abundance of unexceptionable evidence, if we only look for it, and I would look with suspicion on the evidence derived from the public exhibitions of those who make a trade of such exhibitions, and use paid subjects.

These remarks apply more particularly to clairvoyance, but I have alluded to the subject here, because they apply also to the lower phenomena. I shall have to refer to the matter again, briefly, when treating of the higher phenomena.

20. I have not yet fully noticed another fact, namely, that the operator finds much greater difficulty in producing the sleep at first, than he does after it has been produced several times. It often happens, and has several times occurred to myself, that in a subject, in whom the sleep could not at first be induced in less than from half an hour to an hour or more, with constant and laborious exertion in making passes, or gazing with an intent volition and the most complete concentration of the mind on the subject, the same person may, in a day or two, or a week, or a month, be put to sleep, and that far more deeply than at first, in

five minutes, or one minute, or half a minute, or a quarter of a minute. Nay, some subjects are entranced by a single rapid pass, or by a look. Many subjects, however, never reach this degree of susceptibility; but in all, it becomes easier, after some practice, to induce the sleep, than it was at first.

It is often observed, that those who are slowly and gradually brought up to a high degree of susceptibility make the best subjects. At all events, we should never be discouraged by want of complete success, or even by failure, in our first trials.

Cases have been recorded, in which the sleep never occurred till after hundreds of operations, and yet became very deep, and exhibited beautiful phenomena. I believe, from what I have seen, that every one possesses the power to mesmerise others, though in variable degrees; and further, that every one may be himself mesmerised, with patience and perseverance, on the part of the operator. It must be borne in mind, too, that the sleep is not essential, either to the relief of suffering, the cure of disease, or the production of many beautiful results, which, we have seen, occur in the conscious state, as will be more particularly described hereafter. Patience and perseverance, with a strong resolution to succeed, should be the mesmerist's motto. They are the most powerful aids to mesmerism.

It would appear, that persons of a very marked temperament, most readily affect those of the opposite temperament. Thus, a person of a strongly marked, nervous bilious temperament, will succeed best with subjects who are sanguine lymphatic. A large brain and active temperament are favourable to mesmeric power. A powerful and very active intellect, in the subject, is not exactly opposed to his being mesmerised, but renders it often more difficult, because the constant activity of the mind opposes the concentration of the thoughts on the object of being mesmerised, which is so desirable, and also counteracts the attempt to attain that passive state which may be called essential to the result. In experiments made in public, on parties never before mesmerised, this passive state of mind is almost unattainable by them. They are excited by the desire of seeing and perhaps of explaining strange facts; they are also nervous before so many people; they are afraid of being made ridiculous, or of having secrets extracted from them; and, finally, they often resist the influence to the utmost of their power; that is, they keep up an active state, not aware that, to be acted on, they must be passive. This is one reason why experiments made in strict privacy succeed in a proportion of cases so much larger.

One reason why so many susceptible subjects are found, especially in public exhibitions, among the less educated classes, is, that their intellectual powers are not in so constant activity as is the case with men, for example, engaged in business or in professional and scientific or literary pursuits. They become, therefore, more readily passive.

I have already observed that the Hindoos, and the natives of India generally, are more uniformly susceptible, even to men of their own nation, than Europeans. This depends on the temperament. It would appear, that negroes also are both highly susceptible subjects, and very powerful mesmerists. The Obi of the West Indies and of Africa, depends for its influence on their susceptibility; and the distinguished negro mesmerist, Mr. H. E. Lewis, possesses, in a very rare degree, the power of mesmerising others. I embrace with pleasure this opportunity of testifying, not only to that gentleman's qualifications as a mesmerist, but to his great abilities, his pure and disinterested love of science, his gentlemanlike manners and amiable character, his great readiness to assist, in every possible way, those who desire to investigate the subject with the single object of discovering truth, and his intimate practical knowledge of the subject in every department.

21. After the operator has succeeded in producing the sleep easily and in a short time, he can, in many cases, produce it by the silent exertion of the will, without any passes, or other process of any kind. This I have myself done, and in one case, where the subject was deeply engaged in conversation without any idea that I intended any thing of the kind, as I had taken, up to that moment, an active share in the conversation, I put him into a sound mesmeric sleep in 25 seconds (his eyes having been directed to other persons present), by the silent power of the will. I sat about four or five feet from him, to one side.

In doing this, it is therefore, at least in some cases, quite immaterial, whether the subject be aware of the intention of the operator or not. In this case, after sleeping an hour exactly, as I commanded him to do, he woke suddenly at the appointed time, and his first remark was, when I asked if he had had a pleasant sleep, "Oh, yes! but you did not tell me you were going to mesmerise me." Similar facts are of daily occurrence.

22. In such instances as that just mentioned, the subject is put to sleep by the operator, when the latter is in the same room, or near him. But this also may be dispensed with. I have often seen persons put to sleep, both when aware of the intention, and when that has been concealed, by the operator from the next room, or the floor below or above. The fact is, that with a susceptible subject, distance is a matter of little or no moment. The influence, whatever it be, seems to travel to any distance, like light. Many facts of this kind, at distances much greater than I have now mentioned, have been recorded. I shall here give an instance, the details of which I can testify to, as having occurred in my own family.

Mr. Lewis met a party of fifty ladies and gentlemen in my house, one evening in the end of November or beginning of December, 1850. He acted on the company *en masse*, and affected several, among them a lady, a member of my family, who was susceptible, and had frequently been mesmerised by others. This lady, when mesmerised, loses the power of her arms, her eyes are closed, and the sensations she experiences are very marked and well known to her. Mr. Lewis, not being told how strongly she had been affected by him, did not do anything to remove the effect, and the consequence was a headache, to which she is naturally very subject. This she ascribes to her not having been demesmerised, and it continued next morning. When I saw Mr. Lewis, after my lecture, at 11 A.M., he asked me how the lady was. I mentioned the headache, as well as her idea of the cause of it. Mr. Lewis then said, "Oh! never mind the headache, I shall think of her some time during the day, and dismiss her headache." This I begged him to do, as I knew that such things could be done. He then left me. When I returned home, at 5 P.M., I had quite forgotten this conversation, when the lady in question recalled it by saying, as I entered the room, "What do you think of this? I have been mesmerised in your absence." "Indeed? by whom?" "By nobody. I was sitting at the pianoforte, playing, at half-past three, when I felt as if strongly mesmerised; my arms lost their power; I could no longer play, and had all the usual sensations. In a few minutes I was compelled to lie down on the sofa, and fell into a short mesmeric sleep. When I woke, my headache was quite gone." "Did you mention this to anyone at the time?" "I was alone, but, just as I woke, a lady, who was here last night, called, and I told her of it, adding, that I felt sure that Mr. Lewis was mesmerising me." I then said that he had undertaken to do so, but that I did not know whether he had done it or not. In the evening I saw Mr. Lewis again, at a large party, and, in the presence of Dr. W. F. Cumming, who felt much interested in the case, I asked him whether he had kept his promise about the lady's headache. He said he had. Dr. Cumming then asked him at what time, when he at once answered, "at half-past three, when I returned to my lodgings. I could not do it sooner."

It appears to me, that everything was here combined to make the case a good one. It was accidental. The subject had no idea either that she was to be mesmerised, nor of the time; and a lady came opportunely to attest the fact before my return, while a gentleman heard Mr. Lewis' answers to my questions and his own. I may add that the lodgings of Mr. Lewis are in South St. Andrew-street, while my house is at 114, Prince's-street, a distance of nearly four divisions of Prince's-street, or, I should suppose, 500 or 600 yards. I may further state, that on two other occasions, Mr. Lewis affected the same lady, at the same and at a greater distance, without her knowing that he was to do so.

It appears, from this, and other facts of the same kind, that, in susceptible subjects, distance forms no obstacle to the action of the mesmeric influence of the operator, although it may possibly retard or weaken it to a certain extent. When we first hear of such a thing, we are naturally incredulous, but when we have seen it, or produced it, several times, we are not only compelled to accept the fact, but to feel that it must depend on a natural cause, which it is our business to investigate.

23. Not only may the subject be put to sleep by the silent will, but he may be made, also by the silent will, to exhibit all the phenomena already described as producible by the expressed volition of the operator.

He may be made, in this way, to come to the operator, or to sit down in any place, or to perform any act, which the mesmerist may will him to do. It is unnecessary here to repeat details; it suffices to say, that, in many cases, everything that can be done by the expressed will, may be done also by the silent will, of 'the latter. This, too, occurs also in the conscious state.

24. Another remarkable fact, is a kind of attraction felt towards the mesmerist, and which he, by willing, can exert in many cases. The subject then feels an irresistible desire to approach him, and if prevented, will exert great force to overcome the obstacle. He cannot explain it farther than by saying, that he is drawn somehow towards him; some, however, speak of fine filament or threads, often luminous, by which they are gently drawn to him.

This strange attraction may also be exhibited at a distance. I have been informed, on the best authority, of a case where it was exerted at a distance of 100 yards or more, and where the subject moved towards the operator, till stopped by the wall of the house in which she was, in spite of the resistance offered by a strong man. This may also be shown in the conscious state.

25. In some cases, there is observed a permanent liking for the mesmerist, in the ordinary waking state of the subject. I have not had opportunities of seeing this, but it is, I believe, a well authenticated fact.

26. This leads me to another very curious phenomenon, namely, that the sleeper, if commanded, in the sleep, to do a certain thing, after waking, and at a certain hour, will do so, and however absurd or ridiculous the act, he cannot, in many cases, refrain from doing it, if he has promised it in the sleep.

He may have been ordered to go to a certain person's house, at a certain hour, and ask some trifling or useless question. As the time approaches, he is seen to be restless, till he sets out for his destination. He pays no attention to the people he may meet, and if they purposely arrest him, he forces his way onwards, asks his question, and can only say, that he felt that he *must* do so. He is often much hurt at the ridicule excited by his action, and, therefore, should not be made to do any thing that may excite ridicule, as, if that be persevered in, he will refuse compliance with the order or request, when made. This, at least, often happens.

This power, of influencing the waking actions by a promise

made in the sleep, may be most usefully applied. I lately saw a person, who had been induced by Mr. Lewis to promise, while in the sleep, to abstain from fermented liquors, and had, in his ordinary state, steadily adhered to that promise, ever since it was made, three or four months before ; nor had he the slightest desire to break it. I do not know whether he was aware of having made the promise, but that is not at all essential. The desire is extinguished, even when the subject has no recollection of the promise, and has not been told of it in his waking state. Mr. Lewis informs me, that he has broken many persons of the habit of drinking, as well as of other bad habits in this way. From what I have seen, I am satisfied, that a pledge given in the mesmeric sleep, will be found more binding than one given in the ordinary waking state.

I have now described, briefly, the most obvious and remarkable of what are called the lower phenomena, although it will be seen that they pass, insensibly, into the higher.

Proceeding further, we shall find that the higher phenomena develop themselves. The subject, in many cases, after a time, exhibits the highest degree of sympathy with the operator, or with those placed *en rapport* with himself; or he acquires the power of clairvoyance in some one or more of its varied forms.

These subjects we shall go on to investigate in the next chapter. Meantime, let me remark, that the occurrence of these phenomena rests on the very same testimony as that of those already described ; that this testimony is frequently of the highest possible character; and that, in truth, the so-called lower phenomena, which we have seen to shade into the higher, are not in the least more easily explained or understood than the latter.

It appears to me certain, that both classes of phenomena depend essentially on the same cause, and that, a natural cause. There is nothing supernatural or miraculous about sympathy or clairvoyance, if we will only examine them. They occur, as we shall see, spontaneously, and have been observed from the earliest ages. It is probable that the ancients were well acquainted with them, that this knowledge, being kept secret, and perhaps used for bad, certainly for interested, objects, by those who had the exclusive possession of it, had been lost, and that it was necessary to recover it, which was first effectually done, in great part at least, by Mesmer, although Van Helmont, and many others before Mesmer, ad obtained glimpses of the truth.

Chapter III.

SYMPATHY—COMMUNITY OF SENSATIONS: OF EMOTIONS—DANGER OF RASH EXPERIMENTS—PUBLIC EXHIBITIONS OF DOUBTFUL ADVANTAGE—SYMPATHY WITH THE BYSTANDERS—THOUGHT-READING—SOURCES OF ERROR—MEDICAL INTUITION—SYMPATHETIC WARNINGS—SYMPATHIES AND ANTIPATHES—EXISTENCE OF A PECULIAR FORCE OR INFLUENCE.

WE now come to what are called the higher phenomena, namely, Sympathy and Clairvoyance. It has been shown that these are connected, by insensible gradations, with the lower or more usual phenomena, and that, as both classes of facts depend, so far as we know, on the same cause, and both are equally inexplicable on ordinary principles, or rather equally explicable, the one class cannot, with strict propriety, be called higher or lower than the other. Nevertheless, as the effects of Sympathy and Clairvoyance have a peculiar character, which would partake of the supernatural, were we not convinced that they depend on natural causes, and have nothing miraculous about them, it may be convenient to use the term Higher Phenomena, on the understanding that this does not imply a difference in nature, but only in degree, for those to which we now proceed, in contradistinction to those already treated of.

And first, of Sympathy. This power, as we have seen, begins to appear in the earlier stages, and is shown in the form of an attraction towards the mesmerist, or in that of obedience to his silent will. But as we advance, it is further developed, so as to become the chief characteristic of a certain stage of the mesmeric sleep. The sleeper acquires the power of perceiving every sensation, bodily and mental, of his mesmerist. Nay, he often exhibits a like power in reference to all with whom he is placed *en rapport*, especially when this is done by contact. These sensations are so vividly felt by the sleeper, that he cannot distinguish them from the same sensations produced by direct external impressions on his own frame. Indeed, there appears to be no difference whatever between the two. He feels what is felt by the person *en rapport* with him, as truly as if the original impressions were made upon himself. He forms, for the time, a part of the person on whom the direct impressions are made, and all sensations, or many sensations, are common to both parties.

1. There is Community of Taste. If the operator, or any person, *en rapport* with the subject, takes any kind of food or drink into his mouth the sleeper, in many cases, instantly begins

going through the pantomime of eating or drinking; and, if asked, he declares he is eating bread, or an orange, or sweetmeats, or drinking water, wine, milk, beer, syrup, or lemonade, or infusion of wormwood, or brandy, or whisky, according as the operator takes each of them, or any other substance. When the thing taken is bitter or disagreeable, the countenance of the sleeper at once indicates this, while his eyes, as usual, are closed, and the mesmerist or friend may stand behind him, so that he cannot see what is taken. Minute details would be tedious. Suffice it to say, that I have seen and tested the fact in so many cases, that I regard it as firmly established. Moreover, no one who has had opportunities of observing this beautiful phenomenon, can long hesitate as to its entire truth, such is the expression of genuine sensation on the face and gesture, besides the distinct statements made by the sleeper. Like all the other phenomena, this one varies in extent and intensity in different cases. But it is very frequent with advanced subjects.

2. The same thing occurs with regard to Smell. If the person *en rapport* with the sleeper, smell a rose, the latter at once begins to inhale the delightful perfume. If he smell assafœtida, the sleeper expresses disgust; and if he place strong hartshorn under his own nostrils, the sleeper starts back, complaining of its pungency. The Community of Smelling is just as perfect as that of taste, in many cases; but I cannot affirm, that where one is present, the other always occurs. This is probable, but I have not strictly examined it. Of course, as before explained, the sleeper may be rendered dead to either or to both, by the will of the mesmerist, in many instances.

3. There is Community of Touch. Whatever touches the person *en rapport*, is felt by the sleeper, in precisely the same part. If the former shake hands with any one, the latter instantly grasps a visionary hand. If a pin be driven into the back of the mesmerist's hand, the sleeper hastily withdraws his hand, rubs the part, and complains loudly of the injury. This may be tried in all forms with perfect success in very many subjects. These are never deceived. Many most interesting experiments may thus be made, and I have, as in the preceding facts, often seen and minutely tested the phenomenon.

4. I cannot with certainty state whether the same community extends to the Sight. Possibly, the fact that the eyes are closed, and usually turned up, as well as insensible to light, which, in fact, constitutes a leading feature of the sleep, may serve to explain why this is not in general tried. The sleeper's eye may be quite dead to all *external* impressions, even where he possesses internal vision; and the sensations above treated of are the results of external impressions, conveyed to him by sympathy. The question is, can he, internally, see what the mesmerist sees externally, as he tastes, smells, and feels? If in a state of clairvoyance,

no doubt he can, for he then sees all surrounding objects, with his eyes closed. But that state is not necessary to the above mentioned community of sensations, and we must distinguish true clairvoyance, as we shall soon see, from that which depends on sympathy. On the whole, I am inclined to think, that the state of the eye is opposed to the making of such experiments in a satisfactory manner.

5. In regard to Hearing, I have not seen experiments tried in this way. I have already said, that the sleeper is usually deaf to all but the voice of the mesmerist, or of the person *en rapport* with him. Does he hear what is said *to* that person by others? I have no doubt that, in many cases, he does; and that in this way, he often becomes acquainted with matters intended to be kept secret from him. This should be carefully attended to in making experiments.

6. There is often, but perhaps less so than in regard to the senses, a Community of Emotion. In these cases, whatever mental emotion occurs in the mesmerist, or in others placed *en rapport* with the sleeper, is also experienced by him. I have not yet examined this phenomenon so minutely or fully as the others, on account of the difficulty of calling up at pleasure, a genuine and marked emotion. On this account, the observations are commonly accidental. Thus, I have seen some patients smile and laugh when their mesmerist happened to do so; and I have also seen, what has been very often described by other observers, the sleeper painfully affected by nervousness and alarm on the part of the operator.

This, indeed, is the chief cause of all the unpleasant results which occasionally arise when persons, who have no experience or knowledge of mesmerism, try, for amusement, or out of curiosity, to produce mesmeric effects. They succeed better than they had supposed possible, merely by imitating the gestures of some mesmerist whom they have seen operate, without, perhaps, having attended minutely either to his operations or to the cautions and directions he may have given. The subject operated on, probably a young person, or even a young lady, falls into a deep sleep, and hears nothing that is addressed to her, perhaps by her father, mother, or other near relations. These persons become alarmed, never having before seen any thing of the kind, and not being aware that this deafness is a common character of the sleep, and that the sleep is not only harmless, but beneficial. They ask the luckless operator, with much agitation, perhaps with some anger, to relieve her, and while he hesitates and becomes infected with their fears, not knowing how to proceed, they seize her hands, and their own influence, unknown to them, crosses his, producing uneasiness, which appears in the countenance of the sufferer, and almost invariably out of all proportion more intensely pictured there, than truly corresponds, by the patient's own subsequent statement, to the actual suffering. This again frightens them still more; they call on her, they weep, they rage against the mesmerist, and overwhelm him with reproaches. At last, goaded almost to madness, he tries to undo the charm. He takes the patient's hands, perhaps while several other persons are in contact with her, or acting on her, and by sympathy with him, she becomes instantly and seriously worse. This continues for a time, varied only by every sort of useless and hurtful interference on the part of the bystanders, not one of whom, perhaps, knows what ought to be done, and the unhappy victim of ignorance and temerity falls into a fainting fit, and possibly into severe convulsions. I need not pursue the unpleasant picture further, but I may suggest that it is only aggravated by the proceedings of the medical man finally summoned, if, as is too often the case, he have, either accidentally, or acting on a firm resolution, declined making himself familiar with these phenomena, or the laws which regulate them. Then, when it is too late, he regrets that carelessness or prejudice have led him to neglect facts, often presented to his notice; then the parents discover that an able and estimable physician has been induced to commit a grievous error, namely, to shut his eyes to some of the most wonderful and practically important phenomena. But there is an end of their scepticism, if they had any; possibly of the doctor's too. Yet even in such circumstances, where it has been quite evident that the slightest acquaintance with the matter would have enabled us to avoid all that suffering and danger, I have heard of physicians who drew no other moral from the occurrence than this, that mesmerism was dangerous; and they then shut their eyes to it as resolutely and as closely as before.

True, mesmerism is dangerous. But it is not the study of it, nor the knowledge of it, but ignorance of it, and the rash experiments of those who are ignorant of it, that are dangerous.

In the hands of qualified experimenters, I have never seen one unpleasant accident. I have heard of several in the circumstances above sketched, and on the authority of both the operators and of their subjects. But I can go farther. For I have never yet seen a case in which the mesmeric sleep was produced in the proper way, in which the sleeper did not declare, not only that he sustained no injury, but also, that he always felt better, stronger, and more fit for work of any kind after the sleep, than before it. And, in very many cases, the general health, if in any way bad, has been improved, or a complete cure effected, by a course of mesmerism. I do not mean to say, that it never can prove injurious to any one; because I have not sufficient experience to justify me in drawing such a conclusion. But this I can say, that in all the cases I have seen in the hands of others, and in all that have been in my own hands, including in both categories many nervous persons, affected with various maladies, some of them precisely of that kind, such as heart complaints, which would appear the most likely to suffer from any undue excitement, the effect of the mesmeric process in general, and of the sleep in particular, has always been soothing, and never, in any one instance, unpleasant to the patient, besides, as I have said, acting beneficially on the health. I regard it as equally safe and more beneficial to impaired health than ordinary sleep ; that is, as far as my experience goes.

Of course, I do not here speak of exciting exhibitions of striking phenomena ; of causing excessive laughter, or rousing violent passions or emotions. That is a kind of experiment of which I entirely disapprove, as I do likewise of all those in which strong and false impressions, especially of a disagreeable nature, are made in the mind ; as when a man is made to believe he is ruined, or that he is a wild beast. Not that these are always hurtful, but that they may, in very susceptible temperaments, become so. Such experiments, especially in the form of exhibition, are not justifiable, and are at most permissible in private, with a view to the ascertaining a fact, necessary to complete our knowledge of the phenomena, and to enable us usefully to apply it.

Public exhibitions of the phenomena of mesmerism are not, in my humble opinion, good things. I have already given some reasons against them, and I would here add, that to employ these wonderful and beautiful facts merely to excite wonder and produce amusement, is a great abuse of our powers. Mesmerism is not a play-thing ; it is a serious, I would say a sacred, thing, which ought to be studied with reverence, and not degraded to minister to the idle curiosity of those who regard it merely as an exhibition, to be forgotten the hour after it has served to gratify our love of novelty, or to raise a laugh.—In private alone can it be properly studied. No one in a public hall, save, perhaps, one or two close to the subject, can see the phenomena as they ought to be seen, or judge aright of their truth, and of the beautiful evidence of that truth afforded by the countenance and manner of the sleeper. I have seen many persons who came from a public lecture quite unsatisfied, convinced in five minutes in a private room, where they could really see what was done and hear what was said.

I have been led into this digression by considering the effects of sympathy in sometimes giving rise to unpleasant accidents, when mesmerism is tried by the ignorant and inexperienced. I would here repeat what I have formerly stated, that, when such accidents unfortunately occur, the safest plan is to let the patient sleep out his sleep. The mesmerist, if he can recover his own composure, may, in general, easily awake him by upward or reversed passes. But it ought to be an invariable rule, never to try such experiments without the presence of one experienced mesmerist.

7. Returning to the topic of sympathy, in the form of community of emotion, I would further point out the strong effects often produced on the sleeper by the bystanders. Many sleepers

do not require to be placed *en rapport* with others, in order to be very strongly affected by their emotions. Thus sceptical, unreasonable, prejudiced, uncandid persons often excite, by their approach, the most unpleasant, often distressing sensations; while the approach of the rational and kind may easily, in many cases, be traced in the expression of the sleeper. But, above all, the approach of many different persons, all probably much excited, produces, in very susceptible cases, a great confusion of feelings, arising, in part, from sympathy with their mind, in part from the crossing of so many streams of mesmeric influence.

This is one of the chief causes of failure in public exhibitions of the more delicate phenomena, as I have already explained. It ought always to be avoided, as far as possible. Fortunately for exhibitors, there are cases which are not much affected by this source of error and confusion. But some are so quick, that they will detect, by sympathy alone, the presence of one individual, and even discover his state of mind.

8. This brings us to that peculiar form of sympathy which consists in reading the thoughts of others, especially of those *en rapport* with the sleeper. This is quite a distinct phenomenon from feeling their emotions; it extends to tracing all the intellectual processes or images in their minds, and it thus constitutes a kind of clairvoyance, which may be called *sympathetic clairvoyance* or *thought-reading*. It is, as we shall see, a very beautiful and interesting phenomenon. But before describing it, I would remark, that many persons, who are extremely averse to admit the existence of clairvoyance at all, are apt to suppose that they get rid of it, when the facts are forced on their attention so that they can no longer be denied, by ascribing them to thought-reading, as if thought-reading, the power of seeing into another man's soul, (and through his body too,) were at all less wonderful than the power of seeing through a stone wall, or a floor.

To my apprehension, thought-reading is still more wonderful and incomprehensible than that kind of clairvoyance, which takes note of material things at a distance. In the latter case, we can imagine some subtle, rare medium by which impressions may be conveyed to us, as light or sound are. But how do we perceive thoughts, not yet expressed, in the mind of another?

It would appear, then, that those who would explain all clairvoyance by thought-reading, only fall from the frying-pan into the fire. They account for an apparently unaccountable phenomenon, by one still more incomprehensible.

Yet both the phenomena are true; and, as far as we know, both depend, essentially, on the same cause. After the discussion of thought-reading, we shall be prepared to enter on that of true, that is, direct, clairvoyance.

Thought-reading presents itself in every possible variety of form. The sleeper, being placed *en rapport* with any person, can often describe, with the greatest accuracy, the subject that occupies the thoughts of that person. It may be an absent friend, or his own house, or that of another, or his drawing-room, bed-room, study, &c., &c. All these things the sleeper perceives, as they pass through the mind of the experimenter, and describes with great minuteness and accuracy, so as to excite our astonishment.

Or he goes further; he not only perceives the present, but the past thoughts of the person *en rapport* with him; he shares his memory. Thus he will mention facts, no longer so existing, but remembered by the experimenter.

Nay, he goes still further even than this; for he perceives things once known to, and now forgotten by, the experimenter, who very often contradicts the sleeper, and persists in maintaining his own opinion, until, on further inquiry, he not only finds him to be right, but himself is enabled to recall the fact, which had, as we say, escaped his memory.

We all know that we are apt, at times, to forget facts, which subsequently recur to the memory. But here, it would seem that the sleeper so sympathises with our past thoughts, as to read what we ourselves are for the moment blind to. At least, this must be admitted by those who ascribe all clairvoyance to sympathy; but it is difficult, in many cases, to distinguish between sympathetic and direct clairvoyance, if we admit the possibility of the latter.

For example, the sleeper describes a room, at the request of the experimenter. He details the form, size, doors, window, bookcases, tables, carpet, fire-place, sofas, chairs, pianoforte, &c., &c., and, as he goes on, every statement is confirmed by the proprietor, who sees the whole in his mind's eye, as when he left it. But all at once, perhaps, the sleeper speaks of the hangings, or pictures, and says he sees the picture of a dog, a horse, or a man, in such or such a position, with reference to another object. This is denied; but the sleeper is firm. So is the other, and after a long dispute, each retains his opinion. But on returning home, the experimenter finds that he has been mistaken and the sleeper right. He now remembers, that up to a certain period, the picture hung where he had said, but that he himself, or some one else, had changed its position to that described by the sleeper, as he himself formerly knew, but had forgotten. Similar occurrences are very common. But they admit of two explanations.

In the first place, the sleeper, in many cases, declares that he reads the thought of the other party, and certainly does so in some cases, even where he is not aware of doing so, but thinks he is looking directly at the objects described. Now, it may be, and this is the phenomenon properly under consideration, that he reads the past thought of the experimenter, and, in some obscure manner, discovers that it is true, while the present thought is erroneous.

Or it may be, that when asked to describe the room, the sleeper, finding the trace in the questioner's mind, follows it up until he comes into direct communication with the object, by direct clairvoyance. That this often happens, I think cannot be doubted, and we shall see further on, that the experiment may be so made as to prove it; but I am also inclined to believe, that the former explanation applies in some cases, and that, in some instances, the mediate and immediate modes of perception of distant objects are mixed or combined.

One frequent form of thought-reading is that of perceiving the contents of a closed letter, or of a sealed packet, or of a sealed box. Some sleepers can do this readily, if *en rapport* with a person who knows these contents, but not otherwise. But here it must be noted, that, in some cases, the rapport is established without contact, so that it suffices for the sleeper, that one person who knows the contents of the closed objects should be present. And lastly, it appears, that some subjects, who at one time possess the power of direct or immediate clairvoyance, at other times are destitute of this, and have only that of thought-reading.

Of course, when it is done by thought-reading, failure will take place, when no one who knows the object to be described is present, while, on the entrance of such a person, the sleeper succeeds. All these things must be carefully attended to in our experiments, otherwise our results have no value whatever, and only lead to confusion. When a sleeper, of the sympathetic class, fails at one moment and succeeds in the next, after a person at first absent has arrived, the idea of collusion arises in many minds; whereas, if we were well acquainted with all the facts, and with their infinite variety, we should see, in that result, a new proof of the truth of the fact, and the integrity of the sleeper.

Those who meet with cases, in which thought-reading is found to be the true explanation of the phenomena, should reflect, that thought-reading is, in itself, a beautiful and most wonderful fact; and should beware, also, of drawing the conclusion, that, because it applies to one case, or to many cases, of apparent vision, at a distance, it is therefore sufficient to explain all cases, or the only explanation, if it can be called such, that is admissible.

It often happens, as I have explained, that the sleeper or thought-reader is found right as to present facts, where he has been supposed to be wrong. But it also often happens, that he is supposed to be wrong, and is not afterwards shown to have been right. There are, in fact, many sources of error on both sides, which are often difficult of detection.

Thus, the thought-reader may be dwelling on some past event, and be persuaded that it is present. The impressions of past and of present events are of equal vividness in his mind, being, in both cases, internal and indirect impressions. Hence he cannot readily distinguish between them, and may be quite correct, if we could discover the precise time to which his sensations refer. This must be carefully attended to, and our experiments regulated accordingly.

Or he may receive erroneous impressions from suggested ideas. So powerful is his sympathy with other minds, that an idea, directly suggested or indirectly introduced, as, for example, by a leading question, may often produce on his mind an impression as vivid as that caused by the thoughts or memory of his questioner, and all three may become mixed together.

For this reason, all suggestions and leading questions should be carefully avoided, and the sleeper encouraged to tell his own story. Nor is the danger of error equally great in all cases. Many can readily distinguish the different kinds of impression, and steadily reject suggested ideas, even while some of them have a difficulty in distinguishing present from past events. Some, again, can do the latter also, and these, of course, are the best subjects.

It often happens, that in early experiments, the operator is so excited by the novelty and interest of the facts, that he does not calmly examine, and involuntarily suggests, by silent sympathy, his own ideas to his subject. But after a few sittings he becomes more collected; he has only the pure desire to hear what the sleeper says; the sleeper is not disturbed by involuntary suggestions, and his sensations come out more pure and less confused.

Besides, his powers improve, and, by practice a sleeper, at first confused and often mistaken, may become, if properly treated, a very valuable one.

Under the head of direct or immediate clairvoyance, I shall have to return more fully to these considerations.

9. Sympathy extends often to the bodily state of the operator, or other person *en rapport* with the sleeper. The latter will feel and describe every pain or ache felt by the former, and will even in some cases feel, or intuitively perceive, the morbid state of certain parts. He will say, that the other has a headache, or a pain in the side, or difficulty in breathing; he will declare that the brain, or lungs, or liver, or stomach, or heart, &c., &c., are deranged in such or such a manner. And in many cases he will be right. I do not speak here of his *seeing* the state of these organs, which shall be hereafter considered. But the intuitive perception of health and disease, here alluded to, is very often found.

10. This sympathetic intuitive sense of the state of body of another, may be exerted in the absence of the individual, provided a communication be established. For this purpose, a lock of hair, or any object that has been in contact with the person, even a recent specimen of handwriting, will suffice. The sleeper, aided by this, will enter into sympathy with the absent person, as if he were present, and will often be found quite accurate. This I have seen and tested repeatedly.

11. It has been said, that subjects, having this degree of sym-

pathy, often also possess the power of indicating the appropriate remedy. I am always averse to speak decidedly on that which I have not thoroughly examined. I shall not, therefore, say, that such power is impossible, nor that it has never occurred. But I can state, that in all such cases as I have seen, the sleeper uses or recommends the class of remedies, or the plan of treatment, which he has picked up, from having been treated himself, or from having been taught by some practitioner. Thus, one prescribes hydropathy, another homœopathic remedies, a third mesmeric treatment, and so on. There are some cases, however, in which a peculiar sympathy would seem to exist, inasmuch as the sleeper, being shown certain remedies, of the name or nature of which he is ignorant, after feeling them, will often select that which is really most appropriate to the case. But I have not yet had sufficient opportunities of inquiring into this matter, which *may*, possibly, depend on sympathy with the person who exhibits the drugs to the sleeper, and who has probably an opinion as to that which is best adapted for the disease in question.

It will be seen that thought-reading induces many very curious and beautiful phenomena, which have often been called clairvoyance, and which really are so, but are not the only phenomena of clairvoyance. They are sympathetic, indirect, or mediate clairvoyance, the perceptions being derived, not from the objects themselves, but from their images in the mind or thought of others. Some observers, having only met with this form of clairvoyance, have rashly concluded that there is no other; but we shall see, that sympathy and thought-reading cannot explain all the phenomena of clairvoyance, and that we must admit another form of it. We shall find, that it may be *possible* to trace both forms to one common source, and that a kind of sympathy; but in the meantime, there are two distinct classes of phenomena, which must not be confounded together.

12. Sympathy is widely diffused, as a natural, spontaneous occurrence. There are few people who have not experienced something of it, and none who are not, by nature, susceptible to it, although their susceptibility may be masked under ordinary circumstances.

How often does an inexplicable something warn certain persons that an absent and dearly beloved friend or relation is in danger, or dying! This is an effect of sympathy. Every one has heard, in his own circle, of numerous instances of it. I am informed, for example, by a lady nearly related to me, that her mother always had such a warning at the time when any near and dear friend died. This occurred so often as to leave no doubt whatever of the fact. It happened, that this lady more than once made the voyage to and from India, and that during the voyage she, on several occasions, said to her daughter and to others, "I feel certain that such a person is dead." On reaching port, these perceptions were always found to be true.

It often happens, that this spontaneous sympathy goes so far as to produce the vision or appearance of the dying person. But this phenomenon belongs more properly to direct clairvoyance.

The remarkable case of Zschokke, the Swiss romance-writer, who possessed, at times, and quite spontaneously, the power of reading, in the minds of others, the whole of their past history, is a proof of the spontaneous occurrence of thought-reading, in reference to past events.

13. Sympathy often produces a strong attraction between two persons, who see each other for the first time. Neither of them can explain it, but both feel it, and thus, love at first sight is no fancy, but a reality. It arises, in part, from a pleasing correspondence between the mesmeric influences of the parties, and, when this is the case, it is as durable as strong. Nay, it is well known, that there are many persons who frequently quarrel after being long together, yet are quite wretched if separated, and infallibly come together, till a new quarrel again forces them asunder, again to feel miserable apart.

14. Not only do such sympathies exist, but there are antipathies equally strong. Every one must have seen or felt the repulsion exercised on himself or others by certain individuals, which, even in spite of reason, often continues for life. But antipathy is still more strongly exhibited by persons in the mesmeric sleep. In some instances, they cannot bear the approach of persons, who, in their ordinary state, produce no such effect upon them.

Antipathy is frequently very strikingly seen in regard to animals or inanimate objects. Thus, many waking persons cannot endure the presence of a cat, others of a dog, or of a mouse, or of a spider, or of a toad. Many such persons will detect the presence of their *bête d'antipathie* or *bête noire*, when it has been carefully concealed from them, and if it be not removed, will be affected with sickness, fainting, or even convulsions.

Again, many persons are thus painfully and disagreeably affected by inanimate objects, even by such as are pleasing, or indifferent, to people in general. Some cannot endure a rose, others an apple, pear, melon, or strawberry. Some object to sealing-wax, others to resin, some to salt, some to bread, many to less common articles of food, or to objects apparently the most innocent. All these antipathies are so strong, that the mere presence of the objects is sufficient, and cannot be reasoned away, nor overpowered by any volition. The smell of the objects is sometimes considered offensive, but frequently it is not regarded; it is the object itself, from which an influence felt by the victim of antipathy proceeds. Reichenbach has observed a connection between Antipathy and Mesmerism, inasmuch as those who have marked and strange antipathies are usually very sensitive to mesmeric influence. A careful study of the phenomena would soon enable us to discover the laws which regulate them.

We must conclude, therefore, that there resides in bodies, animate and inanimate, a certain force or influence which is felt by certain individuals, who, again, are more or less strongly, and in different ways, affected by it. That this force or influence is the same which, in a peculiar form, gives rise to the mesmeric phenomena, to the mesmeric sleep, and to mesmeric sympathy, as well as antipathy, is the highest degree probable. But I shall not here enter on theoretical questions; I wish merely to point out and establish the facts.

Chapter IV.

DIRECT CLAIRVOYANCE OR LUCID VISION, WITHOUT THE EYES—VISION OF NEAR OBJECTS: THROUGH OPAQUE BODIES: AT A DISTANCE—SYMPATHY AND CLAIRVOYANCE IN REGARD TO ABSENT PERSONS—RETROVISION—INTROVISION.

CLAIRVOYANCE, or what I have defined as direct or immediate clairvoyance, or the direct and immediate perception of absent or distant objects, without the use of the eyes, has been noticed by the earliest authors on mesmerism in modern times. But I shall not here dwell on its history; I shall proceed at once to describe it, in the various forms in which it occurs, as described by those who have seen it, and as I have had opportunities of observing it.

1. The first form in which the power of clairvoyance, or direct perception, called, by the sleeper, *seeing*, but without the use of the external eye, occurs, is usually that of noticing the hand of the operator, while the eyes are firmly closed. This is often observed in the very beginning, when the mesmeric sleep is first induced, and is commonly pointed out by the sleeper, who, without questions being asked, calls out, that he sees the hand, and very frequently describes light as flowing from the tips of the fingers. He sees the hand either when held before his firmly closed eyes, or, in many cases, when held to the side of his head, or above it, or behind it; and it is quite easy to assure ourselves that he does not, and cannot, use his eyes, without the clumsy expedient of blindfolding him, which, to judge by the manipulations sometimes resorted to, would seem to be a very difficult matter, and often causes great annoyance and loss of power to the clairvoyant. The truth is, nature adopts a far more effectual, and indeed a double process of blindfolding; since, as in those cases of spontaneous sleep-walking in which the eyes are open, the pupil is usually found to be fixed, and insensible to light, when, by forcing open the eyelids, we can see it; while, in a large proportion of cases, the pupil is not only fixed and insensible, but is also turned upwards, so that it cannot be seen at all, when the eyelids are forcibly opened. In addition to all this, we can place the hand above or behind the head, positions in which the most sensitive and moveable eye cannot possibly see anything.

The sleeper makes, at first great efforts to look at the hand, and instinctively exhibits the pantomime of intent looking, but with closed eyes. He has the appearance of looking (if his eyes were open) before him, in all cases, even when the hand is above or behind. It is evidently the "mind's eye," the internal vision, which is strained to catch the image of the hand; and it is often described, especially at first, as dim, enveloped in a thick mist, or faintly visible. By degrees the mist clears off, that is, when a deeper or higher stage of sleep is reached, and the hand is seen, without an effort, plain and distinct, in its natural colour. At first, it is often grey, or devoid of colour.

2. When the sleeper has reached a certain stage, he will often notice the objects which happen to be placed behind him, and in such positions that, were his eyes open, he could not see them without turning round; and he will describe them, without ceasing for a moment to look straight forward, or downward, towards his knees, as he very often does, his eyes being firmly closed. If anything be done, however quietly, in any part of the room, he will, if not absorbed in internal contemplation, detect it. This fact, as well as those concerning the hand of the operator, or of others, I have very often seen. They are of daily occurrence.

Now we have here, distinctly, the fact of perception, in the shape of vision, without the use of the external organs of vision. This is the essential fact, and it is as difficult to understand or explain, as any other form of clairvoyance whatever. The question is, by what means is the image of the object conveyed to the internal organ of vision, and to the sensorium? That the brain is vitally concerned in this species of vision, no one can doubt; but, by what kind of influence or emanation is it affected? Common vision, by ordinary light, it cannot be, for the eyes are closed, and the rays from the object do not, in most cases, fall even on the closed eyelids, while, as before stated, the pupil is, or may be, insensible to light.

We must, therefore, admit the existence of some other force or influence, exerted by bodies, and capable of reaching the brain without passing through the eye. When the sleeper finds his vision not clear, or misty, as he calls it, he will, very often, in order to see better an object which is shown to him, apply it to his forehead, to the coronal region, or, in some rare cases, to the occiput, and forthwith perceive it more distinctly. It would appear, then, that this influence, like heat, can traverse the cranium and its membranes, and reach the brain. And it is probable that, when the sleeper perceives objects not in contact with his head, their emanations still fall on, and pass through, his cranium.

Some who admit, as all who examine the matter for themselves must do, that the eyes are not used, have endeavoured to account for this kind of perception by assuming an extreme degree of acuteness of all the other senses; and refer to the case of the blind, who, they tell us, perceive and avoid objects by means of an extreme degree of acuteness of hearing, touch, and smell.

They point also to sleep-walkers, who are supposed to be warned of the presence or proximity of solid bodies by the action of these on the air, perceived by their sense of touch, morbidly acute as it is said to be.

But the latter is not only an assumption, but a complete *petitio principii ;* since it is much more probable, and at all events possible, that sleep-walkers are awake to impressions which are so far new, that they are not usually noticed in the waking state; and the assumed acuteness of touch has never been proved. As to hearing, it is enough to say, that clairvoyants are far more frequently deaf to the loudest sounds, save the voice of the operator, or of those with whom they sympathise, than they are possessed of extreme acuteness of hearing. And in regard to smell, not only have the objects frequently no smell, but their colour, form, transparency, &c., as well as, in most cases, luminous emanations from them, are minutely described; matters which are surely out of the province of the olfactory nerves ; or, at least, if they have come within it, this amounts to a transference of a sense, one of the most striking phenomena of mesmerism.

As to the precise nature of the influence, or force, or emanation, by which this kind of interior visual perception is caused, this is not the place for the discussion of that question. But I may here point to the very frequent, in some cases universal occurrence, of luminous emanations from all objects thus seen by the sleeper, nay, often seen by him while awake (as in the case of the light from the tips of the fingers of the operator, or of other parties present), first, as indicating that an emanation of some kind, whether it be called a force, a motion among the particles of air, or of a supposed ether, or a fluid, as we speak, vaguely, of the electric or mesmeric fluids, really does proceed from bodies in general; and, secondly, as confirming the results of the researches of Baron von Reichenbach, who has proved, independently of all experiments in the mesmeric sleep, the existence of a peculiar influence (force, fluid, or imponderable agent), in all forms of matter, and pervading the universe, the action of which is perceived, in various forms, by a large proportion of mankind, and always very distinctly by spontaneous somnambulists. It is in this direction, that we shall most probably find the explanation so eagerly sought after.

When we have seen, in several independent cases, these first manifestations of the power of clairvoyance, which embrace the essential parts of the phenomenon, and compel us to submit some new mode of access to the interior vision, we are no longer astounded at the further manifestations of the same power. We feel, that he who can see an object behind him, while his eyes are closed, and who sees it best when applied to his head, has some means of perceiving objects, which is either not possessed in the ordinary waking state, or, if possessed, is not attended to, but its impressions are overpowered by the stronger impressions of the ordinary senses. We easily conceive that, in highly susceptible cases, distance may be a matter of no moment; that our new force or influence may, like light, traverse the universe without difficulty, while, like heat, it may be able to penetrate through all objects, even through walls of brick or stone. And such, precisely, is the character of Baron von Reichenbach's Odyle, save that it moves with less velocity than light, and passes through solid bodies much more easily than heat, as we shall explain hereafter.

3. The next observation is, that the clairvoyant can often perceive objects which are wrapped up in paper, or enclosed in boxes or other opaque receptacles. Thus, I have seen objects described, as to form, colour, surface, markings, down to minute flaws and chipped edges, when enclosed in paper, cotton, pasteboard boxes, wooden boxes, boxes of papier mâché, and of metal. I have further known letters minutely described, the address, post marks, seal, and even the contents, read off when the letters were enclosed in thick envelopes or boxes. No fact is better attested than this; Major Buckley, who would seem to possess, in an unusual degree, the power of producing in his subjects this peculiar form of clairvoyance, has brought, I believe, upwards of 140 persons, many of them of high character and education, and 89 of these even in the conscious state, to the point of reading, with almost invariable accuracy, although with occasional mistakes, printed mottoes, enclosed in boxes or in nutshells. He causes some friends, who wish to see and test the fact, to purchase a number of these nuts, in different shops, and to seal them up in a bag, from whence they are taken by the clairvoyant by chance, read, noted, and opened. It is physically impossible for any of the parties concerned to know the contents of any one nut; at the utmost, a clairvoyant, who has had much practice, may, after reading the first few words, guess the remainder of a motto previously deciphered. But out of a certain number, the majority, sometimes all, have been found new, and besides, new clairvoyants constantly succeed in this well-devised experiment.

In regard to this particular form of clairvoyance, I would observe, first; that only a certain proportion of sleepers possess this power, so that any one may very probably not have it. Secondly; that the same clairvoyant may succeed at one time, and fail at another, from causes explained in the first section. Thirdly; that it occurs more frequently in the experience of some mesmerists than in that of others. Major Buckley, for example, is very successful, while there are some mesmerisers who never produce it at all, but who call forth, perhaps, other equally wonderful phenomena. No one, therefore, is entitled to deny the fact, because he does not meet with it in his own cases, or in any given case, or on any one given occasion,

I have already mentioned that there are some clairvoyants who cannot thus read or see, unless some one be present who knows that which is to be read and seen, and that in these cases, it is performed by sympathetic clairvoyance, or thought-reading. I would here add, that it would appear that some clairvoyants can thus read at one time directly, at another only sympathetically, and at a third, perhaps not at all. This is an additional reason, why a failure, or even a few failures, do not entitle us to reject the fact, save as a speciality in the individual cases in which the failure occurs.

I do not think it necessary to say more than I have already said on the subject of attempted impostures in such cases. Nothing is easier than to render all deceit impossible; and I regard Major Buckley's method as entirely satisfactory; but we must never forget that even a genuine and good clairvoyant, if exposed to the close proximity and the involuntary mesmeric influence of a vigorous person, or of several such persons, more vigorous, perhaps, than his mesmerist, penetrated with the conviction that deceit is practised by him, and eagerly bent on detecting it, may easily be deprived of all, or the greater part of his power.

4. Proceeding onwards, we next find our sleeper perceiving objects in the next room, or in that overhead, or in that below. This is a frequent phenomenon, requiring no special preparation, and usually brought to light by the sleeper, of his own accord, remarking what takes place there. It cannot be referred to ordinary sympathy with the operator, for the latter often does not know the room described; and when he does, changes or events in it, taking place at the moment, and unknown to the operator, are noticed. I have myself seen it frequently.

It often happens, that when the operator knows the room described, he and the clairvoyant will dispute about some trifling matter, in which he declares the clairvoyant to be wrong, just as I have explained under the head of sympathy. But it also frequently occurs, that the clairvoyant proves to be right, some change having been made since the questioner last saw it, or even during the experiment. It is needless to quote cases, for this is one of the very commonest phenomena, and there is none which more forcibly impresses us with the fact, that the clairvoyant really *sees* that which he describes. But again it must be borne in mind, that only some clairvoyants exhibit this power, and that those who possess it, do not possess it at all times, nor, when they do, at all times in the same degree.

5. The next step is, that the sleeper can see into another house, as it were, mentally, and describe it in every part, just as he does in the case of the house in which he is. This I have often seen, and have had opportunities of satisfying myself, that his vision, at least in some cases, is not determined by sympathy with the operator's mind.

In the first place, his description is that of a person seeing and examining, for the first time, what he describes. He attends to the minutest details of the objects he looks at, but often omits, till his attention is called to them, objects on which the mind of the operator is dwelling at the time. Secondly, he observes the persons who may be in the room, and what they are doing, the whole of which may be, and often is, unknown to his questioner, but is afterwards ascertained to be correct. Thus, in one case, which I examined, the clairvoyant found my house, which he had never seen nor heard of, and the situation of which he was not told. He first noticed the number of steps at the door (which I could only have guessed at) correctly; he then entered, described the lobby-table and the piece of furniture beyond it, on which coats and hats are hung; but omitted to notice, till I urged him to look, a pillar in the middle of the lobby. He then described the situation, direction, and shape of the stair, but stopped to scrutinise the stair-carpet and "queer brass *fenders*" (carpet rods) which were new to him, but of which I never thought for a moment; then entered the drawing-room, minutely and accurately described the furniture and ornaments, as far as time was given him to do so; noticed various striking peculiarities; saw, at one time, a man sitting in the room (the man-servant at the hour of prayers), and, at a later time, only a lady, in a particular arm-chair, engaged in reading a new book (which turned out quite correct). I give this as a recent specimen of a fact often observed.

In another recent case, a lady, who became a clairvoyante at the first sitting, having been mesmerised by Mr. Lewis, described the same house, and also saw, in another room, a certain lady and gentleman, quite unknown to her, as was the house also; the lady, in a particular dress and head-dress, sitting on a sofa, with other persons, the gentleman standing at a large round table, on which he leaned his hand; on the little finger of the hand was a ring, and he was conversing with a short dark-haired gentleman. All of which was correct at that moment, save that she called the hand on which she saw the ring the right, instead of the left. This is a frequent, but by no means an uniform blunder of clairvoyants. Some of them always put right for left, south for north, east for west, and, in the sleep, adhere immoveably to this error, which is an additional proof that they do not, in these cases, sympathise with the mind of the operator, who of course knows the truth. Thus, in a third clairvoyant, mesmerised by myself, I was at first much puzzled by his apparent blunders of this kind; but on comparing my notes, and asking a few questions, I discovered that the error, in his case, was invariable. He told me, for example, that the houses in Prince's-street, which he saw, looked north; that the Castle was to the north of Prince's-street; that the fire-place in my drawing-room was on the east, and the door on the west side, whereas the former is on the west, the latter on the east, side of the room; and he disdainfully rejected all suggestion to the contrary, declaring that he saw it quite plain, and that I must surely be making a fool of him. I cannot pretend to explain this, which, as I have said, is only an occasional, although I am told by experienced mesmerists, a tolerably frequent phenomenon.

6. The next fact is, that the sleeper, at the request of the operator, and frequently of his own accord, visits distant places and countries, and describes them, as well as the persons in them. This may, as I have already said, be done in some cases, by sympathy, but there are many cases in which ordinary sympathy will not explain it.

Thus, the clairvoyant will often see and describe accurately, as is subsequently ascertained, places, objects, and people, totally unknown to the operator, or to any one present; and he will likewise, in describing such as are known to the operator, notice details and changes which could not be known to him.

The clairvoyant appears, as it were, mentally to go to the place named. He often finds himself, first, in *no place*, but floating, as it were, on air, or in space, and in a very short time exclaims, "Now I am there." The place named is the first, as a general rule, that presents itself to him. But whether it be so, or whether he sees, first, some other place, a certain internal feeling tells him when he is right. If it be a distant town, and no house be specified to him, he will either see a general panoramic view of it, as from a neighbouring hill, or from a height in the air, and describe this as he would a map or bird's-eye view, or he will find himself in some street, place, square, or promenade, which, although not specified to him, is at once recognised from his account of it. He sees and describes the trees, roads, streets, houses, churches, fountains, and walks, and the people moving in them, and his expressions of delight and surprise are unceasing. If *sent* thither, to use his almost invariable phrase, a second or a third time, the sleeper will see the same objects, but remarks the change on the living part of the picture.

For example, Mr. D., a clairvoyant, mesmerised by myself, when in an early and imperfect stage of lucidity, was asked by me to go to Aix-la-Chapelle, he never having left Scotland. He agreed, and after a very short, apparently an aerial voyage, said he was there. He was in a beautiful walk, bordered with trees, saw green turf, and the walk stretched on both sides, till lost, at either end, by a turning, not sharp, but gradual. This was evidently the boulevard. Another time, I specified the Friedrich Wilhelmsplatz, where he saw houses on one side, and at both ends, some much higher than others, the place itself of irregular oblong form, wider at one end than the other, and partly shrouded in a mist, of which he long complained; on the other side a long building, not a house. In the middle, a road, with small trees, having no branches till the stem rose rather higher than a man, and then a number but the top obscured by mist. Another time, he saw the door of Nuellen's Hotel, large enough, he thought, to allow a carriage to enter, but not more, if that; people were going in and out; and a man stood at the door, with a white neckcloth and vest, and no hat; as he thought, a waiter. In the saloon, he saw tables, all brown, no one there. Another time, some tables were white, and people sat at them eating, while others moved about. According to the hours of experiment, he was most likely right both times, although their dinner hour differs so much from ours. One day, I sent him to Cologne. There he noticed, from a bird's-eye position, a large building, seen rather misty, but much higher than the houses. He got into a street near it, and described its long pointed windows, showing with his fingers their form, and its buttresses, which he described, but could not name. In the street he saw people, indistinctly, moving; but he saw, pretty clearly, one "old boy," as he called him, fat and comfortable, standing in his shop-door, and idling. He had no hat, and wore an apron. Mr. D. was much surprised, without any question being asked, at the fact that about half of the men he saw, both in Aix and Cologne, wore beards, and he described different fashions of beards and moustaches. One time, when I sent him to Bonn, he gave a beautiful account of the view from the hills to the west of it, of the town, and the Rhine, stretching out and winding through the plain, with the rising grounds on the other side, such as the Ennertz. But it was remarkable that he stoutly maintained that the hill on which he stood was to the east of the town, the town to the east of the Rhine, between the hill and the river, and the Rhine running towards the south; whereas I knew every one of these directions to be reversed.

The same subject has often spontaneously visited other places, unknown to me, but has given such minute and graphic accounts of the localities, the people, houses, dress, occupations and topography of these places, that I should recognise them at once, were I to see them. I intend to give, further on, some instances of this.

It often happens, that a clairvoyant, who can see and describe very well all that is in the same room, or the next room, or even in the same house, cannot thus travel to a distance, without passing into a new stage. This generally occurs spontaneously, but may sometimes be affected by passes, or by the will of the operator.

The new or travelling stage, in such cases, is marked by peculiar characteristics. Thus, in one very fine case, which I had the opportunity of studying, the clairvoyante, in her first lucid state, could tell all that passed behind her, or in the next room, and could, by contact, perceive, and accurately describe, the state of body of other persons. She could hear, and she very readily answered, every question put to her by any one present, but could not go to a distant place. Yet, as I saw, she would often spontaneously pass from that state or stage into another, in which she was deaf to all sounds, even to the voice of the mesmerist, unless he spoke with his mouth touching the tips of the fingers of her right hand. Any one else might also converse with her in this way, but when first addressed, she invariably started. And now, not only could she go to a distance, and see very plainly what passed, but she was already in some distant place, and much occupied with it. She called this going away, or, when it was done by her mesmerist, being taken away, and when tired, would ask him to bring her back, which he did by some trifling manipulations. She then remembered (in her first state, to which she came back) what she had seen on her travels. I shall have to return to this case hereafter, and some of her visions will also be given further on, in order not to interrupt the current of the description of the phenomena. I shall designate her as E.

7. Allied to the preceding manifestation, is that of seeing any person asked for, by the operator or others. The sleeper, in certain stages, sees the person asked for, not in any locality, but, like himself, floating in air, or in space. He describes the figure, face, hair, complexion, eyes, gesture, dress, &c., with great accuracy, and that when the operator has never seen the individual described. In other stages, the sleeper sees the person in his house, or in the street, or in a road or walk, or at his occupation, whatever that may be. It may happen, that he is seen, either as he is at the moment or as he has been at some former period. The clairvoyant generally expresses a decided liking or dislike for the person seen. If the person be at a great distance, and his actual locality not known to the inquirer, it may often be ascertained by the clairvoyant's description; and it is a very general remark, that clairvoyants have a great difficulty in naming persons or places known to them, and are often very adverse to doing so, while they willingly describe them, and, of course, in regard to unknown persons, places, and objects, we must be confined to description alone.

Experiments of this kind are among the most beautiful and interesting in mesmerism. They may be varied in a thousand ways, and admit, very frequently, of the easiest and most complete verification. Some instances will be given hereafter.

8. Some clairvoyants possess the power of sympathising with the absent or unknown persons seen and described by them, so as to read their thoughts, to know their past actions, at least in part, and even to perceive their intentions. E. often exhibits this power, and I shall have occasion to mention several instances of it in some of the subsequent sections.

9. The powers mentioned in the two preceding sections, are not only exercised when the person asked for is named by the operator or other inquirer, but also, when some object, formerly in contact with the person, such as a ring, a piece of dress, a lock of hair, or even a letter, or piece of writing without the name, all information

being withheld. This power is very highly developed in E., and I have very frequently tested it. A lock of hair, or the handwriting of the person, the more recent the better, seem to answer best, and are, at least, the most convenient means for doing so. I shall give several instances further on, and I shall here describe her mode of proceeding, and the results, in a general form. She crushes the letter or envelope, or hair, in her hands; and if the person do not at once appear, she will often lay it on her head, on the upper part or coronal surface, and this she calls putting it straight before her eyes, which are closed. She then describes the person, and his occupation, which is often, not always, that of the moment of the experiment, sometimes that of the time when the letter was written, or of some former time. This must be attended to, or else E. may be supposed to fail, when, in truth, she succeeds perfectly; and she is one of those who cannot at all times distinguish the past from the present, in the images which present themselves, although she is often quite clear on this point. E. will often trace the history of the person for a considerable period, down to the present moment. This she did, for example, in the case of Mr. Willey, and of another gentleman who had gone from Liverpool to California, and were then in San Francisco, most correctly, as attested by Mr. Willey, publicly, on his return to England. She has often done the same thing in other cases.

E. sympathises to a remarkable extent with the persons thus seen, so as to read their thoughts. But what is most remarkable, is this, that she holds conversations with them, asking questions, and receiving (of course inaudible) answers, as appears from her remarks, which are chiefly addressed, in an easy, familiar tone, to the individuals. She will often scold them, for example, because they have not written to their friends, and will listen to, and admit or reject their excuses. She maintains that she speaks *to* them, and that she can suggest thoughts or even dreams to their minds. Nay, she told one person, seen from his handwriting that at a certain period, when confined to bed from illness, he had had a vision of his wife coming to see him. This she spoke of, as if he had just informed her of it; and it appeared, subsequently, that he had such a vision as she described. She added numerous other details, given in another chapter. She takes likings to some of her new acquaintances thus formed, and dislikings to others, and is much annoyed, and rendered angry, by anything mean or bad. In one case, where she traced a stolen watch, and described the thief, who was not habitually such, she spoke to the thief of that thief's feelings, fears, and intentions, and scolded her severely for her theft and hypocrisy, stating that she was afraid of what she had done, and intended to return the watch, and say that she had found it. "But you took it! you know you did!" said she with angry energy. Before the account of this sitting reached the proprietor of the watch, it had been returned by the person indicated, who said she had found it.

E. has also frequently discovered lost and stolen property, when put in communication with it, or with its proprietor. She has also frequently discovered missing papers of value in the same way. Another clairvoyante lately recovered fifteen bales of cotton, which had been stolen from a ship in New Orleans, and traced it thence in another ship to Havre, where it was found; and the fact is attested by the captain of the former vessel, who thus escaped the loss of a large sum of money. I have myself tested E. with various handwritings and other objects, and found her power to be most remarkable.

10. Another very striking fact is, that some clairvoyants will accurately indicate the time of the places they thus mentally visit. E., who is one of these, does this with great accuracy. It is sometimes said by the clairvoyant to be done by observing the appearance and position of the sun, and is probably, a guess more or less accurate, as in the case of many of us in the ordinary state. But E. declares that she does it by reading the clocks of a place, or the watch of the individual seen by her. And it is found that she will give, for the same place, at any hour of the day, invariably the same difference of time, so that, where both watches are correct, the difference of longitude may be ascertained. Some very well devised and satisfactory experiments were made with her at the request of my honoured friend, Sir W. C. Trevelyan, on this interesting point.

11. Not only do clairvoyants see persons asked for, and such as are not asked for, but whose hair, writing, &c., are put into their hands, but they also see, in the same way, persons who are no longer alive. Mr. D. has described to me, possibly by thought-reading, persons long dead, of whose death, and even of whose existence, he was not aware, when I have asked for them by name. They usually appeared to him as if alive, or, as he expressed it, *like us;* while he saw his own brother, dead five years, *not like us,* but quite different. E. has the same power of seeing dead people, but she also will not speak of them as dead; they are *shelled.* In one or two very curious experiments, she was, spontaneously, or by some obscure inducement, while on her way to visit, mentally, a lady in a distant town, led to enter another house, where she saw a lady, who turned out to be shelled, and rather frightened her at first, till she found that out. She is never frightened, nor are clairvoyants in general, by seeing those who are dead. They rather like to see them. Mr. D. delights in contemplating his dead brother, although he is moved and saddened at the sight. E. also likes to see shelled persons. Both instinctively feel a difference, but never use the word death or dead, and will use the most ingenious circumlocutions to avoid it, till they hit on some peculiar term.

12. Clairvoyants can also see, not only dead persons, but those

of former ages, and the events in which they are concerned. I have heard of some very striking instances of this, in reference to historical personages, which I may afterwards mention, and in which, all that could be verified was found to be correct. One clairvoyant, for example, traced the history of a ring for about 300 years, and was found to be accurate for 70 or 80 years back or more. The shelled lady, seen by E., as mentioned in the preceding paragraph, was in the costume, and the room had the furniture, of 280 or 300 years ago. She saw various events connected with this shelled lady, and when asked what she had died of, started back in surprise, and with a very significant gesture, said, that she died of having her head cut off. Altogether, I have some reason to think, that, by means of very lucid clairvoyance, many obscure points in history might be cleared up, and that, by the discovery of documentary evidence.

This power of seeing the past is truly remarkable, and deeply interesting. It would appear to indicate, that what has once existed, or happened, leaves a trace of some kind, perceptible to the inner vision and soul of man, when no longer obscured or overpowered by the coarser impressions conveyed to the sensorium by the external senses. This idea, which has often been entertained by philosophers and thinkers, we shall again refer to, in its proper place.

13. Another power, exhibited by the clairvoyant, is that of seeing the structure and interior of his own frame. The most eloquent descriptions ever given of the wonders of the human body, never produced half the effect on the mind which is caused by the simple, but graphic words of the clairvoyant, who is perhaps altogether ignorant of anatomy, and yet sees, in all their beauty and marvellous perfection, the muscles, vessels, bones, nerves, glands, brain, lungs, and other viscera, and describes the minutest ramifications of nerves and vessels, with an accuracy surpassing that of the most skilful anatomist. He will trace any vessel or nerve in its most complex distribution ; the whole, to him, is transparent, bathed in delicate light, and full of life and motion. Some, at first, are terrified on seeing these wonders, but soon learn to admire and delight in them. But it is only a certain proportion of clairvoyants who pass into that particular state, and as experiments are most frequently made on the uneducated, or half educated, they are often at a loss for words to describe what they see. I cannot doubt that, when intelligent medical men shall be themselves rendered clairvoyant, some useful information will be derived from the exercise of this power.

It is easy to understand, that when the sleeper sees his frame in this perfect way, he can detect disorder and disorganisation in it. This, indeed, he very readily does, and his diagnosis is often confirmed by that of the physician who attends him when he is suffering from illness.

14. The clairvoyant, in some cases, possesses the same power in reference to the bodies of those *en rapport* with him. He describes their structure, and its derangements; and I have good reason to believe, that in some instances, when the disease is of an obscure nature, his diagnosis has proved, and has been acknowledged to be, correct.

The clairvoyant who possesses this power, can often exercise it at a distance, with the help of the hair or of the handwriting. I have seen it done both ways, and repeatedly, with very great minuteness and accuracy. The observations of the clairvoyant have always corresponded to the opinion of the physician who knew and treated the case, but have often gone further, and in the subsequent opinion of the physician, correctly, in the cases I have studied.

I have already, in treating of sympathy, alluded to this, and stated my opinion, that the treatment recommended by the clairvoyant is almost always a reflection of that which he has himself experienced, or learned from some medical man. It has generally, in each prescribing clairvoyant, one unchanging character. It is homœopathic, or hydropathic, or mesmeric, &c., &c. But some clairvoyants do appear to have an instinctive power of selecting unknown remedies, although I have had no opportunities of seeing this done.

It is much to be regretted, that some persons, not at all qualified for the task, use genuine or possibly spurious clairvoyants, who are made to examine and prescribe, for the object, exclusively, of pecuniary profit to their employers. This ought to be discouraged; but on the other hand, where a well-qualified medical man, of good character, is fortunate enough to meet with a good clairvoyant, he does rightly in availing himself of the power, to assist his diagnosis. I rejoice to know that this is done, in more than one instance, by medical gentlemen of character and standing.

I shall here close this chapter, and, in the next, I shall go on to some other phenomena connected with the subject of direct or true clairvoyance.

Now ready, price 2s. 6d., paper covers.

LIFE & WORKS OF DR. JUSTINUS KERNER,
With Memorials of Mesmer.
By MRS. HOWITT WATTS.

(Being Part II. of Vol. I. of "Pioneers of the Spiritual Reformation.")

CHAPTER I.—BIRTH AND PARENTAGE.
CHAPTER II.—AT COLLEGE.
CHAPTER III.
IN THE BLACK FOREST AND IN THE WELZHEIM FOREST.
CHAPTER IV.—AT WEINSBERG.
CHAPTER V.
THE LAST DAYS OF JUSTINUS KERNER.
CHAPTER VI.—KERNER'S HOME.
CHAPTER VII.
SPECIMENS FROM THE WORKS ON PSYCHOLOGY OF DR. KERNER.—*History of the Two Somnambules. The Seeress of Prevorst.*

No. I.
PERIODICALS EDITED BY DR. KERNER, WITH OTHER LATER WORKS ON PSYCHOLOGY.—*Leaves from Prevorst. Magikon. An Appearance from the Night Realms of Nature: proved legally by a series of witnesses. History of Modern Possession. Magdalene, the Maid of Orlach.*

No. II.
SOME RESEARCHES AFTER MEMORIALS OF MESMER IN THE PLACE OF HIS BIRTH.—Experiences of Mesmer in Hungary, together with translation of the documents discovered. Discovery of Mesmer's Ring, &c. Something about Mesmer's birthplace and childhood.

No. III.
MESMER'S FIRST PRACTICAL CAREER AS A PHYSICIAN TOGETHER WITH STRANGE EXPERIENCES IN HUNGARY.—Projection of Magnetism through music. Impressions produced on patients. The *Baquet*, or magnetic apparatus. Remarks of Kerner on previous narrative.

No. IV.
STATEMENT MADE BY THE FATHER OF THE BLIND GIRL FRAULEIN PARADIS REGARDING WHOM MESMER SUFFERED PERSECUTION.—Fraulein Paradis placed under the care of Mesmer. Her new-born power of vision. Newly awakened sensations, Mesmer's observations on the cases of persons born blind. Further accounts of Fraulein Paradis in Mesmer's words. Misconception of hostile persons. Persecution of Mesmer. An uproar.

No. V.
MESMER'S TWENTY-SEVEN APHORISMS, CONTAINING IN BRIEF THE SUBSTANCE OF HIS DISCOVERIES.—Mesmer re-examines his ideas. Ardent search after Truth. Thinks for three months without words.

No. VI.
MESMER'S DEPARTURE FROM VIENNA. JOURNEY TO MUNICH, AND SOJOURN IN PARIS.—The Academy of Sciences rejects his discovery. Uproar in Paris.

No. VII.
REGARDING MESMER'S FOLLOWERS AND OPPONENTS IN GERMANY. Also regarding the gradual development of Animal Magnetism, and the Publication of Mesmer's Collected Writings. Animal Magnetism tabooed. Co-labourers in Mesmer's Field, Mesmer in seclusion.

No. VIII.
MESMER'S LAST YEARS.—His pet canary. His musical gift of improvisation. His mode of life. His peaceful death. His canary found dead. His funeral and grave.

THE PSYCHOLOGICAL PRESS ASSOCIATION.

The PSYCHOLOGICAL PRESS ASSOCIATION beg respectfully to announce that they now offer for publication by Subscription,

"PRESENT DAY PROBLEMS,"

By JOHN S. FARMER,

AUTHOR OF

"*A New Basis of Belief in Immortality;*" "*How to Investigate Spiritualism;*" "*Hints on Mesmerism, Practical and Theoretical;*" "*Ex Oriente Lux,*" &c.

This work, first announced a year ago, has been unavoidably delayed, owing to the Author's numerous engagements. It is now, however, ready for press, **as soon as a sufficient number of copies have been subscribed for.** The plan of the work has been considerably enlarged; its scope may be gleaned from the following draft synopsis of the sections into which it is divided.

I.—Introductory: Giving brief résumé of ground to be traversed, and present position of Psychological Science, embracing—(a) What is known based on personal observation; (b) What is believed on reasonable grounds; (c) What is speculation only; (d) The Tendency of Material Science towards the Realm of Spirit.
II.—Methods and modes of investigation, with suggestions.
III.—General difficulties experienced by investigators (a) on Scientific grounds, (b) on Religious grounds.
IV.—The Present Day Problems and their general bearing on Modern Thought.
V.—Mesmerism. Its Rise, Progress, and Present Position. Recent Investigations, Comparison and Analyses of Results, &c.
VI.—Thought Transference. VII.—Clairvoyance.
VIII.—Reichenbach's Researches and the Luminosity of the Magnetic Field.
IX.—Apparitions, Hauntings, &c. X.—Spiritual Phenomena.
XI.—Summary.

This book is intended to present to the student of Psychological Science a succinct and bird's-eye view of the subjects enumerated, in each case narrating and discussing the results of recent research, and attempting to shew how each new development of science is bringing us nearer, step by step, to the Unseen Realm of Spirit. It advocates the existence of the Counterparts of Natural Laws in the Spiritual world, and proves by scientific methods that the Spiritual is not the projection upwards of the Natural; but that the Natural is the projection downwards of the Spiritual,—in short, that the Unseen World is the world of Causes, and this the world of Effects. The Author also endeavours to trace out some of the laws which appear to govern the abnormal phenomena with which he is concerned in this volume.

To Subscribers only :—Single Copies, **7s. 6d.**, *or Three Copies for* **£1.**

THE BOOK WILL BE PUBLISHED AT 10s. 6d.

ORDERS AND REMITTANCES TO BE ADDRESSED TO

THE PSYCHOLOGICAL PRESS ASSOCIATION.

Chapter V.

LUCID PREVISION—DURATION OF SLEEP, ETC., PREDICTED—PREDICTION OF CHANGES IN THE HEALTH OR STATE OF THE SEER—PREDICTION OF ACCIDENTS, AND OF EVENTS AFFECTING OTHERS—SPONTANEOUS CLAIRVOYANCE—STRIKING CASE OF IT—SPONTANEOUS RETROVISION AND REVISION—PECULIARITIES OF SPEECH AND OF CONSCIOUSNESS IN MESMERISED PERSONS—TRANSFERENCE OF SENSES AND OF PAIN.

WE now come to a part of the subject which, to some, is the most interesting; nay, which, in some minds, swallows up all the rest, so that when Mesmerism, but especially Clairvoyance, is spoken of, this phenomenon, namely, Clairvoyant Prevision, or the power of predicting future events, is alone understood.

I would, in the first place, remark, that whether Prevision exist or not, we have now what I consider sufficient evidence that clairvoyants do possess the power of seeing contemporary or present events, as well as that of seeing past events. And even if it should turn out, that all alleged cases of prevision are founded on some fallacy, this would not affect these other phenomena, which must rest on their own evidence. I premise this, because I have often heard the alleged impossibility of prevision, or its absence in a particular case, employed as an argument, or rather a proof, against the possibility of clairvoyant vision, introvision, and retrovision. Now, I cannot consent that these should be, in any way, made to depend on the other.

But, on the other hand, these things, vision, introvision, and retrovision, being, as I think, established as facts, though not explained or understood as occurring under any known laws, furnish undoubtedly an additional argument for the possibility of prevision. If, in some way, to us at present unaccountable, present and past events are presented to the mind's eye, may not future events be also thus perceptible? If past occurrences leave a trace behind them, may not "coming events cast a shadow before?" If the latter is inconceivable, the former, had we not seen them, would be equally so, and, whether conceivable or not, the one is as easy, or rather as difficult, if not impossible, to explain as the other.

1. We have already seen, that many somnambulists can predict, to the second, how long they have to sleep, and they will do so, however often they may be asked, at different intervals of the same sleep, and always name the same minute. I do not mean to

say that they are never wrong, for some known interferences may derange the phenomena, while some unknown ones seem also, at times, to operate. But in those who possess the power, failure is the exception. Thus, in a set of observations I made on this point in one case, where this power was early developed, out of thirty-five sittings, the time of waking was precisely predicted in thirty-one, in many of these three or four times, at different intervals; in all, more than once. Of the remaining four, in two, the time was not asked at all; in the other two, disturbance occurred, as I shall explain below.

The form of prediction varies. Some subjects name the hour and minute, usually by the watch of the operator, without, however, its being shown to them; or by some clock, to which they are accustomed to refer. They say, for example, "I shall sleep till eight," or, "I shall wake at thirty-four minutes past nine," &c. Others, among whom is the person referred to in the preceding paragraph, name the number of hours or minutes they have to sleep. That person, for example, would say, "I have fifty-three minutes to sleep" (I never caused him to sleep more than an hour in these trials); and, if I asked again, after twenty-one minutes by my watch, he would say, after a moment's attention, "I have still thirty-two minutes to sleep." When I again inquired, after an interval of fourteen and a half minutes, he said, "I have to sleep eighteen minutes, no! only seventeen and a half," and so on.

It would appear that these two modes of fixing the time of waking, depend on the form of that which presents itself to the interior vision. The former class see the hour on some imagined clock or watch, or possibly look by their lucidity at the house clock, or at the operator's watch. I know that some have said that they saw a watch or timepiece of some kind before them, when asked the question, and that some internal, inexplicable feeling, showed them the point at which the hands would be at the moment of their awaking. The individual above referred to, on the other hand, spontaneously told me of, and most minutely described, a sort of apparatus that enabled him to answer the question. He saw a kind of scale, or measuring rod, which accompanied him, and seemed to pass before his eyes, moving from left to right, slowly, so that while certain marks or divisions, at one time to the left of his eye, when they seemed to advance out of a dark cloud or mist, moved towards his eye, others came into view on that part of the rod formerly shrouded from his sight. On the right, the portion of rod (the ends of which he never saw) which had passed his eye, at a certain distance entered a similar cloud. He was very intelligent, and compared it to an endless revolving tape, of which only a small portion, and that straight, was at any one time visible to him, extending to a variable distance on each side of his eye, and constantly advancing. It had marks or divisions, representing minutes, and at every ten there was either a longer mark, or the mark was broader, so that he could instantly distinguish it, but he said this was not necessary. Indeed, I think he occasionally spoke as if the marks were all alike, and at all events he counted them by an instinctive and instantaneous process. He did not always notice this scale, that is, attend to it, while looking at other objects; but he had a sense of its constant presence, even when not attending to it. When asked how long he had to sleep, he had only to look, and there it was. There were *no numbers* on the scale, but he could always see as far as was required (in his case never more than sixty minutes or degrees of the scale, to the left of his eyes). The degree right in front of his eye he knew to be that of the present minute, and a peculiar feeling, which he spontaneously compared to the consciousness of right and wrong, told him, on looking towards the cloud on the left, the degree which, at waking, would be found opposite his eye. He was also quite sensible that the degrees to the right represented past minutes, as those to the left represented future minutes. He could apparently recognise, after it had passed on towards the right, the degree which was present when he first fixed the time, and found it easiest to count how many degrees it had moved to the right, and deduct this number from that originally fixed, when he was again asked. But he could answer either way. The whole of this description was given, as nearly as possible in the words and order here set down; and I asked no questions until he had finished his account of it, which he had spontaneously offered to give me.

I should observe, that in about one-half of the trials made with him, I first commanded him to sleep 30, 40, 45, 50, 55, or 60 minutes; and in the remainder I allowed him, when first asked, to fix his own time, which he did instantly, by looking at his scale, and which was not copied from my experiments, inasmuch as he varied from 7 or 8 to 12, 14, 15, 20, 22, 34, 35, 40, 41, 43, 47, 50, and 52 minutes. One reason why he never exceeded the hour, any more than myself, was doubtless the fact, that our time, on both sides, was limited, and that he retained this impression in his sleep.

From the variation in the times fixed by himself from those fixed by me, and from his invariably, as he told me, looking first at his scale, before he could specify the time, I conclude, that the length of the sleep was not, in those instances in which he fixed it, the result of suggestion. But, granting that it was so, surely the fact, that he could see the moment fixed upon, when asked, visible as a future point, advancing to the present, and subsequently fading into the past, is truly remarkable. I am not aware that this curious phenomenon has been, in other cases, so minutely inquired into. I had the great advantage of having a subject highly educated, of great natural ability, and able to express his feelings and observations in good and precise language. But I have no doubt, that if other cases were fully investigated, we should arrive at very interesting results in reference to this point.

In the two instances in which the prediction of the time of waking was not precisely fulfilled, in this case, the subject slept about fifteen minutes longer than the allotted period. The first time I observed some symptoms of change in his face and manner, but was not aware, till he awoke, that the time had been exceeded. Next day, I observed him more closely, and saw the same signs of something unusual. He became silent, and, after a time, told me, that he was in no place, but, as it were, in air, and then he said, he was in a different world,—not meaning by this, as I ascertained, more than that he was in a different country, and among strange people. He also saw better. At the time he was in a progressive state, the power of predicting the time of waking having appeared to him very early, before he saw plainly. This change occurred about seven or eight minutes before the time he had named for waking. He continued to describe the new scene for about fifteen minutes, which, as I was busy noting what he said, seemed to be a shorter time. All at once, he became again silent, and then spoke of the things he had been alluding to before the change. I then asked him how long he had to sleep, and he said seven minutes, which proved correct. Here, being in a progressive state, he had spontaneously passed into another stage, the fifteen minutes passed in which did not count in his first state, but had been, as it were, interpolated. It is highly probable, that many instances of inaccuracy in the time fixed as the duration of the sleep, especially in the early stages, may admit of a similar explanation. I think it also very probable that, occasionally, the interference of a third party, especially if he should touch the sleeper, may derange the result, either by confusing his sensations, or by temporarily inducing a different stage of sleep.

2. The next form in which prevision appears, is that in which the somnambulist predicts changes in his own state of health. This, of course, is observed in those affected with diseases, and chiefly in such as suffer from attacks of a spasmodic nature, or from fits of neuralgic pain, migraine, fainting, &c. They will often, quite spontaneously, predict the precise time of one or more attacks; they will describe their intensity, and specify their duration; and they frequently do so long before their occurrence, so that the necessary precautions may be taken.

They further announce, and not unfrequently, especially when under mesmeric treatment, that the first, second, third, or other attack, to take place on a certain day, at an hour and minute named, will be the last. And all these predictions are very frequently fulfilled, quite independently of any regularity, nay, along with the utmost regularity, in the recurrence of the attacks. It is also to be observed, that the patients are very often, perhaps always, in perfect ignorance of their own prediction, having no recollection of what occurs in their sleep, so that it is often difficult, without causing alarm, to keep them under our eyes, and within reach of assistance, when the predicted fit approaches. Such cases have been recorded, in great numbers, but I have no personal experience of them, my experiments having been chiefly made with healthy subjects.

3. The somnambulist often predicts, in his own case, the precise period when lucidity is to appear, or when it is to reach its highest degree. He says, on such a day, I shall be light, or I shall see, or I shall be able to see such a person, or such a place, or to fix the time when I shall become so, or the time when I shall have a fit. He will often prescribe the peculiar form of mesmerisation, which will bring about the result, whether gazing, or laying the hand on the head, or passes downward in front, or behind, or passes round the head, or breathing over the head or forehead, or over the heart or epigastrium, or holding the hands, &c. He tells us how many times he must be mesmerised, and for how long each time. And when he predicts the result, and prescribes a method, it is generally, perhaps always, found that he is right.

In the case above referred to, when I asked the subject, who had begun to show imperfect lucidity, but could predict the duration of his sleep, whether he should become lucid, he said he should; but must first be mesmerised many times. When I inquired how many, he looked intently, and then said that he could not specify the number of times; that he saw the figures (Arabic) so dimly, that he could only say there were two of them, and that these were not only shrouded in a thick mist, of which at this period and long after he bitterly complained, but were also in constant motion, so that he could not fix them. The next time I asked, he saw two figures, no longer obscured by mist, but of a red colour, dazzling his eyes, and whirling round with such rapidity, that he could only distinguish that the first of them was a 6 or a 9, and the second a 6, a 9, or possibly a 0. I had thus, at this period, the agreeable choice of the numbers 60, 66, 69, 90, 96, 99. This referred to his highest lucidity; but I soon found that he improved rapidly in clearness of vision, as far as his power extended, and that the high degree or stage to which he looked forward, was one in which he expected to possess powers at this time not observable to all. I shall hereafter give more details of this case, which is that already referred to as Mr. D. Many instances of the same kind of prevision are on record, and while I write, I am awaiting, in the case of E. formerly mentioned, the fulfilment of a very remarkable prediction concerning trance or extasis, the time of which she has fixed.

I should here add, that Mr. D., in his waking state, has no idea either of what he predicted as to his attaining a high lucidity, nor the slightest recollection of his remarkable scale of time, of which he will be informed, for the first time, by reading these chapters. I believe that E. is equally unaware of having predicted a trance. And this happens generally, if not uniformly, in such cases.

4. The somnambulist will often predict the course of disease and its termination, in those persons with whom he is *en rapport,* or sympathy. This phenomenon has been frequently recorded, but I have not hitherto had an opportunity of personally examining or verifying it. I may allude, however, to one remarkable case of this kind, in which *Adèle,* the somnambulist of M. Cahagnet, predicted his death at the end of six years; and I am informed, on good authority, that he died at the time specified, from natural causes. There exists also a very striking case of a similar prediction, made by a sorceress or divineress in Venice, concerning the death, at different periods, fixed by her, of three gentlemen, friends, who together consulted her. They all died at the time predicted: one from an accident; the other from acute disease; and the last, who died of fever, was so far from labouring under the depressing influence of the prediction, that he was, not long before his death, full of his hopes of recovery, and of his plans for the enjoyment of fêtes about to occur. I would not allude to this case, were it not that I have it on the best authority, and that it must be regarded as perfectly attested. The divineress, it is in the highest degree probable, was in the state, either spontaneous or artificially produced, of waking clairvoyance.

Another case illustrates the power of predicting illness, and shows that clairvoyants do really perceive changes in the state of health of others, before the persons examined do so themselves. A gentleman, highly distinguished by his acquirements, his vast abilities, and his position, happened, when, as far as he knew, in perfect health, to visit a remarkable clairvoyant, who told him he felt a chill in his limbs, and a severe pain in his side. As he felt nothing of the kind, he regarded the statement as a mere blunder. But not many hours after, he was attacked by violent pain in the side indicated, and soon afterwards felt a chill in his limbs; he then recollected that, just before seeing the clairvoyant, he had gone out in a biting wind, with very thin trousers, and had felt much chilled at the time, but had forgotten the fact till he perceived the invasion of illness. It appears to me clear, that the clairvoyant here perceived a change in the part, and its nature, before it had advanced so far as to be sensible to the patient. And thus the acute observation of the present fact amounted to a prediction, unless we suppose that the future state of the patient appeared to the clairvoyant to be present, in which case it was a case of real prevision.

5. Other cases are recorded, and not unfrequently, in which the clairvoyant has predicted an accident which was to befall him, and perhaps to cause a fit of fainting or of epilepsy. This I have not

seen, but it appears to be well authenticated, and is remarkable for the prevision of something external to the subject, and with which he has no direct sympathy, or means of communication, save the unknown or obscure one to which these phenomena are usually referred The accident is often predicted vaguely as to its nature, but precisely as to the time of occurrence, and the effects produced by it. It may be a fright from a rat or mouse, or other cause, or it may be a fall or stumble, &c.

6. In other cases, again, the clairvoyant is said to have accurately predicted events, altogether unconnected with himself. He has told the operator of a letter to be received next day, or several days, or even weeks after, and the name of the writer as well as the contents of his letter. I know of one very remarkable case, in which a clairvoyant, whom I afterwards saw, and found to possess considerable lucidity (although he was, when I examined him, in a different, and lower stage of lucidity, which had spontaneously supervened), predicted to his mesmerist the arrival of a letter from a distance, on a particular day, with other details of a private nature, which were found to be correct. Not having any personal experience of this form of prevision, I shall not dwell further upon it.

The same degree of prevision has been alleged, as occurring in reference, not only to letters, but to events of various kinds; but, for the reason above given, I shall not here enter into further details, as enough has been said to illustrate the nature of the alleged facts. And it appears to me, that as the facts of prevision, in regard to the duration of the sleep, to the course of the somnambulist's illness, or of his fits, and in regard also to the time when certain powers or stages of lucidity are to occur, must be admitted, we must pause before rejecting those phenomena of prevision, in reference to persons and things unconnected with the sleeper, which, often on the very same authority, and that good authority, are likewise recorded.

Those who have the opportunity, which I have not at present, should particularly investigate this part of the subject; and as cases of this kind, though somewhat less frequent than others, are yet said to occur pretty often, there is every reason to hope that the question will, if this be done, be very soon and satisfactorily settled.

Does clairvoyance, using the term in a general sense, occur spontaneously? When we consider, that the state of natural or spontaneous somnambulism is, in all probability, or rather certainty, identical with the artificial mesmeric sleep, we are prepared to expect that, as in some cases of the latter sympathy and clairvoyance occur, so also will these phenomena be sometimes present in the former. And when we further reflect, that sympathy and even clairvoyance may and do occur, when produced artificially, in the ordinary waking or conscious state, that is, in a state the consciousness of which is continuous with our ordinary consciousness, we may look for their occurrence, naturally, in the ordinary state. Indeed, as to sympathy, that is a well-known fact, although there is good reason to believe that the state of reverie or abstraction, as might be expected, is the most favourable to their production.

Now when we inquire, we find, that facts are known, and have been recorded, which prove the occurrence of spontaneous clairvoyance in reference to present or passing events. Every one has heard of such examples, which, however, are generally noticed as freaks of imagination and strange coincidences. If they be coincidences, they are indeed most strange, for the chances against them as such, must have been almost infinitely great.

I have been informed, on what I consider perfectly good authority, of the following case, which is not generally known. A lady was subject, occasionally, not to illness, but to a certain state of mind, possibly connected with a tendency to reverie or abstraction, in which she became aware of what the persons she thought of were doing at the moment, even at very considerable distances. She did not know, at least so far as I am informed, any cause which produced this state. On one occasion, residing at some distance from town, she had, in this way, while fully awake, but late in the evening, a vision of her son's chambers in town, such as I believe she had often had before. She saw the porter of the chambers leave his own room, with a candle in one hand and a knife in the other, and proceed to her son's bedroom, which he entered softly, and, going to the bedside, ascertained that his master was asleep. He then took from the clothes of that gentleman a key or set of keys, went to the other end of the room, opened a trunk or box, took out a pocket-book, and from it a £50 Bank of England note. The thief then returned to the bed, replaced the key, and once more looking, to ascertain that the sleeper still slept, retired to his own room. The lady was naturally much alarmed, and next day drove to town, and saw her son. Without letting him know her vision, she contrived to ascertain that he had placed in his box a bank-note for £50, and begged him to look whether it was still there. The note was gone; the lock uninjured. She now told the story, and after consultation with her son, who agreed with her that he could not, on this evidence, accuse any one of the theft, the note, the number of which was known, was stopped at the bank, and the fact advertised. It was never presented for payment. The porter soon left the chambers, and there the matter would have ended; but some time afterwards he was taken up for some other robbery. And when his lodgings were searched, there was found the very identical note, rolled up hard into a small bulk, and at the bottom of the criminal's purse.

This remarkable vision was told me by a gentleman of great

acquirements and of the highest character, who himself had it from the lady, and told me that I might entirely rely on its authenticity. It was no dream, but a waking vision. And had it been a dream, this would only prove that the clairvoyant state had this time occurred during sleep, which had often occurred in the waking state to the same lady. The fact would not be less striking nor less valuable. I shall return to this case.

I cannot entertain a doubt, that many dreams, which are found afterwards to have been true and exact, depend on the same cause; nay, I think it probable, *à priori*, that the state of spontaneous clairvoyance, like natural somnambulism, occurs much more frequently in the sleeping than in the waking state. We all know how heterogeneous dreams often are; but it is very far from being impossible, or even improbable, that, in certain persons, many of their dreams are the result of true clairvoyance. Baron von Reichenbach has observed, that not only those who walk in their sleep, but those who talk much in their sleep, are uniformly sensitive to the odylic influence; and we know that the highly sensitive, in this sense, are found to be most easily thrown into mesmeric sleep and clairvoyance.

The striking case of Zschokke, the celebrated and amiable Swiss novelist, proves that sympathetic retrovision is also a spontaneous occurrence. He frequently found himself, as he has described it in his works, possessed of a perfect memory of the past life of the person he was speaking to; and, on one occasion, he confounded a sceptic who defied him, by declaring to him certain passages of his past life, known to himself alone, and such as he could not have wished to be known to others. This was done in a large company.

With regard to clairvoyant prevision, I have already spoken of its occurrence in the Venetian sorceress, as possibly induced by artificial means. But I have no information on this point, and it is at least possible, that, in her case, it may have been spontaneous.

But there are numerous recorded instances of spontaneous prevision, and among them, that of M. de Cazotte, which, as it is far from being so well known as it deserves, I shall give further on. In his case, as in that of the lady above-mentioned, the occurrence of the peculiar state was frequent, and it was always observed that he was, when lucid, in a peculiar sleep, not ordinary sleep. There was, very probably, in his case, divided consciousness. Persons of the highest character are, or lately were, yet alive, who heard his remarkable prophecy spoken of and often ridiculed, before the events to which it refers took place.

Moreover, we have the recorded, and in many cases, well-attested instances of second sight, to prove, if not the existence of spontaneous prevision of coming events, at least the firm conviction of its existence, impressed on the popular mind in many countries. I am disposed to think that no such general belief ever prevailed, without a natural truth for its origin or foundation, apart altogether from the precise nature of that truth, and of the true explanation, which is matter for inquiry. I shall return hereafter to the subject of popular predictions.

I have now to mention a few circumstances, connected with the mesmeric sleep, which I have either not hitherto noticed, or only briefly alluded to, because they do not occur so frequently or uniformly as those which I have discussed. It is possible, and, in some instances, probable, that these also, when the matter is carefully investigated, may be found to be present in most or in all cases, and to present themselves in particular stages, so that they may often have been overlooked. Observers should therefore attend to them.

1. The first is, the occurrence of a very great unwillingness to name any person, place, or object. The sleeper will often take a minute or two to describe, rather than use the appropriate name. He seems often to labour under a difficulty in finding the name, but still oftener, his manner indicates that he will not use the name. Thus, while he often, if urged, gives himself a wrong name, very often that of his mesmerist, he will not address his mesmerist, or speak of him, under his name, but will use a circumlocution. He will not speak of being lucid, or clairvoyant, in some cases, but will say he is light, or bright, or warm, or sent or taken away, and so on. And very many clairvoyants will not speak of death, but will use the most laborious circumlocutions rather than do so, whether it be, that the dead do not appear as dead to them, or that the idea of death is repugnant to them ; and I have not had, as yet, sufficient opportunities to enable me to speak confidently on this matter. When they adopt a word or phrase, they usually adhere to it, for the person or object to which they have applied it; and hardly any very lucid subject is without some peculiar form of expression. Thus, E. always speaks of a dead person as *shelled*, and of being mesmerised as being *warmed*; and I have heard of various similar examples. In other respects, the language of clairvoyants is generally improved, and often remarkably distinct and energetic in its character. This point is well worthy of study. Many cases, however, at least in the stages observed, either exhibit no peculiarities, or such as have been overlooked.

2. I have already alluded to the difficulty, in some cases, of ascertaining, at the moment, whether a clairvoyant is describing a past or a passing scene. It would appear that the impressions of both, being alike peculiar, are of so equal a vividness as to be liable to be confounded together. I think it right here again to point out, that this often causes an appearance of doubt and of failure, where all is true, and is found so, when we are enabled to trace the impressions to their source. It often happens, that the

subject, if his attention be called to the point, finds the means of distinguishing past from present things, but this appears not to be always the case.

3. Although the clairvoyant has, naturally, no recollection whatever of his mesmeric sleep, yet, in many cases, he may be made to remember the whole, or part, of what passes, by the will, and at the command of the mesmerist. In many cases, in which this has not been observed, I believe it has not been tried. But I have myself often desired the subject to remember, when awake, any fact or statement, or the whole, of what he has seen and described; and, on waking, he has done so, just as we sometimes remember the whole, or part of a dream, and at other times remember only that we have dreamt, but not the subject of our dream, which yet was very vivid at the time.

The influence of the command of the mesmerist is so great, that we can even affect the feelings by a command given in the sleep. We can cause the sleeper to awake with a pleasant sensation, even when he has seemed like one under a nightmare; and although I should be sorry to do it, I am sure that in some cases I could cause the sleeper, when awake, to feel uneasy and wretched by a command given in the sleep. I have already mentioned the power we have of impressing the sleeper's mind so that, when awake, or long after, he shall, without knowing why, feel compelled to perform some act, although it may have no object, and can only cause him to be laughed at. But I repeat, that we do not find it, as a matter of fact, easy to do this, when the act required appears to the sleeper improper or wrong.

4. I would here mention a fact, which has several times presented itself to me, and which I suspect is much more frequent than is generally supposed. I mean, the occurrences, during the mesmeric sleep, or rather during the existence of a certain consciousness separate from the ordinary one, of a third consciousness separate from both, which spontaneously occurs, and is, as it were, interpolated. If this happen on several successive occasions, I observe that the sleeper does not mix up any of the three consciousnesses, but when his ordinary mesmeric state is present, has no more recollection of what has passed in the more unusual one, than he has of the former in his waking state; nor is there any connection between the waking state and the second mesmeric consciousness, any more than the first. Whichever of the three states he is in, he remembers the previous periods passed in that state alone. In one case, that in which I first noticed it (the fact has, I believe, been often recorded, but I had forgotten it, when it was forced on my attention), the sleeper all at once changed his manner, and spoke of a new vision, which was clearer than any previous one, and was preceded by a comparatively long journey through space or air, to the scene of it. He was much delighted with the distinctness of what he saw, and, although he had never

been within perhaps thousands of miles of the place, he described it, down to the minutest details, exactly as if he were on the spot, pointing out the features of a landscape, and attracted by a thousand minute beauties. Nay, he identified himself with the scene to such an extent, that he spoke of every hill, tree, house, man, and beast, as if they were old friends, and insisted that he was born there, and had lived there a long time when a boy. I thought he had passed, permanently, into a higher stage of lucidity than before, and that thenceforward I should always obtain the same state; but I was much mistaken, for when I asked him how long he had to sleep, the time he had fixed being past, he strenuously denied that he was asleep, and when I told him his eyes were shut, declared they were wide open, even after feeling them at my request. As this had never happened before, I saw he was in a new stage, and I was calculating on its permanence, when I observed a change come over him, and when I again spoke to him, he had returned to the place in which he had been before his first change. He now not only admitted being asleep, but fixed the time of waking, when it appeared, that the new state, which had lasted about fifteen minutes, had been interpolated merely. He had now not the slightest recollection of that new state, or of his erroneous idea of having been born and having spent his early years in that country; and when awake, he had quite forgotten both the mesmeric states.

It may appear to many, that the remarkable vision above alluded to, was simply a vivid dream, and I am not disposed to quarrel with the name. But then, what is a dream? The truth is, we know so little of the matter, that it is quite possible, nay, more than possible, that all, or at least many dreams are clairvoyant visions, the fine and subtle influences by which clairvoyance, if it exist at all, must be produced, being noticed in their impressions on us, better during sleep, natural or mesmeric, because the disturbances caused through our external senses are then shut out. And when we find the dreamer describing not only localities, but the occupations of people in those localities, correctly, as then passing events, it is easy to see how dreams may often be true, and to understand how it is, that we sometimes remember, and sometimes forget them. When our dreams are obviously only imaginary, as concerns ourselves, we may be mixing up various impressions, and sympathising with others, in their actions, or in their thoughts, and the sudden transitions of our dreams are matters of daily experience in mesmerism.

5. It sometimes happens, strange as it may appear, that a person is seen to be, at one and the same time, in two distinct states, and to possess, at once two consciousnesses. Thus, while he is conversing rationally with you, and observing all that goes on, he will suddenly, and without ceasing to follow the current of talk, see before him scenes and objects and persons of which he

has no recollection, but which his mesmerist instantly recognises from the description. Thus E., formerly alluded to, while awake and perfectly collected, will suddenly see persons whom she has formerly seen and described in the sleep, but does not remember, the mesmeric consciousness, which, in the full sleep, would enable her to remember them as seen in a former sleep, being apparently overpowered by the stronger ordinary consciousness. When told that the persons she sees are not present, but only visions, she is puzzled and alarmed.

I am not given to accounting for facts, especially when they have not been fully studied; I am satisfied to observe and to verify them; but I cannot refrain from here saying, that this strange phenomenon of double and simultaneous consciousness may not, after all, be so very strange. It seems to be caused by the circumstance, that the two halves of the brain, which are equal and alike, so that we have two brains, as we have two hands, or two eyes, are not acting together, so that while the one half is in the ordinary waking state (and, like one eye, it suffices for most purposes), the other, probably by some derangement of health, has passed into the mesmeric state, spontaneously. It recalls the visions of a former mesmeric sleep, or rather, it repeats the act of clairvoyance on objects formerly seen.

Every one has felt the strange sensation of knowing beforehand what is about to be said or done. I have often perceived it myself, so that I felt as if I could say, "Now you, Mr. A., are going to say so and so, to which Mr. B. will answer so and so, and then you will reply so and so," &c. But I never *could* actually force myself to say this, till too late, nor, so far as I know, has any one else been able to do so. I am inclined to believe, that this sensation also depends on the unequal action of the two brains; that while we are, with the one, in a reverie, which is a close approach to the mesmeric state, we are, with the other, drinking in what passes, but not attending to it at the time, and that on suddenly coming out of our abstraction, we notice, at a glance, and a lightning glance, all that has passed, which, by some confusion of two consciousnesses, appears future; but it may also be, that with one brain, we have unconsciously exercised a certain degree of clairvoyant prevision. This confusion is always of short duration. I do not pretend that I have here explained it; I merely wish to show, that it *may* be connected with unequal action of the two brains, and that a careful study of similar facts in mesmerism, will probably enable us to explain it.

6. I have not hitherto noticed, save in passing, a phenomenon which occasionally presents itself, but which is not by any means uniformly present in a marked form; I mean, transference of the senses to some special part of the body. I have already stated, that the clairvoyant sees, without his eyes, by means of peculiar emanations, we shall say, which reach his sensorium by another

path. I have also mentioned, that he will often place an object on his head to see it better.

But it sometimes happens, that the power of seeing, not the ordinary sense of sight, but the clairvoyant power, is located in some special part. It has been observed to be located in the pit of the stomach, in the tips of the fingers, in the occiput as well as in the forehead, or on the top of the head, and in one case which I heard of from a scientific gentleman who tested it, in the soles of the feet. The books and journals which treat of mesmerism teem with similar facts; and the head, hand, and epigastrium, seem to be the usually selected parts, probably from the proximity to the brain in the first, the great development of the nerves of touch in the second, and the presence of the great sympathetic plexus of nerves in the third. The fact itself is beyond all doubt, and it is quite unnecessary to accumulate cases. In one form or other, the power of dispensing with the eyes, and yet perceiving colour, &c., quite plainly, is found in every good subject.

The same thing frequently happens with hearing. Thus E., when in her travelling state or stage, is utterly deaf to all sounds, save those which are addressed to her by speaking with the mouth in contact with the tips of her fingers. This fact I have myself verified. I believe she would not hear a pistol fired at her ear, in that state.

Cases are also recorded, in which the sense of taste was transferred to the epigastrium; and, if I am not mistaken, although I cannot find the reference to it, the sense of smell has also been located there. As for touch, being already present in all parts, it cannot of course be transferred.

In all these instances, it is not that the part acquires the peculiar properties of the regular external organ of the sense transferred; but that the nerves of the part serve as conductors of the subtle (probably odylic) influence to the cerebral organ of the internal sense. The fingers do not collect and transmit the rays of light, so that they shall fall on the retina, and the image there formed, according to the laws of optics, be conveyed by the optic nerve to the sensorium; but the nerves of the fingers convey to the sensorium, directly, an influence, which there produces an image of the object. At least, so far as I have been able to trace it, such appears to be the process.

7. Another curious phenomenon, frequently witnessed, is the apparent transference of pain from the subject to the mesmerist. If the subject have a headache, or a toothache, or a neuralgic or rheumatic pain, it often happens that the mesmerist, who does not perhaps know of its existence, finds himself affected with it. And, at the same time, the patient is partially or entirely relieved. I am not satisfied that the two facts are related to each other necessarily, as cause and effect; for a pain is often relieved, when the mesmerist perceives no pain himself; and he may suffer, with-

out sensibly relieving the patient. But it does occur, especially where the object has not been to relieve the pain, that he, as it were, catches it ; and, in many instances, it is also relieved in the patient, not, however, as I conceive, so much because the mesmerist has got it, as because he has mesmerised the patient.

I had once, in my own experience, a striking proof that the mesmerist may suffer. I was, however, in that case the patient. I have for years suffered from weakness, swelling and pain of the limbs, arising from a chronic affection of the lymphatics. Of late the pain has been nearly altogether removed, so as only to appear when I walk or stand too long, and the limbs have become much stronger than before, by the use of the mineral waters of Aix-la-Chapelle. On one occasion, Mr. Lewis tried to act on my right leg, with the view of showing, to an audience, that he could paralyse the muscles by his influence, in those who were susceptible to it. I was but slightly susceptible. this being the first trial ; and Mr. Lewis, before he succeeded, which he did, in paralysing the limb so that I could not move it from the floor, had to exert all his power for a considerable time. He also made passes over the limb, but not with the intention of relieving pain, of which I had not complained, as at the time I had little or none. Next day, and even the day following, Mr. Lewis suffered so much, although otherwise perfectly well, from weakness, swelling, and some little pain, but especially weakness of his leg and ankle, that he was compelled to put on a bandage. He informed me, that similar things had often occurred to him, as he is of an exceedingly susceptible and sensitive temperament ; but that, had he known, or thought of, the state of my limb, he could have prevented the effect on himself. All this, from what I have seen, I believe to be true.

I may here allude to a fact, which I cannot, however, vouch for as having seen it, or tested it, but which I have on what I regard as the unexceptional testimony of a gentleman much experienced in these matters, and of a truly philosophical turn of mind, that a mesmerist may thus be affected with the aches of his patient, even at a great distance, if some object, such as a glove or a handkerchief, be used ; which, being placed on the suffering part, is then sent to him, and will often produce in him the same pain. I may return to this, when treating of the mesmerisation of inanimate objects.

It also occasionally happens, if the mesmerist be suffering from pain, and if he mesmerise a healthy but susceptible person, that the pain is transferred from the former to the latter. This does not, of course, occur frequently, because it is a general rule, not to mesmerise, unless the mesmerist be in good health. But I have seen a headache thus transferred, so that the patient had it for the remainder of the day, while the mesmerist was instantly and entirely relieved. Many such cases have been recorded, and have led to the adoption of the rule above mentioned.

In the next chapter, I shall consider the phenomena producible in the conscious state, that is, without going so far to induce the mesmeric sleep, and which have often been supposed to be radically distinct from those of mesmerism, and have made much noise, both in America and in this country, under various new names, such as Electro-biology, Electro-psychology, &c., &c. I shall also bring before you that peculiar method of inducing the mesmeric sleep, or at least, a mesmeric sleep, which is preferred by Mr. Braid, and by him called Hypnotism.

Chapter VI.

MESMERISM, ELECTRO-BIOLOGY, ELECTRO-PSYCHOLOGY, AND HYPNOTISM, ESSENTIALLY THE SAME—PHENOMENA OF SUGGESTION IN THE CONSCIOUS OR WAKING STATE—DR. DARLING'S METHOD AND ITS EFFECTS—MR. LEWIS'S METHOD AND ITS RESULTS—THE IMPRESSIBLE STATE—CONTROL EXERCISED BY THE OPERATOR—GAZING—MR. BRAID'S HYPNOTISM—THE AUTHOR'S EXPERIENCE—IMPORTANCE OF PERSEVERANCE—THE SUBJECT MUST BE STUDIED.

As I am now about to discuss matters to which various names have been given, I think it best to remind you that I use the term mesmerism, not as theoretically perfect, nor as quite satisfactory, but because it is established, known, and generally used in Europe; and is, theoretically, at least as good as any other that has been at all generally employed. Mesmerism has come, in this country, to have a meaning limited, in some instances, to the sleep, and its phenomena, in others, to the curative agency. Electro-biology signifies vital electricity, or the electric theory of life; electro-psychology signifies the electric theory of thought or of mind. Both proceed on the assumption, now generally regarded as fallacious, that that which we call life, or vitality and thought, mind or soul, are essentially electrical, or depend on electricity as one of their conditions. Now it is impossible to deny, that the view which regards the nerves, whether of motion or sensation, as wires conducting electric currents, is very tempting to the speculative inquirer, and can appeal to some very striking analogies. Mr. Smee's work will explain what I mean. Nay more, Matteucci and Dubois Reymond have proved that electric currents do take place in the body, and especially during muscular action. But then, there are many points which electricity cannot clear up; the analogies are partial only; and the presence of electric currents does not prove these to be the causes of muscular action. They may be its effects. Indeed, since we know that at every moment, in every motion, in every sensation, in every thought, a chemical change is concerned, we might deduce from this, *à priori*, the probability of electric currents in the body. But we are not entitled, on that account, to say that the vital phenomena are electrical, more than we are, perhaps not so much as we are, to say that they are chemical. Chemico-biology is a better founded name than electro-biology; but yet it is not to be recommended. It may be urged, and truly, that mesmerism places mesmerism where the other names place electricity and chemical action. But then this name is established; and the analogies between vital and mesmeric phenomena are quite as strong as in the other cases. Besides, the magnet, in addition to Ferro-mesmerism, does actually possess a force capable of producing the phenomena in question, such as the mesmeric sleep. And if we use the term mesmerism, as distinguished from Ferro-mesmerism, to designate simply this unknown force or influence, we do not thereby attribute life to Ferro-mesmerism as its cause, as the name might possibly seem to indicate, if not explained. In this sense, then, mesmerism is synonymous with the odyle of Baron von Reichenbach; it is generally received; and it embraces the whole series of phenomena. When I come to treat of the cause or of the explanation of these phenomena, I shall endeavour to show, that, if we are to have a new name, that of odyle is unexceptionable.

I have already stated that many remarkable phenomena may be, and daily are, produced on persons in the ordinary conscious, or waking state, by the usual mesmeric processes of gazing, with or without contact, or passes, when not pushed so far as to cause the mesmeric sleep, or when the operator wills that the sleep shall not be produced.

These phenomena are chiefly such as exhibit the control acquired by the mesmerist over his subject's movements, sensations, perceptions, memory, will, &c., &c. I need not here recapitulate them in detail, because they have been already mentioned, each in its place, and because I shall presently have to describe them, as produced in a different way. I shall merely remind you, that every one of them may be produced *in the sleep*, as well as *in the waking state, by the ordinary processes*, and that I have myself often, by that process, produced them in both states.

But they may also be produced, or rather the state favourable to their production may be induced, in a manner somewhat different, without the necessity of the influence which is exerted by the mesmerist, in the usual method, over his patient, and rather by the subject's action on himself; and this is what has been called, in America, electro-biology or electro-psychology.

I have recently had, and most fully availed myself of, the best opportunities of witnessing and studying these phenomena, both as produced in the usual method, by Mr. Lewis, and as produced in the other way, by Dr. Darling. I have also, as above stated, produced them myself in both ways. I cannot allude to these gentlemen without recording my grateful sense of their extreme kindness in enabling me to see and study their operation, both in public, and on very numerous occasions in private. Both of them have been honourably anxious to promote the progress of science, and have spared no trouble to assist me in my investigations. It is quite impossible for any one to be more ready and willing than they are, to explain and communicate all they know; and their extended experience renders their communications exceedingly valuable.

MODES OF OPERATING.

1. The first observation I would make is, that there is not the slightest shade of difference between the phenomena produced by Dr. Darling, who is a singularly neat operator, on persons in the conscious, waking state, by the peculiar process he employs, and those produced by Mr. Lewis, on the same persons, in the same state, by the usual process.

It may happen, and it does happen, that at one time Dr. Darling exhibits certain forms of experiment, which Mr. Lewis has not on a given occasion exhibited, but which he could, if he tried, produce. And I have also seen, in the same cases, things done by Mr. Lewis, which Dr. Darling had not tried in these cases, but could have shown, had he tried, and which he has shown perhaps in other cases. This is the only difference I have been able to detect, and it depends on the circumstance, that the time is limited, and that commonly, some particular form of experiment, which comes out well, is dwelt upon so long, that no time is left to try others. But I have seen no one form of experiment tried by either, which has not, at some time, been also exhibited by the other, in the same ordinary, waking state of consciousness.

2. The process followed by Dr. Darling, which, he informs me, he has never made a secret, is to cause a certain number of persons, willing to try, to gaze for ten or fifteen minutes steadily at a small coin, or double convex mass of zinc with a small centre of copper, placed in the palm of the left hand. The other conditions are, perfect stillness, entire concentration of the mind on the object, and a perfectly passive will, or state of mind. Dr. Darling does not profess to affect those who sit down with an active determination to resist; nor such as come with an eager desire to detect the imposture which they politely attribute to him; nor such as gaze, not on the coin, but on their neighbours, to see how they get on; nor such persons as an ingenious gentleman, who, after descending from the platform, declared, in the hearing of my informant, that he had given the thing a fair trial; for that he had looked steadily at the coin, keeping his eyes shut (!?), and had, besides, occupied himself, the whole time, in solving a problem ! Truly it would have been wonderful indeed had he been affected.

Of the persons tried as above described, a certain proportion, and of those who really fulfil the conditions, and are not agitated or alarmed, a much larger proportion, are found on examination, to be more or less subject to Dr. Darling's will. He ascertains, in the first instance, which of them have been affected, by desiring them, singly, to close the eyes, when he touches the forehead with his finger, makes a few passes over the eyes, or rather presses the eyelids down with a rapid sideward motion, and then tells them that they cannot open their eyes. If, in spite of him, they can do so, he generally takes hold of one hand, and desires them to gaze at him intently for a moment, he also gazing at them, and then

2 F

repeats the trial. If it fail, he tries no further at that time, but goes on to the next case. In me, he succeeded in this on the second trial—I could not open my eyes, Seeing this, he said, "Now you can," and I could instantly do it.

I have seen, especially in private, a considerable proportion found to be thus affected, and I have never seen the experiment tried on even a small number, without at least one being affected to that extent. Those thus discovered to be susceptible are requested to remain and to keep their eyes shut, the others are dismissed.

3. He now takes one of them, and, having repeated the trial with the eyelids, to make sure that the effect continues, tells him to close his mouth; and then, after pressing the lips together with his hands, and making a pass under the jaw, tells him he cannot open it, which in many instances proves to be true, but was not so in my case. He then, perhaps, causes the subject to stretch out his hands and place palm to palm, presses the hands strongly together by a rapid motion of his own, and defies him to separate them. This also he cannot do. Or he makes him place one or both hands on his (the subject's own) head, strikes them rapidly down on the head, and defies him to remove them, which, again, he finds it impossible to do, till, as all in these cases, Dr. Darling says, "Now you can," or "All right."

In the same way, Dr. Darling proceeds to show his power over the sensations of his subject. For example, he deprives one hand, or one arm, of all feeling, and renders it utterly insensible to the most acute pain; or he makes his subject feel a cold pencil-case burning hot, or himself freeze with cold, or taste water as milk, brandy, or any other liquid, as I shall illustrate by a case or two, further on.

In like manner, he controls the will, so that the subject is either compelled to perform a certain act, to fall asleep in a minute, or to whistle, &c., &c., or is rendered unable to perform any act, as to jump on a handkerchief, which if he tries to do, he is sure, according to the volition of Dr. Darling, either to come down straddling over it, or to come down on one or other side of it; or he may hit out straight at Dr. Darling's face, but cannot touch it, &c., &c.

Dr. Darling further controls the memory. He causes the subject to forget his own name, or that of any other individual; or to be unable to name a single letter of the alphabet, &c., &c.

Moreover, he causes him to take any object to be what Dr. Darling says it is, a watch for a snuff-box, a chair for a dog, &c., &c., or to see an object named, where nothing really is, as a book in Dr. Darling's empty hand, or a bird in the room, where none is. The illusion is often absolutely perfect.

Again, he will cause the subject to imagine himself another person, such as Dr. Darling, Father Matthew, Prince Albert, or the Duke of Wellington, and to act the character to the life; to lecture on biology or on temperance, &c., or, if he imagined himself an officer, to drill imaginary troops, and so on *ad infinitum.*

Lastly, Dr. Darling can control, perfectly, the emotions. If the subject be laughing, he causes him first to stop laughing, then to feel serious, sad, and miserable, and to burst out in tears and lamentation; or, if that appear, as it often does, too painful, he will make him feel intensely happy, or laugh incessantly, without being able to assign a cause for his mirth.

Every one of these forms of influencing the subject I have seen, varied in a hundred details. The effect is usually, but not always, instantly produced, and as instantly removed by the operator's simple word. And there is no mystery, no secret, nothing supernatural in it. It is a perfectly natural phenomenon, and any one who tries, may do it, not indeed so well or so successfully, at first, as Dr. Darling, who, as I said before, operates with extreme neatness, and has vast experience to aid him. But, with practice, even this may be attained; and Lord Eglinton, Col. Gore Browne, and other gentlemen, as well as myself, have found no difficulty, when we lighted on a susceptible subject.

Good subjects are easily found, if we only make the trial. Dr. Darling showed his power, on three different occasions, to large parties at my house. On the first, he was entirely and most beautifully successful with a gentleman whom he had never before seen, but whom Col. Gore Browne had just ascertained to be susceptible. On the second, he was equally successful with a gentleman whom he had himself discovered to be susceptible, on the preceding day, at the house of a lady well known in literature. On the third, he was again equally successful with a young gentleman, who, at my request, consented to be tried, and who had never been even tried by any one; and on a fourth gentleman, the secretary of a public institution, whom Dr. Darling had himself discovered, at his lecture the day before, to be susceptible. These cases, all utterly indisputable, and which were seen by many persons of high standing, both in society generally, in literature, in art, and in science, will be given further on. I can testify to the exactness of all the details. I could multiply similar cases without end, but that would be superfluous. Every one who saw the facts exhibited in these cases, was thoroughly satisfied of their genuineness.

4. Now, when we inquire into the cause or the explanation of these facts, the first point to be borne in mind is, that the subjects, in order to be successfully operated on, must not only be susceptible, but must be brought into a certain state. This, in Dr. Darling's process, is done chiefly by themselves, by steadily gazing at the coin, which, according to Dr. Darling, has not, as some imagine, a direct electric or galvanic action, but simply assists in enabling the subject to concentrate his thoughts, and thus to bring

himself into a state of abstraction favourable to the further operations.

Mr. Lewis produces the same state, by gazing for five minutes only, with extreme earnestness and concentration, at the subject, while the latter gazes either at him, or at an object in the same direction. The other conditions are the same as those of Dr. D. He adds certain gestures and passes, all of which are most deeply imbued with that energetic concentration of will, which I have never seen so strongly developed, nor so beautifully exhibited in the natural language, as in Mr. Lewis.

The same state, I say, is produced; that is, the same in this, that the subject is now, if susceptible, under the control of the operator. But I conceive that there may, nay, must be, a difference in the two states, inasmuch as we can hardly suppose the effect to be the same when the powerful and often strongly felt influence of another is added, and is indeed the chief agent in disturbing the equilibrium of influence, as when the subject acts on himself, without external aid.

I have not observed that a greater proportion of persons is affected one way than the other. Sometimes one method appears more successful, at other times the other. But this, so far as I have seen, depends on the more or less strict fulfilment of the conditions. I believe that if ten persons were tried singly tête-à-tête with the operator, and with a sincere wish to fulfil, fairly, the conditions, seven or eight would be affected in either way; and if more time were allowed, I have little doubt that all would be, sooner or later, influenced in some degree.

It must be at once obvious to every person acquainted with physiology, that the peculiar phenomena now under consideration, and which occur in the conscious ordinary waking state, depend on the principle of suggestion. This principle has often been noticed; but it was reserved for modern times, and for the cultivators of mesmerism, to show how the phenomena could easily be produced, in the utmost perfection, in a very large proportion of mankind, and thus to compel the universal admission of the truth.

If we try to produce these effects on any one, by suggestion alone, we shall in all probability fail, unless we happen to light on a singularly susceptible subject. There are some, who, especially after having been once operated on, as above, can be, at any subsequent time, and without preparation, influenced by the same operator. I am not aware that any are, or have been, so influenced for the first time, without the preliminary process; but I consider it very probable, in certain cases. As a general rule, however, the preparation or preliminary process is required, at all events the first time, although it may now and then be afterwards dispensed with.

The cause which produces the state in which suggestion becomes efficient is, I think, identical, in the ordinary process, with the mesmeric influence; for, if pursued a little longer, it will cause the sleep, with its phenomena. I shall have, by and by, to treat of this cause, and shall not here dwell on it.

In Dr. Darling's preliminary process, the chief part of the work is done by the subject himself, through intense gazing at an object. Now we know, that in Mr. Braid's process, even the sleep is produced by the gazing of the subject at an object rather above and a little before his eyes. This gazing, therefore, since it produces the greater effect, naturally also produces the lesser, and thus Dr. Darling's process also is essentially the same as the ordinary one.

But there is this difference, that both in Dr. Darling's and Mr. Braid's method, the operator does not, in producing the state in which suggestion acts, in the former, and the sleep in the latter case, direct his own mesmeric or vital influence on the subject, as is done in the ordinary method. I speak here only of *producing* an impressible state; for Dr. Darling also uses to a certain extent, passes, touch, and gazing, the ordinary means, apparently to heighten the impression at first produced.

Now, if we suppose, hypothetically, the peculiar state in any degree to consist in a disturbance of the natural equilibrium in the distribution of the nervous, vital, mesmeric, or odylic force or influence in the patient's system, it plainly cannot be a matter of difference *how* that equilibrium is disturbed. When the amount of force, natural to the individual, is by his own act otherwise distributed than usually happens, if more be sent to the brain, or to any part of it, and less to the muscles or to the other viscera, or to the skin, the equilibrium is indeed disturbed, but no force is added. Whereas, when a foreign influence is thrown into the brain or any part of that organ, the equilibrium is also disturbed, but the other parts, while having, relatively, less of the influence, have, absolutely, as much as before.

Hence the two states are not identical, although they agree in this, that in both, suggestion has the force of fact. And we shall see, that in the self-produced sleep of Mr. Braid there are very marked differences from that of ordinary mesmerism.

The phenomena of suggestion, then, whether produced by the usual method of mesmerism, or by that of what is called electro-biology, are the same; while there is, in all probability, a considerable difference in the state of the subject, according to the method employed; which difference becomes more marked in the sleep and higher stages.

The subject having been brought into the state above mentioned, is found to be under the control of the operator. He is accessible to, and so deeply influenced by, any suggestion made by the latter, that he finds it impossible to resist or counteract it. He is told that he cannot perform a certain act, and he forthwith loses the power of doing so. The muscles are so far, and no farther, paralysed, as is necessary for the act. If that be, for example, to open the clenched fist, and drop an object which is grasped, he can move the arm up or down, backwards, forwards, and laterally; he can bend or extend it, &c.; but his will no longer acts on the extensors of the fingers, and is powerless on these alone. Or, if he cannot tell his own name, it is not that he cannot speak, nor yet that he cannot tell any other name, but that he has lost the power of recalling that one object of memory. When he drinks water, and tastes it as if it were the strongest brandy, it is that the suggested impression quite overpowers the real natural one. Just as he can, at certain times, recall, by memory, the taste of brandy, or any other taste, although his mouth be empty, and many can do this very vividly, so, when brandy is suggested to him, a similar, but still more vivid, secondary impression is excited, and overpowers the evidence of his palate. As I have already stated, the singular power of suggestion has long been known, as a fact rarely occurring, and presented, no doubt, when some patient had spontaneously fallen into the impressible state. But it is only of late years that we know, that this state may be produced at will, in a few minutes, on a large proportion of mankind; and it is highly probable that all persons, with a little patience and perseverance, may be brought into that state. We have, in this phenomenon, an additional proof of what I formerly stated, namely, that every leading fact in mesmerism has occurred spontaneously, just because these facts depend on natural causes.

5. It would appear, that many persons may be brought into the impressible conscious state, who cannot, or who cannot without great difficulty, be brought into the mesmeric sleep. If this be so, then the process followed by Dr. Darling and others, becomes at once of great practical value. For it is highly probable that the curative agency of mesmerism, even in the sleep, depends on the impressibility, which, as we have seen, is common to the sleep, and to the ordinary state of consciousness. This explains the often recorded fact, that many mesmeric cures are performed without the occurrence of the sleep, or indeed of any very marked or unusual sensations. The patient is, to use a most barbarous expression, which I only do, in order to protest against it, *biologised;* that is, he is only so far mesmerised, as to be thrown into the impressible state.

When we have seen the soundest natural sleep thus produced, in one minute, in persons who were actually, at the beginning of the minute, in convulsions of laughter, and in others who did their utmost to keep themselves awake; when we have seen, in a person so perfectly conscious as to direct and dictate the experiments, one arm rendered, instantaneously, absolutely insensible to the most severe pain; nay, when we have ourselves obtained these results, as I have done, we cannot possibly hesitate as to the practical value of the facts above described.

6. Let us now attend, for a moment, to the hypnotism of Mr. Braid. I have had the pleasure of seeing that gentleman operate, and I most willingly bear testimony to the accuracy of his description, and to the very striking results which he produces.

Mr. Braid causes his patients or subjects to gaze steadily at an object, such as the knob of a pencil case, held a little above the eyes, and in front of the upper part of the forehead. It would seem, that gazing in this strained position, very soon and easily produces the necessary disturbance of equilibrium in the peculiar influence, vital or mesmeric, concerned in the result (to use the merely hypothetical term above employed). In a short, but variable time, a large proportion of the persons tried are not only affected, but put to sleep. Nay, there is, as I have proved on my own person, no plan so effectual in producing sleep when we find ourselves disposed, in spite of our wish to sleep, to remain awake in bed. Some persons have found reading, especially the reading of certain tomes, the contents of which have a ponderous character, to possess a powerful soporific agency; and, in addition to the narcotic influence of the style or matter, it is highly probable that the concentration of mind necessary to penetrate through the *copia verborum* to the meaning, if any, contained in these literary anodynes, tends to produce a sleep, very likely mesmeric. But let these persons try the experiment of placing a small bright object, seen by the reflection of a safe and distant light, in such a position that the eyes are strained a little upwards or backwards, and at such a distance as to give a tendency to squinting, and they will probably never again have recourse to the venerable authors above alluded to. A sweet and refreshing slumber steals over the senses; indeed, the sensation of falling asleep under these circumstances, as I have often experienced, is quite delightful; and the sleep is calm and undisturbed, though often accompanied by dreams of an agreeable kind. Sir David Brewster, who, with more than youthful ardour, never fails to investigate any curious fact connected with the eye, has not only seen Mr. Braid operate, but has also himself often adopted this method of inducing sleep, and compares it to the feeling we have, when, after severe and long-continued bodily exertion, we sit or lie down, and fall asleep, being overcome, in a most agreeable manner, by the solicitations of Morpheus, to which, at such times, we have a positive pleasure in yielding, however inappropriate the scene of our slumbers.

To return to Mr. Braid. His subjects, as I have seen, fall indeed into sleep; but if tried, are found to be in a state of somnambulism or mesmeric sleep. This would probably also be found to be the case, when we put ourselves to sleep as above explained; but we do not have it tried, our wish being to sleep. It is not, in fact, that the mesmeric sleep differs from ordinary sleep, as far as mere sleeping or restoration of the machine is concerned; but that the internal senses are awake, while the external senses, and the bodily frame, are drowned in oblivion. It is totally unnecessary to recapitulate here all the phenomena which are observed in the hypnotic sleep. Up to a certain point, they are the same as those of the ordinary mesmeric sleep, so often already referred to. There is a divided consciousness, closing of some of the senses, and, above all, subjection to the will of the operator, or the impressible state, in which suggestion or command from him are omnipotent. Questions are readily answered, without the sleeper being awakened; and finally, the curative effects are strongly manifested. So great is the power exercised by Mr. Braid, that feeble women, who were, moreover, in a great measure deprived of the use of their limbs, can be made, while in the sleep, readily to walk, and, by frequent repetitions, are often restored to activity. In some cases, aided, I doubt not, by the energetic volition of Mr. Braid, the effects produced in the first sitting continue, more or less, in the waking state. In other cases, slender, delicate men, of small natural power, are made, in the sleep, to exert a muscular force superior to that of strong men; to raise, with one finger, weights which, if awake, they could not move with both hands, &c., &c.

7. But it is remarkable, that Mr. Braid has not produced, in his subject, what are called the higher phenomena, especially clairvoyance, so often met with by those who employ the ordinary mesmeric process.

Two hypotheses may be proposed to explain this fact. The fact is, that Mr. Braid may be one of those persons who cannot produce these phenomena. There are mesmerists, very successful in the treatment of disease, who produce profound sleep, yet never see clairvoyance in their own subjects. I have heard of some mesmerists who are in this predicament, although they have produced other of the higher phenomena. There is a great variety in the influence of different mesmerists, insomuch, as I have already stated, that the impression caused by some, on certain patients, is distressing and intolerable, while that of others is pleasant and soothing. In like manner, it often happens that one operator can produce effects which another cannot succeed in producing, and *vice versâ*.

The second is, whether Mr. Braid's process, being one of self-mesmerisation, or auto-mesmerism, as it has been called, may not produce certain powers, which would perhaps appear in the same cases under the influence of a different operation, even if performed by Mr. Braid himself. A series of comparative trials can alone determine this point, and I hope that such trials may, ere long, be made.

Mr. Braid, not having produced or seen clairvoyance, has gone so far as to deny its existence. I entertain the highest respect for

Mr. Braid; but I cannot help thinking he has here been too hasty in his conclusion. It was long before I myself saw the higher phenomena, and, on more than one occasion, I have mentioned this, when writing on the subject. But I did not feel warranted in rejecting the prodigious mass of evidence, much of it, to all appearance, unexceptionable, of their occurrence. My opportunities were limited, as I never had it in my power at all to study this subject either in London or in Paris; while in Aberdeen and Edinburgh, in which cities I have lived since 1839, no cases of clairvoyance occurred, for some years, within my reach. I was further under the impression, that I possessed little or no mesmeric power, although I had frequently, in persons who had been already mesmerised, and in some very susceptible subjects who had not previously been operated on, produced the sleep, insensibility to pain, and various forms of sympathy. I did not then know, practically, how much depended on patience and perseverance in these matters, and, not at once meeting with the higher phenomena, I hastily concluded that I could not produce them. I am now convinced, that had I persevered for a short time, some of these cases would have exhibited the finest phenomena, and 1 can only regret that I lost, through ignorance, opportunities so valuable.

More recently, however, I have not only been enabled to see and study these phenomena as produced by others, but also to produce them myself; and I can only urge once more, on all inquirers, the importance of time, patience and perseverance in these researches. Without these, few will be fortunate enough to succeed; with them, no one, of average power, need fail (unless by nature incapable, as a few are, of producing certain results), in obtaining evidence of very wonderful facts, and in acquiring a conviction of their interest and value, and of the necessity of thoroughly investigating them.

That Mr. Braid has not met with clairvoyance is to be regretted; but I entertain the confident hope, that even if he should not succeed in producing it himself, he will yet be enabled to see it produced by others. He has, I believe, produced the state of trance, and has, at all events, written a very interesting work on that subject. Now trance or extasis is, in the opinion of all writers, so far as I know, a higher stage of the phenomena than clairvoyance, and many have not met with it in their own experience. But we must not forget that it is possible, that Mr. Braid's method may not induce the state of clairvoyance at all, a point which can only be decided by experiment. For my own part, I find it difficult to conceive that a person, who, when put into the mesmeric sleep in the usual way, becomes clairvoyant, will not present that power, if put to sleep by hypnotism. But an experiment of this kind, on a subject previously mesmerised in the old way, would not be perfectly satisfactory, since many of them can be put into the full sleep by a variety of means, in a few seconds, without any notable exertion of influence on the part of the operator, beyond telling them to go to sleep in a certain way and in a certain time, and thus we should run the risk of producing the accustomed state, when we rather wish to see the new one.

In fact, the facility with which many subjects, after having been brought fully under the influence, may be sent into the deepest sleep, is one of the most striking facts of the whole subject ; and must be borne in mind when we would make such experiments.

It appears, then, from what has been stated, that Electro-biology, Electro-psychology, and Hypnotism, are essentially the same with Mesmerism, although there is probably some difference in the precise characters of the states produced. The former may, indeed, be regarded more properly as parts of the latter, than in any other way. But it is nevertheless probable, that each may have some advantages and disadvantages, peculiar to itself. All of them should be diligently and carefully studied and investigated, with the firm conviction, that, like all other natural truths, they must prove beneficial to mankind ; and the more so, the better they are known. The danger, if danger there be, and I cannot, for my part, conceive the existence of a dangerous *truth*, lies, we may be assured, in ignorance, not in knowledge. " A little knowledge" has been said to be "a dangerous thing ; " but why ? because it is *little*. Make it more, and the danger diminishes ; if we could make it perfect, no danger could possibly exist.

You would do me great injustice, if you supposed that I propounded the facts contained in these chapters, as truths fully ascertained, or duly investigated, so as to be understood. On the contrary, I give them, simply, as facts, so attested and authenticated, that we cannot disregard them, least of all on the ground that they are incredible, or that they cannot be explained. They never can be either understood as facts, or explained in the way and degree in which other natural facts are explained or understood, unless they are thoroughly and scientifically investigated. When this shall have been done,—and it is no easy task, no matter for an idle hour, or for an evening's entertainment, but a serious, important, and, above all, laborious work,—we shall find that, in proportion as we advance in knowledge of these phenomena, they will lose the character of strangeness and supernaturality which to the ignorant they exhibit. They will arrange themselves under natural laws, whether known, or yet to be discovered, as the law of gravitation, which had acted from the creation, yet was discovered only about 200 years ago by Newton. And they will be found, like all other natural facts, even those at first sight most unpromising in this respect, to admit of a multitude of useful applications. Man will benefit by this, as he has done by all other knowledge ; but we cannot expect, in this, the empirical stage of the inquiry, when we are groping in the dark to find the facts, and can as yet discover no order or beauty in them, to be able to appreciate, worthily, the purpose of the Creator in giving to us that power of influencing each other, which is the essence of Mesmerism.

My sole object is to convince those who still entertain doubts on the subject, that certain facts exist, which are worthy of the best and most earnest study we can bestow on them. My own observations have been directed solely to the ascertaining of some of these facts, and I make no pretensions to account for them. My desire is to promote scientific inquiry into the subject, not to present it as already exhausted. And I shall feel amply rewarded for my labour, if one qualified person shall be induced, by what I have said, to devote his energies to the scientific prosecution of the inquiry into Mesmerism.

I have still to allude to the state of Trance or Extasis, one of the most striking, but of the rarer phenomena connected with this subject. I have not hitherto done this, because I have had no opportunity of personally examining the fact, and I wished, in the first instance, to confine myself to such phenomena as I had been enabled for the most part to see, and in many instances to produce. But many phenomena remain, which I have not yet had the good fortune to meet with; and no doubt can be entertained, if we may judge from the results of other scientific investigations, such as those of Astronomy, Geology, Physiology, Optics, and Chemistry, that a rich harvest of new observations will reward those who devote themselves to the cultivation of this boundless field of inquiry, with the means and appliances of scientific training, with the genuine and sincere desire for truth, and, a matter of almost equal importance, the leisure necessary for the full investigation of any branch of so extensive a subject. While I rejoice in having been early trained to habits of scientific study, and while I have endeavoured, to the best of my power, to look at the subject of Mesmerism in the light of scientific research; while, moreover, I may claim to have approached the subject with a due sense of its importance, and as in other sciences, with, I trust, an earnest longing after the truth; yet, occupied as I have ever been with the cultivation and the teaching of a science which I chose from preference, and to which I daily feel more and more attached, I have not, and cannot expect to have, command of the requisite leisure for such an investigation as this. Had it been otherwise, I should long ago have done my utmost to prosecute the inquiry; and it is only by means of observations, made chiefly during the vacations, or when opportunities presented themselves occasionally at other times, that I have been able even to jot down these imperfect sketches. All I can hope to accomplish, is to aid in stirring up to active research those who possess, and in a far higher degree than myself, the necessary qualifications; and perhaps to assist younger observers in their efforts to advance; to enable them to know what to look for, and how to recognise the different phenomena.

It is very gratifying to me to be able to say, that men of the highest ability, and already distinguished in various difficult branches of science, are now turning their attention to this hitherto neglected subject; and I have had the greatest pleasure in placing it in the power of some of these gentlemen to see phenomena which I was sure, once seen, would never cease to interest them. And as Austria has produced a Reichenbach, who, by five years' incessant labour, has shed a flood of light on the phenomena of that influence which we must regard as the cause of vital mesmeric effects, so we may hope, that Scotland, the country which first adopted and taught the doctrines of Newton, when he had no adherents in Oxford and Cambridge, may also produce men who shall raise the veil which conceals the truths of Mesmerism.

When men such as Sir David Brewster, Sir W. C. Trevelyan, Sir W. Hamilton, Dr. Simpson, Professor Forbes, Professor Bennett, and Professor Goodsir,—when men like these, veterans in science, though some of them are young in years, besides many others, have not only seen the facts, more or less extensively, but admit their importance, and have personally investigated into some of them, the time cannot be distant, when the subject of Mesmerism shall assume a truly scientific form. If I can contribute, in any degree, however small, to hasten that most desirable consummation, I shall ever feel grateful that I was led to devote a part of my spare time to the subject.

Chapter VII.

TRANCE, NATURAL AND ACCIDENTAL; MESMERIC—TRANCE PRODUCED AT WILL BY THE SUBJECTS—COL. TOWNSEND—FAKEERS—EXTASIS—EXTATICS NOT ALL IMPOSTORS — LUMINOUS EMANATIONS — EXTASIS OFTEN PREDICTED — M. CAHAGNET'S EXTATICS—VISIONS OF THE SPIRITUAL WORLD.

I NOW proceed to describe, briefly, that state, or rather those states, which are often included under the name of Trance, or Extasis. I shall not attempt to go very fully into them, because I have not had an opportunity, such as I have had in regard to most of the phenomena hitherto described, either of seeing them, as produced by others, or of producing them myself; and, consequently, have not been enabled to compare with my own observations, the accounts given of the phenomena by those who have seen them. But as, in all cases in which I have thus been able to test the published accounts of mesmeric phenomena, I have found a very great degree of accuracy and truthfulness in the accounts given of the facts by the best observers and authors on the subject, I consider it but just to regard their account of the phenomena as accurate, until it shall be shown to be otherwise.

1. The first observation I would make is, that it is necessary to distinguish two states; one in which we have the appearances of death, and which may be compared to the hybernation or torpid winter sleep of some animals; the other, in which the subject enters, apparently, into a higher state or phase of existence, and is deeply interested, nay, often absorbed, by his contemplation of visions, or scenes of beauty and happiness, so perfect, that, in comparison, the world, with all its luxuries, appears utterly worthless and insignificant. These two states have been confounded by some writers, and the term trance applied to both, indiscriminately. And it cannot be denied that there is some analogy between them, and that, in one sense, both may be called trance, because, in both, the subject, as it were, leaves the world, or indeed may be said to leave life, for a time. We shall call the first state Trance, and the other Extasis.

2. Trance, or a torpid, apparently dead state, occurs spontaneously, and has been often recorded. In one remarkable case described in an early volume of the Philosophical Transactions, the patient, a labouring man at Tinsbury, continued in this state, with hardly an interruption, for many weeks. He took a little

food only once or twice during the whole time, and did so mechanically, and, as it were, instinctively, without awaking. In the same way, he occasionally, but only at very long intervals, performed certain bodily functions. In short, he was, for almost the whole period, in a state closely resembling hybernation, in which, as is well-known, the hybernating animal requires, and indeed can take, no food, and the animal heat is kept up, though at a temperature lower than the normal one of the waking state, by the consumption of the fat stored up in its tissues by the waking animal. He was deaf to all sounds, and never spoke, and when at last he woke, he would not believe that he had slept more than usual till he saw the fields, which, when he went to sleep, were green, now ripe and yellow, ready for the sickle. Another similar case is recorded as having occurred lately in France.

Such cases have in all ages been observed, and even recorded; yet only a few years since, when the Tinsbury case was brought forward as a proof that a state of trance, including insensibility to pain, was possible, some of those who seemed resolved not to believe in the possibility of painless surgical operations, performed on persons in the mesmeric sleep, declared that the man must have been an impostor. Yet there was no ground for this assertion, for the case excited much interest at the time, and was examined by several men of science, members of the Royal Society, who could not detect any imposture.

That such an unconscious, torpid state is possible, has, moreover, been proved by the effects of accidents, as in the well-known case of the man, who, falling from the mast-head on the deck of a ship of war in the Mediterranean, fractured his skull, and lay for months in a perfectly unconscious state, eating and performing other necessary functions by a mechanical instinct, until he was trepanned, and the depression of the bone removed, in London, when his memory went back at once to the period of his fall, nor had he the slightest idea that any time had elapsed since the accident.

We may see also, by the occurrence of such cases, either spontaneously from some unobserved affection of the nervous system, or in consequence of accidents, that all the cases of the alleged power to do without food, or to sleep, for a long time, need not be supposed impostures. It is much more probable, that the occurrence of a genuine case, and the curiosity excited by it, as well as the profit derived from its exhibition to wondering crowds, may have led to its simulation, in some instances, with a view to gain.

We may also see, in these facts, the origin of the oriental tales of sleepers, who, falling asleep in some cavern, have found, on awaking, a new world around them. It is evident, that where one man had slept, were it only for a week or a month, and had awoke unconscious of the lapse of time, this would infallibly grow, in the vivid Eastern fancy, to years and centuries of sleep. Here again we see, that the most incredible stories, if they have ever formed an article of general belief, must have had some foundation in natural truth.

3. Now the same state is said to occur, and this, on apparently good testimony, as a result cf mesmeric processes, and, at all events, artificially produced; as, indeed, we might naturally expect, if it occur spontaneously. I need not here enter into details, for the appearances are the same as above described, in the natural trance. I would only say, that they are described, by those who have seen them, as occurring in a high or deep stage of the mesmeric sleep. The mesmeric trance must be carefully dis· tinguished from the ordinary mesmeric sleep, which has usually a short duration, and in which the sleeper is conscious, although not in his ordinary consciousness, and speaks or thinks, or acts accordingly. But in the trance, he is apparently unconscious, and it may last much longer. Some subjects would appear to have a much greater tendency to fall into this trance than others, and this, also, might be anticipated from analogy. Of the fact there can be no rational doubt, but for the reasons above given, I do not enter into a full or minute description of it.

4. It has long been known, but little attended to, that certain persons have had the power of producing, in themselves, at pleasure, this state of trance, or partially suspended animation. Mr. Braid, in an interesting little work, lately published, has collected the most satisfactory evidence on this point. He has quoted the recorded case of Col. Townsend, who often threw himself into this state of apparent death, nay, who did so in the presence of medical and scientific men, who found his pulse and respiration to cease, and were really alarmed, lest they should not return, and real death ensue. But Col. T., after a certain time, gradually awoke to life; the heart began to throb, the lungs to play, and full vitality was soon restored.

Mr. Braid has also given, on the authority of Sir C. Wade, and other gentlemen of the highest character, who had seen the fact in India, several thoroughly attested cases of Fakeers, who made a profession, somewhat religious in its character, of throwing themselves into a perfect trance, and allowing themselves, in this state, to be enclosed in a coffin, and buried for periods of several days or even weeks. These cases leave no room for doubt as to the fact; and the falling asleep, as well as the waking, after certain frictions and bathings had been employed, are described in a manner which is truly natural and convincing.

As we have seen, that the action of a subject on himself, that is, the great concentration of his mind on one object, can produce, not only the impressionable state, necessary for experiments on suggestion, but also mesmeric sleep, as in what are called Electrobiology and Hypnotism, we may reasonably conjecture, that, in such cases as those of Col. Townsend, and of the Fakeers, the trance also is produced by auto-mesmerism, and by a rare degree of the concentrative power. It is possible, that in some instances, as for example in that of Col. T., this may be aided by a peculiar power of checking the heart's action, which hardly any one possesses.

Such experiments, notwithstanding the resuscitation of the Fakeers, cannot be considered devoid of danger; and therefore, while I would gladly embrace any opportunity of studying the phenomenon, should it occur spontaneously in a mesmeric case, I should not think it justifiable to try to produce it. I have seen the heart's action so affected, by a mesmerist, at the earnest desire of the subject, and for the edification of certain sceptical gentlemen, that the pulse rose to 200 beats in a minute, or rather, became so frequent that it could not be counted, while it became, as it rose in frequency, so feeble as hardly to be felt. I have seen sickness and fainting thus produced, and it is possible that the fainting may have been a trance. But I could not bring myself either to try such experiments, or to countenance them again, after seeing the effect produced.

5. With regard to the state of Extasis, it is of comparatively rare occurrence, and this, perhaps, because it is not looked for, or because mesmerists do not seek to produce it. It agrees with trance, in the complete separation from ordinary life, and even, in some cases, in the existence of danger, if pushed too far. M. Cahagnet has stated, that he has seen cases which convinced him, that if prolonged a little longer, death might have ensued. I have never seen this peculiar state, nor tried to produce it; but, as already mentioned, I have seen, in the ordinary mesmeric sleep, a state supervene, obviously different, and apparently higher, in which the subject was intensely happy, and complained bitterly of being brought back to this dull, wretched, every-day life. This is a feature of Extasis, and therefore it is probable, that, if we were to try, such subjects might be rendered ecstatic. Nevertheless, I should hesitate before trying the experiment, since I regard it as not altogether free from danger.

6. Extasis, as well as trance, has often been recorded as a spontaneous occurrence, usually in females, of a highly excitable temperament, and affected with hysteria or other nervous disease, and under the influence of intense religious or devotional excitement. Such a person is called an "Estatica," and many such "Estatice" have been described. These stories have been rejected, offhand, as mere impostures: but we must beware of supposing that all is imposture in them, even where the patients have come into the hands of persons willing to use them, either for purposes of gain, or in order to promote certain religious opinions.

The "Estatice" see visions, of saints or angels, perhaps of heaven, and describe these visions in glowing colours. Now, granting that these are, as is probably often the case, mere dreams, the nature of which has been dictated by the priest, or suggested by reading, there is nothing in this to justify the charge of imposture. A highly susceptible patient, whether in the mesmeric sleep or not, may be made, as in the experiments on suggestion, to see anything that the operator, in this case the priest, suggests or commands. He, on the other hand, is perhaps aware of his power; as some priests are, who, from the study of forgotten books, and by tradition in their monasteries, possess a knowledge of mesmerism, and have practised it in secret, since it is regarded, by the decree of the Church, as allied to magic, if not identical with it. His strong convictions may lead him, innocently, to suggest to the estatica precisely what he wishes to find true; and she sees his patron saint, the holy Virgin, or any other saint, in a form as real to her, as are the fancied objects to Mr. Lewis's or Dr. Darling's subjects. If she is desired to see and describe heaven or hell, she sees that of the priests, or of her books, and, of course, finds in the former all the orthodox, in the latter all heretics, according to his or her own views.

All this, and a great deal more, may occur quite honestly; but it is not wonderful, if an ignorant and superstitious priest, of whatever denomination, engaged, perhaps, in a fierce controversy on some mysterious point, should, now and then, avail himself of his influence over his patient, or of her delusions, in a way that indicates more zeal than honesty.

But the estatica herself is usually sincere. And, to show that in these spontaneous cases we have to do with the same unknown cause as that which produces the more usual phenomena of mesmerism, I may mention, that these patients are generally also somnambulists, or exhibit, spontaneously, the other effects, observed in the lower stages of mesmerism. It is highly probable that their visions are frequently the results of real clairvoyance, which many of them possess, in so far, at least, as regards the visions which refer to natural objects and persons. But such cases, interesting as they undoubtedly are, have not been studied scientifically.

One of the statements most frequently made with regard to such cases, is, not only that the patients see luminous appearances proceeding from objects or from persons present, but that they themselves exhibit a luminous appearance, often described as a halo or glory round the head. Baron von Reichenbach has proved, that luminous phenomena, visible in the dark to the sensitive, who are far from rare, proceed from all objects, more or less brightly, and especially from the *head* and *hands* of human beings. Some are so sensitive in the waking state, as to see these emanations even in daylight; and somnambulists almost always do so, as already mentioned. I am not prepared to reject this as altogether imaginary. Now, if we suppose a greatly excited state of the nervous system to intensify these luminous appearances, they may be observed in the patients, by such among those who approach them as are more or less sensitive, even in daylight, and by many more in the dark. The appearance, once seen, and regarded as miraculous, will not, probably, be very soberly described, and may have been much exaggerated.

Here it must not be forgotten, that it has been recorded, among others, by Sir Henry Marsh, that dying persons often exhibit such a halo; and it may be regarded as an universal belief, that dying persons often acquire the power of seeing what may be called visions, but which are, most probably, the effects of clairvoyance.

Let us not, then, rashly pronounce all ecstatics to be impostors, but rather investigate the phenomena. I would not even rashly decide on the falsehood of the apparent suspension, or rather counteraction, of the law of gravity, asserted to have been observed in such cases, as well as in the celebrated one of the Seeress of Prevorst, and in which the patient is said to have remained for a short period, suspended in the air, without support. There appear to be facts, in artificial mesmerism, which, if confirmed, would warrant us in admitting this to be possible. I allude to the strange attraction exercised on the subject by the mesmeriser, which, in cases where the patient was extremely susceptible, and the mesmeriser very powerful, is said to have occasionally reached the point of raising the former from the floor, in opposition to gravity, and of preventing him from falling, in positions in which he could not otherwise have remained for an instant.

7. With regard to Extasis, as occurring in the course of experiments on mesmerism, it must be regarded in so far as artificial, that it occurs as a consequence of these experiments, and would probably, in most cases, not occur without them. Still, as those in whom they occur, even though healthy, are the most sensitive to all mesmeric or odylic influences, extasis, as we have seen, does occur spontaneously. And even in the course of mesmeric experiments or mesmeric treatment, it usually comes unsought, perhaps always the first time, although the subject may sometimes become able to induce it at will.

When it occurs in a subject, fortunate enough to be in the hands of a judicious mesmeriser, who does not thrust his notions on the subject, but leaves the ecstatic to tell his own story, it certainly offers very remarkable phenomena, whatever interpretation be put upon them.

The patient, or the healthy subject, will often predict with great accuracy, and a long time before the extasis, the day, hour, and minute of its occurrence. E., formerly spoken of, did so two years ago, with regard to one remarkable extasis of hers, and I believe also predicted several less striking since that time. As I write, I am in hourly expectation of hearing the details of a second great or strongly marked extasis, which she has for some time fixed, in the mesmeric sleep, for Jan. 8th ; I shall, if permitted, give some account of it further on. E., in her waking state, is not aware of her own prediction, which of course is not spoken of to her, or indeed to any one, except myself and one or two others, who are much interested in the result.

In the very remarkable work of M. Cahagnet, already alluded to, there is an account ot a most remarkable clairvoyante, who could at pleasure, and with the permission and aid of her mesmeriser, pass into the highest stage of extasis, in which she described herself as ineffably happy, enjoying converse with the whole spiritual world, and herself so entirely detached from this sublunary scene, that she not only had no wish to return to it, but bitterly reproached M. Cahagnet for forcing her back to life. On one occasion, at her urgent request, he allowed her to enjoy that state longer than usual. But he took the precaution of placing another very lucid clairvoyant, a young lad, *en rapport* with her, with strict orders to watch her closely. She seemed at first unconscious, but by degrees her body assumed an alarming aspect, became to appearance dead, that is, was in a torpid trance, like that of the Fakeers, pulseless, cold, and devoid of respiration. The lad, who kept his eye (the internal vision of clairvoyance) on her, at last exclaimed, "She is gone! I see her no longer!" M. Cahagnet then, after much fruitless labour, and not until, as he informs us, he had prayed fervently to be enabled to restore her to life, succeeded in re-establishing warmth and respiration. The girl, on waking, overwhelmed him with reproaches for what he had just done, and could not be pacified till he succeeded in convincing her, she being a young woman of pious character and good feeling, that what she desired amounted to suicide, and was a grievous crime, for which he would be held responsible.

Various other examples of this form of extasis are mentioned in the work of M. Cahagnet, to which I refer the curious reader. M. Cahagnet is since dead, or I should have endeavoured to see his experiments; he was an operative, who seems to have been possessed of excellent abilities, and to have made his observations with great care. His subjects exhibited clairvoyance in its most perfect forms, and most, or all, of them also passed into extasis, in which they described the spiritual world. Indeed, this is the distinguishing feature of extasis; and the extreme form above described, where the body assumes the aspect of death, as in trance, is very rare. In general the ecstatics, as in the cases of M. Cahagnet, and in that of E., describe minutely all they see and feel.

Now there can be no doubt that M. Cahagnet was an enthusiast, in the genuine and good sense of the term, in reference to this subject, and no wonder, when such facts were presented to him. But I can see no reason, in his book, to suppose that his enthusiasm in any way affected his intellect. Many are ready to imagine without inquiry, that the visions of his ecstatics, concerning the spiritual world, are only dreams, the character of which is determined by his views on the subject, and hence the remarkable agreement which in general exists between the statements of his different ecstatics Such was the view which first offered itself to my own mind. But I am very averse to deciding such questions without inquiry, and on reading further and more attentively, I found that this view would not apply to all the facts recorded. Indeed, if on some points the ecstatics expressed views and opinions in accordance with his, in many others they not only differed from him, but pertinaciously held their own opinions, and the result finally was, that he adopted, and says he was compelled to adopt, notions in regard to the spiritual world, entirely opposed to his former views, which seem to have been materialistic.

I do not propose, in such a work as this, the object of which is chiefly to record my own observations, to enter fully into such matters as these, of which I have no experience. But I have thought it right to mention the subject, and to refer to the work of M. Cahagnet, which will be found very interesting by all those who wish to penetrate, as far as is permitted, into the mysteries of the world of spirits. If there be a spiritual world at all, and such is the almost universal belief of mankind, it is at least possible, that the revelations of ecstatics *may* be more or less true, just as they may be supposed to be mere dreams. I confess that what most strongly affects my mind, and deters me from assuming the latter hypothesis, until I shall have been enabled to study the phenomena of Extasis, is the singular harmony between the visions of different ecstatics, between those, for example of E., and those of M. Cahagnet's subjects. I may add, that Dr. Haddock, in whose house E. lives, and who has described the phenomena of her case, is a gentleman of cool and reflecting mind, a good observer, and besides, as I know from experience, a most judicious and prudent mesmerist. And I can only state, having, by his kindness, been permitted to study his interesting clairvoyante, although not in the extatic stage, at my leisure, and in his absence, as well as when he was present, that she is a genuinely honest, truthful, and intelligent girl, although she has not had the advantages of even a common education. I have, therefore, no reason to doubt the facts as described by Dr. Haddock, any more than those given by M. Cahagnet and others, whatever may hereafter prove to be their true nature.

If the visions of mesmeric ecstatics be nothing but dreams, then, as described by the observers of such cases, they must be regarded as dreams of a very remarkable and peculiar character, and they are found, in different cases, very closely to resemble each other in their general or essential peculiarities. The ecstatics find themselves (and this is said by all, whether educated or not, and, so far as I can see, not only without prompting on the part of the mesmerist, but very often to his great surprise, and sometimes contrary to his belief) in communication with the spiritual world. They hold long conversations with spirits, to whom they often give names, and who, in many cases, according to their account, are the spirits of departed friends or relations. The remarks and answers of these visionary beings are reported by the ecstatics. Some of them affirm that every man has an attendant good spirit, perhaps also an evil one of inferior power. Some can summon, either of themselves or with the aid of their attendant spirit, the spirit or vision of any dead relation or friend, and even of persons also dead, whom neither they nor the mesmerist have ever seen, whom perhaps no one present has seen; and the minute descriptions given in all these cases, of the person seen or summoned, is afterwards found to be correct. Many other details, some of them still more astounding, are given, but, for the reason already given, I confine myself here, to a brief and general indication of the strange phenomena of Extasis, which, be it remembered, I have not myself had an opportunity of observing.

Now, certainly such visions as these, whatever be their real nature, are not ordinary dreams. It is idle to reject them as altogether imaginary, and illogical to do so without inquiry. And I repeat, that all those who believe in the existence of a spiritual world, must feel that they may possibly contain revelations of it.

The belief in the existence of the world of spirits is as old as mankind; and the belief that men are, in certain circumstances, capable of entering into communication with it, is not much less venerable. It has been the favourite dream of philosophers, poets, and divines, in all ages, and therefore, without venturing to pronounce dogmatically, I would say to all, observe, study, reflect, and examine, before coming to a decision on this mysterious subject. It is easy to say that Swedenborg was a mad enthusiast; but it is not the less certain that he was a man of prodigious ability and learning, thoroughly familiar with the science of his day; and the most striking circumstance, in my opinion, connected with mesmeric ecstatics is, that they agree in very many points with Swedenborg; and that this agreement is found to occur precisely in regard to those things which we are accustomed to regard in him, as the products of an insane enthusiasm. It is observed, moreover, in ignorant persons, who have never even heard of the name and opinions of the Swedish philosopher.

I do not here refer to the case of the Poughkeepsie Seer, Andrew Jackson Davis. I think there can be no doubt, that his revelations, which present an appalling hotch-poch of all possible metaphysical systems, are essentially the genuine results of a most remarkable degree of mesmeric sympathy with all who approach him, which leads him to retail, as they are imaged in his own mind, the heterogeneous opinions and ideas of such as act upon him, unknown to themselves, and have read and thought upon metaphysical subjects.

There are, I believe, cases of extasis, not disturbed by this kind of sympathy, and such cases are well worthy of and will richly reward the most diligent and attentive investigation.

In my next chapter, I shall go on to those facts connected with the excitement and manifestation of individual mental faculties, which have been called by some, with not a happy selection of terms, phreno-mesmerism. I have already very briefly alluded to the fact that the mental faculties may be roused into action, in a person who is in the mesmeric sleep, in a variety of ways. I now propose to examine this matter somewhat more fully.

Chapter VIII.

PHRENO-MESMERISM—PROGRESS OF PHRENOLOGY—EFFECTS OF TOUCHING THE HEAD IN THE SLEEP—VARIETY IN THE PHENOMENA—SUGGESTION—SYMPATHY—THERE ARE CASES IN WHICH THESE ACT, AND OTHERS IN WHICH THEY DO NOT ACT—PHENOMENA DESCRIBED—THE LOWER ANIMALS SUSCEPTIBLE OF MESMERISM — FASCINATION AMONG ANIMALS — INSTINCT — SYMPATHY OF ANIMALS—SNAIL TELEGRAPH FOUNDED ON IT.

1. WHEN certain subjects are thrown into the mesmeric sleep, it is found, on trial, that by touching certain parts of the head, marked, and sometimes violent, manifestations of certain mental faculties occur. It is further observed, that these manifestations correspond, in their nature, to the part of the head touched, on the principles of Phrenology. This has been proclaimed, by some, as a convincing proof of the truth of Phrenology, and by others, either rejected, *because* it seemed to favour that science; or, it has been ascribed to other causes, entirely independent of the cerebral organs of the Phrenologist. Both parties appear to me to have been hasty in their conclusions. For the phenomena may, and do occur, occasionally, in such a way as not necessarily to prove the truth of the organology of Gall, while, on the other hand, cases are met with, in which we cannot, I think, explain the facts, except on the hypothesis of Gall, and every mental faculty, whether it be a propensity, a sentiment, or an intellectual aptitude, is dependent for its manifestation in this life on a certain portion of the brain.

2. This is not the place for a discussion of the truths of Phrenology, even if these were less generally adopted, and felt to be true, than they are. In spite of the storm of abuse which was showered on Gall and Spurzheim, as quacks and impostors, their anatomy of the brain is now universally admitted to be the best, and their mode of dissecting it the only good one; and their view of the constitution of the mental faculties, considered by itself, as well as their classification, are regarded as eminently practical, and at least equal to those of any other metaphysician. Men see that those who have thus distinguished themselves as investigators in Anatomy, and as thinkers in Mental Philosophy, are not likely to have been quacks, and still less impostors. And even their organology, after so short an interval as barely half a century, is no longer regarded as absurd, but, on the contrary, its ideas pervade our conversation and literature, and even its language is

employed, as singularly precise and convenient, by writers of every class. Phrenology has passed through the first stage of violent opposition, in which it was decried as a mischievous novelty; it has even passed through the second, in which, as invariably happens with new truths, it was declared, by those who had at first denounced it as new, to be old and well known; and it will soon be, generally, not only admitted, but taught as true science. The late Lord Jeffrey, in an article on the subject in the *Edinburgh Review*, boldly scouted it in every aspect, and fulminated the memorable dictum, that "there is not the smallest reason to suppose that the brain is concerned in any mental process, save only the perceptions derived from the external senses!" a statement which, even at that time, must have appeared ludicrously absurd, and a proof of the grossest ignorance, to all acquainted with either physiology or mental philosophy; not to speak of those who could see what daily passed around them, and who saw the effects of a blow on the head, or an apoplexy, on all the mental powers without exception; or who knew that idiots have commonly either singularly small, or else misshapen and diseased heads, and that there is no instance on record, of any one whose head measured less than 14 inches in its greatest circumference, who possessed human faculties and intelligence at all, while yet many such unfortunates have lived, or rather vegetated, possessed of a few animal instincts and of the external senses. Nay, I may go much further, and say that few, if any, are now to be found, who doubt that a well-formed head, that is, brain, is desirable; or who will deny, that a full, high broad forehead indicates superior intellect; that a large base and posterior part of the head indicates powerful animal propensities, and that a full development of the coronal region is somehow connected with refined, virtuous, and religious feelings. These three great regions are admitted by most people; but they hesitate at what they call the small details of the organology; they imagine Gall to have invented these, and to have, of his own fancy, subdivided the three regions, which they suppose to have been generally admitted before his time, into those smaller mappings, which to some are so great a stumbling-block. Such is a very common notion of Phrenology.

But that notion is not only unfounded, it is the direct reverse of the truth. The three great regions of the brain were not admitted before Gall's time. Some authors had done, what he did not; they had arbitrarily mapped out the head, not from observation, but from fancy. Some of these cranial maps are extant, and one exhibits the intellect located in the occiput. And even Gall himself did not at first recognise or admit the intellectual, animal and moral regions. It was only after he had, by observation and comparison, pursued with unwearied diligence for years, detected the connection of the powers of certain isolated feelings, propensities, and talents, and of their absence or deficiency, with the more or less perfect development of certain parts of the brain, indicated externally by the corresponding parts of the head, that he was at last struck with the fact, that the organs of certain allied feelings, &c., were allied in position also; that the groups of the domestic feelings, of the observing powers, of the reflecting faculties, and of the higher sentiments, were represented by groups of organs associated together in place. Hence the idea, which belongs to Gall, of the three great regions. His march was precisely the reverse of what is often attributed to him. It was strictly inductive; and the details of mapping, which are often denied and rejected as purely arbitrary, were actually the first observed, and furnished the evidence for the existence of those great regions which are admitted by those who reject the details, and are, indeed, instinctively felt to be true, when pointed out.

Such being the case, it is not unreasonable to anticipate the universal reception of Phrenology in the course of another half century or another generation. I may safely hazard the prediction, that the *Edinburgh Review* will not venture again to issue the anti-physiological dictum above referred,to. Should it do so, the merest tyro among its readers would laugh it to scorn. The lamented author, who fell into so gross an error (which, by the way, was not, so far as I know, reprehended or even noticed, by any physiologist, except the phrenologists), when he ventured into regions previously unexplored by him, announced, in the same article, his intention, if at the end of ten years from that date, phrenology should still survive, to annihilate it once more. At the end of the ten years, with the inconvenient tenacity of life exhibited by all truths, physiological or metaphysical, or of any other kind, phrenology was more rampant than ever, and as we have seen, it continues to exist.

" The Thane of Cawdor lives; a prosp'rous gentleman;"

and phrenology still awaits the *coup de grace*, for that promise was never fulfilled.

3. As I admit the fundamental doctrines of phrenology, without imagining so new a science to be perfect or complete, I am quite prepared to find it confirmed in the mesmeric sleep, as it is in the waking state. Let us now see what the facts are.

In some mesmeric sleepers, if we touch with the finger any given part of the head, such, for example, as the organ of Tune, or of Self-Esteem, without a word of suggestion, we instantly obtain a corresponding manifestation. It is really, in many cases, like touching the keys of an organ (in the other sense of the word) when the bellows are full of wind, and the sound instantly follows. If Tune be the organ touched, the subject forthwith breaks into song. If it be Self-Esteem, he throws back his head, struts with immense dignity, and declares himself superior to the rest of mankind. Touch the organ of Love of Children, and he dandles an imaginary babe, with most paternal affection. Touch Benevolence, the expression changes to that of compassion; his hand is thrust into his pocket, and held forth with all his store. Touch Acquisitiveness, the griping miser instantly appears, and with appropriate look and speech, the money is restored to its original receptacle; it is well if the nearest object, however bulky, be not also "boned," to use a slang but expressive phrase. If Caution be the stop touched, the music is the most distressing, nay, often appalling pantomime of fear, or of misery. But if Hope be played on, the clouds vanish, and joyous sunshine gilds every feature. Such are a few of the effects produced. I speak of but a small part of what I have often seen, and often produced. It is unnecessary to say that I have done so in cases where no deception was or could be practised. The question is rather, How are these effects produced?

4. There are two theories; the first supposes them to be the mere results of the operator's will, or of sympathy with him; the second believes them to be the genuine results of the effect of the touch in exciting the subjacent cerebral organs.

I believe that both are true; that is, that some cases may be explained on the first theory, but that other cases exist, in which it does not suffice; the second alone can be adopted in these.

There can, I think, be no doubt, that in some cases, the will of the operator is almost omnipotent. Even in the conscious impressible state of Mr. Lewis, or in that of Dr. Darling, precisely similar effects are produced by suggestion. The subject, if told that he is Father Mathew, delivers a lecture on temperance; if desired to sing or spout, he does so; if persuaded that he is ruined, he exhibits in perfection the pantomime of despair, and so on, as I have often seen. And as I have also seen many effects, of all kinds, produced by the silent will, I cannot doubt that it is, in some cases, capable of replacing suggestion. Besides, to a person acquainted with phrenology, the suggestion may be conveyed by touching a particular part of the head. I confess that I think that this last explanation, although certainly possible, can apply, at most, to a very few persons; for on trial, I seldom find any one, not a professed and practical phrenologist, who can point out the position of the phrenological organs, with any degree of accuracy, even among those who admit phrenology to be true.

Further, the manifestations can be often called out not only by suggestion, but by touching other parts of the body than the head, and that, in cases where perhaps touching the head, as often happens, has no effect. Some have endeavoured to show, that touching a particular part of the arm, leg, or trunk, is followed by the same manifestations at all times and in all cases; but I have not seen any good evidence of this. Mr. Lewis informs me, that in cases in which he can produce such effects by the will, he can do so, whatever part he touches, and can call out the same faculty by touching many different parts. His will, as I have already mentioned, is singularly powerful.

As to sympathy, it is necessary in so far, as it is a condition indispensable to the successful exercise of the will. But no one who has ever seen these beautiful manifestations, which are invaluable, from their truth and beauty, to the intelligent artist, can suppose that the state of the subject is a mere reflection of the operator's mind. For while the latter is tranquil, the former may be heaving with emotion; and, as I have seen, accidental emotions in the operator are, very often, not communicated to the subject, who may be an excellent one, and is perhaps acting some passion or feeling to the life, when the operator becomes convulsed with laughter, &c., and yet he is not thereby affected at all.

I therefore admit, nay I maintain, that there are many cases, in which suggestion, or the will of the operator, or sympathy with him, will suffice to explain the occurrence of the facts.

6. But there are other cases, in which this explanation does not apply. And I would again remark, that I have taken all precautions to avoid the possibility of deception.

First, the subject is often unacquainted with the very name of Phrenology, and ignorant of the position of a single organ. Yet he will, if a good case, respond to the touch instantly wherever it may be made, just as where will is the agent.

But secondly, when the operator is himself, as often happens, as ignorant of phrenology as the patient, he is surprised and confounded at the results, because, when touching a part, he knew not its function, and therefore had no volition on the matter at all. Yet here also, as I have seen, the manifestation will often come out as well as before. Nay, the pressure of a chair, or of the wall, on a part of the head, will sometimes, when quite accidental, as well as the accidental touch of a hand or an arm, whether of the operator or not, produce the same effects. Indeed, it often happens, that when an operator knows phrenology, intends to touch one organ, and, turning to speak to some one, touches a wrong organ, with the idea of the first on his mind, or when his hand slips from one organ to another, he is surprised at what appears a wrong result, till he detects the cause of it, and all this in cases where the subject has no idea whatever of phrenology.

Thirdly, it frequently happens, that the operator, when touching an organ, either does not know what manifestation to expect, or possibly expects one, and yet an admirable manifestation is the result, different from any he had imagined. Thus, in trying a number of organs, I had no idea what to expect from that of weight, which is believed to give the sense of resistance, and to aid in preserving the balance. I tried on two subjects, both seated, and I took no time to consider what would be the result. The first drew himself up from a stooping posture, into a perfectly upright one, with a deep sigh. With this in my mind, I tried the other, but he immediately leaned forward, his face assumed an expression of horror, and he screamed out that he was falling into a bottomless abyss. Both manifestations belong to the faculty, but most certainly neither was expected by me. When I tried, in the second case, the organ of Size, the subject instantly spoke of an elephant, 40 feet high, which he saw, " a big black beastie," as he, being an Aberdonian, called it at first, with that intense delight in diminutive terms, which characterises the natives of Bon Accord. I had formerly seen the same faculty manifested in the form of the perception of vast distance, one of its functions being to observe distance, and was expecting that, rather than the "beastie." I might give other examples, but I refrain.

Fourthly, when I tried the combination of two or more organs, touched at once, combined manifestations came out, as quickly as the single ones, and before I could even conjecture what was likely to ensue. Thus, when I touched Acquisitiveness and Benevolence at once, in a subject in whom both came out well separately, he began discoursing to an imaginary beggar, with his hand in his pocket, which, however, he could not prevail on himself to extract from thence, concerning the duty of assisting the poor, by good advice, and by taking trouble about them rather than by giving money. In another subject, I obtained accidentally a very fine combination, indicating at the same time, that the excitement produced by touch lasts for a certain time, which is longer in some cases than in others. When Veneration was excited, the subject exhibited a beautiful picture of devotion. He knelt and prayed, with a fervour and intensity of expression which it would be difficult to surpass. Humility was intensely predominant in his gesture. When Self-esteem was touched, the organ being large, he exhibited pride and hauteur to a most ludicrous degree, and this faculty was kept excited for a time. At this moment a gentleman entered the room, who anxiously desired to witness the manifestation of humble devotion. I therefore touched Veneration, being firmly convinced that I should obtain, as I had frequently done, the former result; but I was disappointed. Devotion indeed came out, but with a totally different character. Instead of kneeling, he stood erect, and his prayer began : "O Lord ! I thank Thee that Thou hast made me so much superior to all other men in knowledge of Thee," &c. The tone of voice was no longer humble, and in short, an artist, wishing to paint a picture of the Pharisee in the Temple, and the Publican afar off, would have found this man, in his last state, a perfect model for the former ; in his first, an equally perfect study for the latter. I have seen many other instances of combined manifestations, where I knew not what to expect.

From the above considerations I think it is evident, that there are cases, in which Sympathy and Will do not suffice for the explanation, and where nothing but the admission of the phrenological organs, and of the influence of the operator on these by contact, can explain the results obtained.

7. I have seen cases, in which I could easily, by touch, excite some organs, but not others, which in other cases I could easily bring into action. In one case, where the organ of Caution could not be excited, my hand happened to touch that of Secretiveness, lying just below it, unknown to me. Instantly a bystander said, "Look! look! what is she doing?" and I saw her secreting under her shawl, some small object taken very cunningly from the table. I was thinking of and desiring a manifestation of fear or of terror.

8. Lastly, I have observed, that in some persons, in their ordinary state, certain faculties not only become unusually active, but are easily excitable by touching their organs. A lady mentioned to me that she was annoyed with spectral illusions, which indicated morbid activity of Form, Colour, and other perceptive faculties. I tried the effect of touching, in succession, all the perceptive organs. Those of Form, Colour, Size, Order, and Number, all responded to the touch, and exhibited beautiful objects of many kinds, singly, or in vast numbers; grouped in disorder, or symmetrically arranged; grey, or splendid in varied hues; and extending to infinite distance, and small or large, according to the organs touched. When Weight was touched, she felt as in a bad dream, as if falling from a precipice, or the ground falling away from her feet.

This case confirms that theory of spectral illusions, which refers some of them to morbid excitement of the perceptive organ, which, however, does not explain all spectral appearances.

On the whole, we must, I think, admit that in the mesmeric sleep, and in susceptible subjects, the mental faculties may be really excited to action, by touching, or pointing at, their organs, as well as by the will of the operator. See further on.

The next subject to which I wish to direct your attention, is the fact, that the lower animals are susceptible of the mesmeric influence. This has been often observed, and it is important, as excluding the action of the imagination, as well as the idea of collusion, and leading to the conclusion that a real influence exists, which passes from the operator to the subject.

1. There is reason to believe that the celebrated horse-tamer, and others who were in the habit of subduing unbroken or savage animals, have used, perhaps without knowing it, some mesmeric process. The great Irish horse-tamer is said to have shut himself up for a short time with the horse, and at the end of the time to have produced him tame. He evidently had a secret, and it would appear that it must have been very simple, or he would not

have been so fearful of its being discovered. It is said, however, that either he, or other tamers, breathed into the animal's nostrils, and certainly this process has been found very powerful by some who have tried it. Now, we know that breathing is one of the processes adopted in mesmerism, and there is reason to believe that the breath is strongly charged with the influence. The influence of the human eye on the lower animals is a familiar fact, and a great part, if not the whole of the feats of Van Amburgh, and others who subdue lions and tigers, depends on the use of the eye. They never, if they can help it, especially in the case of very fierce animals, withdraw their fixed gaze, and so long as that is kept up, and the eye of the animal is fascinated in this way, so long does he attempt nothing against his subduer. It is well known, however, that it is often dangerous to relax the steadiness of the gaze, or to turn away the eye. Now gazing is an extremely powerful means of mesmerising, so much so, that in my experience, I always begin, in a new case, by gazing steadily for five or ten minutes into the eyes of the subject. Mr. Lewis, whose power is so remarkable, operates chiefly by gazing, and those who have seen him operate, can readily understand the fascination of the eye, when used with so intense a power of concentration as is found in Mr. Lewis. Within these few days, Mr. Lewis easily and completely mesmerised a cat, in the presence of several persons. A case was lately published in the *Zoist*, in which the Duke of Marlborough mesmerised a very fierce dog by gazing alone.

Miss Martineau has also recently published a case, in which she not only mesmerised a cow, suffering from acute disease, but cured the animal by mesmeric treatment.

It would appear that the lower animals, being in a more natural state than civilised man, are generally, perhaps always, susceptible to mesmeric influence, as man, in his natural state, probably is also. Experiments on this subject would certainly yield very interesting results.

2. Not only are the lower animals susceptible to the human mesmeric influence, but they can exert a similar influence on each other. The power of the snake to fascinate birds by gazing, is pure mesmerism. Not only is this fact daily observed in America, but snakes are very often seen to fascinate larger animals, such as the domestic cat, and even at incredible distances. Mr. Lewis informs me that he has often seen this. The cat becomes strangely agitated, when those who observe it see no snake; but a snake is always found on looking for it, with its eyes fixed on the cat. The latter is compelled to move towards the reptile, and after a time, falls down apparently unconscious, and quite helpless, unable to use its limbs, when, if not rescued, it falls an easy prey to the snake. It is also observed, that when the snake is frightened away, or killed, and its gaze suddenly removed, the-

cat, in some instances, instantly dies. This Mr. Lewis has seen. This reminds us of the facts formerly mentioned, in regard to the mesmeric trance, or extasis, which may pass into death, and in which it is sometimes difficult for the mesmerist to restore full life.

3. There would appear to exist among the lower animals, some means of communication unknown to us. This has been observed in all animals, and is usually ascribed to instinct. But what is instinct? This is merely giving the fact a name, not explaining its nature. In the dog, this peculiar sagacity has been much noticed, because the dog is so much in contact with man; but there is hardly an animal which does not exhibit it, and in regard to which it has not been recorded. The courts of justice and punishments of rooks, the movements of birds of passage, and hundreds of analogous facts, point to some peculiar influence. How does a dog trace, not only his master, but also any thing which his master has touched, and commands him to seek, even although it be concealed? How does a dog, carried to a distance by sea, or in a bag, find his way home by the direct route? Who can explain the well-attested fact, that a Scotch terrier, having been taken to England, and there cruelly mangled by a large dog, not only found his way home, but immediately again departed for the scene of his ill-usage, not however alone, but with a companion, an old friend, in the shape of a large dog, who, when they arrived at their destination, assisted him to worry his tormentor? If we ascribe to the scent, the dog's power to trace his master, the degree of scent required is so great, that it amounts to a new sense; for he will, after long confinement, often go to where he last saw his master, and not finding him there, will yet trace him, through many places, till he find him. We cannot suppose ordinary scent here to be the agent.

I am rather inclined to ascribe many of the marvels of instinct to mesmeric sympathy, which there is reason to believe is very active and powerful in animals. We know that animals of different *genera*, and even classes or orders, often exhibit an attachment very similar to that sometimes observed between a mesmerist and his subject; and they also show very unaccountable antipathies, both to men and animals.

It has lately been stated, by M. Allix, on the authority of M. Benoit in Paris, and of another discoverer (also, I believe, a Frenchman, who is now in America), both of whom, during the last ten years, have been employed in working out the discovery, which they had severally and independently made, although they are now associated to work it out, that this mesmeric sympathy is remarkably developed in snails; that these animals, after having once been in communication or in contact, continue ever after to sympathise, no matter at what distance they may be, and it has been proposed to found on this fact, a mode of communication between the most distant places. Nay, M. Allix describes, with care and judgment, experiments made in his presence, in which the time having of course been fixed beforehand, words, spelled in Paris by M. Benoit, and also by M. Allix himself, were instantly read in America, and as instantly replied to, by words spelled there and read in Paris. All this was done by means of snails, and although the full details of the apparatus employed, and of all the processes necessary to ensure success, have not yet been published, yet the account given by M. Allix, and also by M. Benoit, goes so far as to enable us to conceive the principle made use of.

It would appear that every letter has a snail belonging to it in Paris, while in America, each letter has also a snail, sympathetic with that of the same letter in Paris, the two snails of each letter having been at some period, and by some process, brought into full sympathy, and then separated and marked. There is, of course, a stock of spare snails for each letter, in case of accident, but it is found that these animals will live for a year without food, should that be necessary. When a word is to be spelt in Paris, the snail belonging to the first letter is brought by some galvanic apparatus, not yet fully described, into a state of disturbance, with which his fellow in America sympathises. But this requires to be ascertained; which is done by approaching, in America, to all the snails successively, a testing apparatus, not described, which however includes a snail. On the approach of this, the snail whose fellow in Paris has been acted on, exhibits some symptom, which is not exhibited by any other, and the corresponding letter is noted down. This is done with each letter, and thus the word is finally spelled.

Now all this may appear, at first sight, very absurd and ridiculous. I confess it appeared so to me, when I first heard of it. But when I recollected all I had seen of sympathy in man, all that was known about sympathy in the lower animals, and when I read the account given by M. Allix, a gentleman well versed in science, of the successful experiments at which he had assisted, I preceived that the only difficulty lay in admitting the fact of the extraordinary sympathy of snails, and that, this being granted, all the rest was not only possible but easy. Now, I know nothing whatever about the habits of snails; and surely I am not entitled to reject facts, thus attested, without some investigation into them. I cannot say that the alleged sympathy is impossible. But an investigation into the matter, so long as the full details of the experiments made by the discoverers are not published, is certain to be a laborious task, and probably a fruitless one. It cost them a long time to ascertain the facts, and they have been, for ten years, engaged in bringing their discovery into a practical form. Till the promised publication appears, we can only admit the possibility of the thing, and wait for the explanation, which shall enable us to verify it for ourselves. It will certainly be very remarkable, if a snail telegraph should come into existence, which, in spite of the proverbial slowness of the animal concerned, should rival in rapidity the electric telegraph, and surpass it in security, inasmuch as there are no wires to be cut by an enemy, besides being infinitely less costly, since no solid, tangible means of communication are required, and all that is needed is the apparatus at either end of the line, and the properly prepared snails.

It appears from the paper of M. Allix, that even this astounding novelty is not new. At least, it would seem, that a long time ago, I cannot fix the period, a secret mode of correspondence was devised, intended chiefly for commmunications with a beleagured fortress, in which it is believed that animal sympathy played the leading part.

Chapter IX.

ACTION OF MAGNETS, CRYSTALS, ETC., ON THE HUMAN FRAME—RESEARCHES OF REICHENBACH—HIS ODYLE IS IDENTICAL WITH THE MESMERIC FLUID OF MESMER, OR WITH THE INFLUENCE WHICH CAUSES THE MESMERIC PHENOMENA—ODYLIC OR MESMERIC LIGHT.—AURORA BOREALIS ARTIFICIALLY PRODUCED—MESMERISED WATER—USEFUL APPLICATIONS OF MESMERISM, PHYSIOLOGICAL, THERAPEUTICAL, ETC.—TREATMENT OF INSANITY, MAGIC, DIVINATION, WITCHCRAFT, ETC., EXPLAINED BY MESMERISM, AND TRACED TO NATURAL CAUSES—APPARITIONS—SECOND SIGHT IS WAKING CLAIRVOYANCE—PREDICTIONS OF VARIOUS KINDS.

IN this chapter, I shall proceed to mention, in a very general way, the action, on the human system, of inanimate objects, such as magnets, crystals, &c. It is the less necessary to dwell at length on this subject, as Baron von Reichenbach's *Researches on Mesmerism*, (a translation of which I published, in so far as they have appeared in Germany, in the month of May, 1850, having previously given, in 1846, an Abstract of Part I.) contains the results of the only truly scientific investigation which has yet been made on that point. It is true that we only possess, at present, a part of these admirable researches, which were continued, with great labour and astonishing perseverance, for five years. The Baron, during that time, collected full materials for a work on the numerous branches of the subject investigated by him ; but he has not, as yet, been able to publish more than a part, sufficient, however, to make us eager to obtain the remainder. The labour and time required for arranging and publishing the details of so many investigations, made on upwards of 100 different persons, is very great, and the remainder of the work can only appear gradually, as it is brought into a state fit for publication.

In a former chapter, however, we have a general summary of the whole investigation, and it is to that that we must refer for the facts which have been ascertained on this matter.

1. Mesmer observed the effects, not only of magnets, but of other things, on the human body ; but he seems to have been, and his followers certainly were, in too great a hurry to apply the power he observed to profitable purposes, and to the cure of disease, so that they made no thorough or scientific examination of it ; and the whole subject fell into discredit.

Nevertheless, it now appears, that the fundamental facts are true. Magnets do act on the human body. When passes are made with them, the same sensations are experienced, as when the operator uses his hand. Here, no doubt, the influence of the hand is combined with that of the magnet; but, by using the magnet without the hand of the operator, or in the hand of a person whose hand, by itself, has no perceptible effect, it is ascertained that the magnet does exert an influence identical with that exerted by the human body.

This influence may go so far as to produce, even at a great distance, unconsciousness, as well as the true mesmeric sleep, and in highly susceptible cases, even cataleptic rigidity and convulsions. In one such case, mentioned by Reichenbach, a large magnet, being disarmed at the greatest distance permitted by the room, instantly struck the patient into complete rigidity and unconsciousness. But Reichenbach has also shown that a large number of perfectly healthy persons are sensitive to the influence from the magnet. The sensitiveness is not a morbid condition, and is found, in different degrees, in one person out of three, on an average, of healthy and diseased people.

2. This influence is conducted, or passes, through all matter, differing in this from the electrical influence, which cannot pass, for example, through glass or resin, but passes easily through metals.

3. Like electricity and ordinary mesmerism, it is polar in its distribution. In the Magnet, this influence, which Reichenbach has named Odyle, is associated with Ferro-mesmerism, that power by which the suspended needle points to the north, and by which the magnet attracts iron-filings. But it is found, as we shall see, unconnected with ferro-mesmerism, as in crystals, or in the human body. But wherever it appears, that is, whether in mesmerism, in crystals, or in the human body, it is polar, like ferro-mesmerism: that is, there is a difference in its manifestations at the two ends or poles of the magnet or other body possessing it. It occurs, also, in amorphous matter, &c., without distinct polarity.

4. The odylic influence is characterised, in its flow out of one body towards all others—for, like heat, light, and electricity, it is sent forth in all directions—by its emanations being luminous, that is, to sensitive persons, in the dark. The light is very faint, so as generally to be overpowered by the faintest glimmer of ordinary light, although very sensitive persons, and most persons when in the mesmeric sleep, can see it in daylight. It presents the rainbow colours, but at the northward pole of magnets the blue, at the southward the red, predominates. For a multitude of very interesting details on the odylic light of magnets, I refer to the work of Reichenbach already mentioned above.

5. Not only is the odylic influence found in magnets, but also in crystals. All bodies, when in distinct and large crystals, possess it, and exhibit, to the sensitive, the same or analogous luminous emanations, often of great beauty. Crystals are also

odylically polar, and produce, though less powerfully, the same effects as magnets, or as the human hand.

6. The human body is found to possess the same influence, and to produce the same effects on the sensitive, as magnets do. I have already spoken of the light seen, by persons in the mesmeric sleep, to issue from the tips of the operator's fingers. This is odylic light, which is seen by the sensitive, at least in the dark, without their being in the mesmeric sleep. The hands are oppositely polar; and the head, eyes and mouth are also foci where the odylic influence appears to be concentrated. This is the reason why passes with the hands, and gazing, are the most powerful means of mesmerising.

7. Besides the sources of odyle above mentioned, Reichenbach has shown that it is present in all material substances, though generally in a less degree than in magnets or crystals. He has found it to be developed by heat, light, electricity, common or galvanic, friction, and every species of chemical action, such as combustion, the solution of a metal or of an alkali in an acid, respiration, and the changes going on in the animal body generally. This explains why the human or animal body is so plentiful a source of odyle. He has also found it in plants; and has detected its presence in the light of the sun, moon, and stars.

8. Another important observation is, that the human body is strongly influenced by the magnetism of the earth. Many very sensitive persons cannot sleep unless their bed lies in a plane parallel to the mesmeric meridian, with the head towards the north. I have had opportunities of seeing several, and hearing of many more, persons who experience this; and many of them had observed it, without being able to account for it, long before Reichenbach's experiments were made. It appears extremely probable, that some diseases may be more easily cured when this position of the bed is observed. To some patients, the position, at right angles to it, is quite intolerable, and this has been noticed long ago, but ascribed to fancy or idiosyncrasy alone.

It is found that people are more readily mesmerised when they sit, with the head towards the north, the face turned, and the feet extended towards the south, than in any other position. I have myself repeatedly experienced this, and probably, if observed, it will be found to be a general fact, although many are easily enough mesmerised in any position. Reichenbach has also found, that to see the odylic light, for example, best, the subject should be in north and south position, with the head towards the north.

9. Reichenbach has also observed many very curious facts concerning the distribution of the mesmeric or odylic influence in the body, at different hours, and before or after meals. On awaking in the morning, or rather with sunrise, it begins to rise, sinks a little before breakfast, from the effect of hunger, then rises steadily, with a sudden increase at dinner-time, continuing to rise till the

evening, or after sunset, when it begins to fall, and falls during the night, till before sunrise it is at the lowest ebb. For many curious and interesting details on this point, and for the application of these facts to the due regulation of our mode of life, with a view to the preservation of health, I refer to the work of Reichenbach, so often mentioned.

10. In all these researches, odyle appeared to be polar; and the negative and positive, northward or southward poles of any object possessing the odylic force, such as a magnet, a crystal, the human body, of which the hands are the chief poles, always produced peculiar effects. The negative or northward pole caused a great coolness, and gave out light in which blue predominated. The sensation caused by the positive or southward pole was a disagreeable warmth; and in its light red prevailed. The right hand is negative and cool, the left positive and warm. The sun's rays are negative, and cause to sensitives a strong but delightful coolness. Nay a hot stove caused, to the very sensitive, until they came so near as to be affected by the radiated heat, a cold feeling amounting to that of frost, due to its negative odylic emanations; and in some, the numerous tapers in a Roman Catholic church caused not only cold, but fainting. The moon, on the contrary, is odylically positive, and her rays excite a feeling of warmth in the sensitive. All the planets, which, like the moon, shine by reflected light, are, like her, odylically positive.

11. In short, odyle is universally diffused throughout the material universe, and in this respect, it agrees with heat, light, and electricity. By a laborious and beautiful investigation, Reichenbach has, in my opinion, demonstrated the existence of a force, influence, or imponderable fluid, whatever name be given to it, which is distinct from all the known forces, influences, or imponderable fluids, such as heat, light, electricity, mesmerism, and from the attractions, such as gravitation or chemical attraction. But it is highly analogous to the other imponderables, and, as we have seen, is found associated with them. All of them may possibly be hereafter reduced to one primary force, but in the meantime odyle must be distinguished from the rest, just as heat, light, and electricity are from each other.

Although Reichenbach has not made his experiments with artificially mesmerised persons, nor of those persons in the mesmeric sleep, or the state of artificial somnambulism, yet he has observed, that persons who are subject to spontaneous somnambulism are almost always very sensitive when in their ordinary state, and that when they fall into somnambulism, their sensitiveness is greatly increased. Now we know that persons in artificial somnambulism or mesmeric sleep are also highly sensitive, so that they see the odylic light from the hand or from other objects, even in daylight.

12. Hence it is hardly to be doubted that the odylic influence,

which exists in magnets and in the hand, and which in magnets produces the sensations formerly described, and even the mesmeric sleep, is identical with the mesmeric influence of the hand, which is usually employed to produce the effects of mesmerism.

Thus, when Mesmer spoke of an influence from magnets, and of a mesmeric fluid, as producing these effects, he was right, in so far, at least, as the existence of the influence is concerned; which, if not a fluid, is as much entitled to the name as the electric fluid. But he was wrong in supposing it to be identical with ferro-magnetism, with which it is only associated. The magnetic *baquet* was merely a mixture of all sorts of things, which, giving rise to a slow chemical action, furnished a slow but constant current of odylic, or mesmeric influence.

We may, therefore, for the present, safely assume the odyle of Reichenbach, discovered by a totally different and independent train of researches, as being the influence or cause to which are due the effects of mesmerism, as above described. The coincidence of the two modes of investigation in this great point, of the existence of an influence, which may be exerted, or pass, from one individual to another, is the best guarantee for the accuracy of both.

We can no longer have a difficulty in conceiving how a susceptible person may be thrown into somnambulism by the influence of another, even without contact. If a magnet can do this, why may not the hand, which has been shown to possess the very same influence as the magnet? It appears to me, that the laborious and truly scientific researches of Reichenbach have for ever settled the question as to the existence of an external and universally diffused influence, different from all known influences, although closely allied with and analogous to several of them, and which is capable of producing the effects of mesmerism.

13. I have, with very imperfect means and very limited leisure, repeated many of Reichenbach's experiments, on magnets, crystals, chemical action, and the human hand, on sensitive persons in the ordinary waking state. And in every such case, with great variations in degree, I have found his statements and descriptions rigorously exact. I have also been informed, by many friends, as well as by many persons unknown to me personally, who have repeated some of these experiments, chiefly those with small magnets and crystals, that their experience, in every point observed by them, confirms the statements of the Baron. Sensitive persons are easily found, if we only look for them, and, according to my own observation, they are not less frequent here than he found them to be in Vienna.

I would here, in recommending the repetition and prosecution of these attractive researches, by enquirers possessing the leisure which is necessary, urge on them the absolute necessity of attending to the conditions minutely laid down by Reichenbach. In order to see the odylic light, for example, not only must the person be sensitive, but the darkness must be absolute, and the sensitive should remain in it for an hour or two, before we can expect the eye to be fully awake to the faint but beautiful luminous emanations of magnets, crystals, the hand, &c. And after the observer and his subject or subjects have entered the darkened chamber, not the smallest gleam of light, even of the dullest daylight or of a candle, must be allowed to enter at chink or cranny, door or window. No one should come in or go out during the experiment; for if the door be opened, the admission, for an instant, of light from the next room, blinds the subject, unless of the highest order of sensitiveness, which is rare, for half an hour, an hour, or even longer, to the feeble beams of odylic light. Another essential precaution is, that no one should be close to the subject or to the object observed. The approach even of the observer often extinguishes the light, visible but a moment before, by changing the odylic state of the magnet, &c., as well as that of the subject himself. Unless all these and other precautions are attended to, failure is the result.

14. Before passing on to the next point to be discussed, I would here refer to two beautiful applications of Reichenbach's discoveries. First, since all chemical action is attended with the emission of odylic light as well as odylic influence, the changes which take place in dead bodies by decay, which are chemical, are sources of odylic light, just as are the changes in the living body, respiration, digestion, &c., &c. Hence sensitive persons see luminous appearances over graves, especially over recent graves. There will be found in the work of Reichenbach several most interesting and instructive cases of this fact, and thus we find, that science, with her torch, dissipates the shades of superstition. Corpse-lights exist, but they are not supernatural; neither are those, who habitually see them, what we call in Scotland "uncanny." The lights are perfectly natural and harmless; and the seers are only sensitive persons. I have been informed of several such cases, in which these lights are always seen at night (if dark) over churchyard graves or burying-vaults, and in which the observation dates many years before Reichenbach made his investigations.

Secondly, as magnets emit beautiful odylic light, so the earth, which is a vast magnet, emits its odylic light; which, in consequence of the great size and enormous power of the magnet concerned, becomes visible to all eyes, perhaps more vividly to the sensitive; but this is not easy to ascertain. This is not a mere hypothesis. It is supported by a series of the most beautiful experiments with which I am acquainted. Reichenbach converted a large iron globe, two or three feet in diameter, into a powerful temporary magnet, by causing an electric current to traverse a wire coiled round a bar of iron passing from pole to pole of the sphere. When the globe was suspended in the air, in an absolutely dark room, the sensitives saw the odylic light in the most exquisite beauty, and with all the peculiar characters of the Aurora Borealis and the Aurora Australis. At each pole appeared a wide circle of light, more blue at the northward, more red at the southward pole, but at both with all the rainbow hues. The equator was marked by a luminous belt, towards which on or close above, the surface of the sphere, lines of light constantly streamed from the polar circles. In the polar circles, as well as in the streaming lines, the colours were arranged so that red predominated in one quarter, the south, blue in the opposite, yellow in the west, and opposite to it, grey, or the absence of colour, while, as in all the odylic rainbows, a narrow strip of red appeared near the grey, at the end of the iris most remote from the great mass of red—a most beautiful confirmation of Sir David Brewster's analysis of the spectrum. The delicate streaming lines or threads of light passed by insensible gradations from one colour to the other, so that any two contiguous lines appeared to have the same colour, yet, on looking a little farther on, the colour gradually changed, and thus the whole of the rainbow hues appeared in their order, red, orange, yellow, green, blue, indigo, violet, and last of all the small red stripe, and the grey. But the passage from red to orange, or from orange to yellow, &c., was not sudden, but slow and gradual, so that all the intermediate tints were seen. Nor was this all, for in the air, above each pole, appeared a splendid crown, of light, more blue at the northward, more red at the southward, but exhibiting also all the colours, and sending towards the equator splendid streamers of many-coloured light, dancing and leaping, lengthening and shortening, just as the finest northern streamers do, to the delighted eye of the observer. I cannot here enter into a full description of this artificial Aurora, the first ever produced; but I may record my conviction, that this experiment gives, to that theory which regards the Aurora as odylic light, a degree of probability far greater than attaches to any other theory of that phenomenon. I may observe also, that the Aurora does not cease to be a magnetic phenomenon; and that it should affect the needle is to be expected, since, in magnets, odylic influence and odylic light are found associated with the ferro-mesmeric influence.

15. It still remains for me to speak of a class of facts, which has not yet been alluded to. I mean the power which we possess of communicating, to certain objects, the mesmeric influence. Mesmer spoke of mesmerised water; but this idea was scouted and rejected as absurd. But every one who has studied mesmerism, and tried the experiment, knows, that water may be so charged with vital mesmerism (with odyle) that a person in the mesmeric sleep, without the slightest knowledge that the experiment is made or intended, instantly and infallibly distinguishes such water from that not mesmerised. It is generally described

as having a peculiar taste, not easily defined, and as producing internal warmth when swallowed, and these peculiarities are very strongly marked. Some subjects describe it as vapid and tasteless, like rain, or distilled water, whereas the same water, if not mesmerised, has to them (if it be good spring water) its usual agreeable sharpness. Of the fact, which I have often tested, there can be no doubt.

This effect may be produced either by the hand, in which case, while the vessel is held resting on the left palm, and grasped by the fingers of the left hand, circular passes are made above it with the right hand, or the fingers of the right hand are held with their points close to the surface of the water or by magnets, held in the same way, or by crystals.

Reichenbach has shown, that sensitive persons, even when not in the mesmeric sleep, often readily distinguished mesmerised water from ordinary water. The effect, in all cases, lasts only for a certain time, which may extend, when the charge is strong, to a good many hours.

Mesmerised water, as I have seen, will often cause the mesmeric sleep, in persons who have been, on former occasion, put to sleep by the operator in the usual way. I have seen it also produce natural sleep, in excitable persons, not formerly mesmerised, and sometimes the sleep has taken place instantly on swallowing the water, and has been sound and refreshing. It is indeed possible, that it may have been mesmeric sleep, at least in some of these persons; but as the object was to produce sleep in those who were restless, no experiments were tried.

Not only water, but any other body, as has been shown by Reichenbach, may be charged with the influence; and it is not uncommon for the mesmeric sleep to be produced, in the absence of the operator, by an object thus charged and sent by him to his patient. The patient will easily detect the attempt to impose on him by an object not charged; at least in many cases, just as he knows mesmerised from unmesmerised water.

I now come to another matter; namely, the applications which may be made of the facts hitherto noticed. And here I would again urge on your attention the consideration, that it is no argument against the study of a fact, or of a series of facts, to say, that it is, useless, and nothing more than a mere curiosity of science. There is no such thing as a mere curiosity of science; that is, there is nothing of which it can be said that it may not, at a moment's notice, become useful, either in reference to some other scientific matter, or in its practical application to the purposes of ordinary life or of the arts. I have already given many instances of this, and one of these was the sudden application, to a most important practical purpose, the alleviation of pain, of a substance long regarded as a scientific curiosity, of no value whatever, namely, chloroform. The same substance supplies us with the example of the application of a discovery to the improvement of the method of investigating another branch of science, which, in its turn, may bear rich fruits of practical utility, especially in reference to the cure of disease ; and to the investigation, again, of another subject, the laws of the nervous system, or of the vital principle.

Every one knows that many important points in physiology have been ascertained by experiments on living animals, the cruelty of which, in most minds, so far exceeded any possible benefit to be derived from them, that they were only undertaken by a few of stronger nerves, and less sensitive to the sufferings of the dumb victims to science, than most people are. Now, by the use of chloroform, all conceivable experiments on living animals may be at once divested of pain. The animals will not suffer, and if their lives are sacrificed, it will be with less suffering than when they die to furnish our tables, or to supply us with sport. No one need now recoil from such experiments ; they will be prosecuted with greater success, because there will be no writhings, no struggles, no cries, to interrupt or dismay the experimenter, whose mind, free from the reproach of his own conscience for cruelty, will be far better fitted to discern the truth. Such is the second, possibly in its consequences to mankind the most important, application of chloroform. And more remain behind.

1. In like manner, if mesmerism had never yet been applied to any useful purpose, this, so far from being a reason for neglecting it, would furnish the strongest reason why it should be more diligently studied, since it is only by a knowledge of all the properties which can be ascertained to belong to any agent, that we can hope to find useful applications of it. It was because the discoverers of chloroform confined their observations to its physical and chemical properties alone, or rather to some of these, and neglected to try its action on the system when inhaled, that it continued so long useless.

But mesmerism, in point of fact, already presents many useful applications. It has been, and daily is, used to produce insensibility to pain, in surgical operations. It is used with very great success, to relieve rheumatic and neuralgic pains. Many cases of severe neuralgia, but not all, yield to its use. It daily removes headaches, and produces refreshing sleep in persons who have long suffered from wakefulness. It relieves, nay, with perseverance it often cures, many diseases of the nervous system, such as paralysis, hysteria, epilepsy, catalepsy, and chorea, or St. Vitus' dance. And all this, from its direct and powerful action on the nervous system, might be anticipated.

2. The effects of mesmerism, however, are not confined to such cases. It acts on the general health, doubtless through the nervous system, in such a way as to produce very often the most

marked improvement, and in many instances to cause, sooner or later, old and very annoying complaints to disappear. Nay, cases occur, in which one operation, especially if it induce the sleep, will be followed by a rapid and permanent cure. This, it is true, is the exception; but, with patience and perseverance, even without ever producing the sleep, we may cure or relieve a large number of cases, provided they be not of that nature which precludes hope of amendment. An immense number of mesmeric cures have been recorded, both by medical and non-medical operators, among them that of a well-marked case of cancer by Dr. Elliotson; and making every allowance for imperfect observation, and for the tendency to exaggerate the merit of any new method of treatment, no doubt can reasonably be entertained, that mesmerism is a very powerful means of cure, and ought to be in the hands of every physician. The absurdity of the idea of an universal medicine, or panacea, is obvious; but that is no reason for rejecting a method which, in many cases, will prove of essential service, and whch is not only manageable but safe; which, therefore, if it do no good, will at least, in good hands, do no harm.

I have had many opportunities of seeing the good effects of mesmerism, even where the object has been only to study the phenomena. It daily happens that persons mesmerised for that purpose, astonish and delight the operator, by telling him that, since mesmerism was used, they have got rid of some obstinate complaint; or that their general health and spirits are strikingly improved.

If it be said, that these effects are due to the imagination alone, I answer, that if so, they are not on that account unreal or imaginary; that it is then our duty to study the power of the imagination, and use it as a most powerful agent for good; that at all events, mesmerism, in that case, has a very great action on the imagination, and is probably the best means of acting on it. But, in many cases, the imagination does not act, because it is not appealed to. Both subject and mesmerist are often taken by surprise, when they find that some distressing complaint, perhaps regarded as hopeless, but which neither of them had thought of curing, has been, as if by magic, relieved or cured.

I would particularly direct the attention of physicians to the value of mesmerism, in reference to insanity, not only as regards treatment, but also for another reason.

There is no doubt that many persons who are subject to attacks of insanity, as in the case of other diseases of the nervous system, are very sensitive, and susceptible to mesmeric or odylic influence. Hence we are prepared to find, as has been recorded in many instances, that mesmeric treatment is sometimes effectual in insanity, where all means have failed. The treatment, and above all, the moral management of the insane, has of late been greatly improved; and for the greater part of this improvement we are indebted to phrenology. Violence, restraint, and cruelty, are banished from our asylums, and means are judiciously and kindly resorted to, in most of them, for employing such faculties as remain capable of being used. The result is, that, in spite of the sad reflections which arise in the mind when visiting an asylum, we feel, if it be well conducted, that, for most, if not all, of the unfortunate inmates, it is a scene of pleasure, and of such happiness as they can enjoy, which often surpasses, by far, the lot of the sane man. I rejoice to think, that the improvement is progressive.

But I am now convinced, that the treatment of the insane will not be so complete or so efficient as it may become, till mesmerism is regularly introduced into the practice of every asylum. Indeed, there can be no doubt that the control exercised, by the aid of the voice and eye, on many of the insane by an experienced physician, depends mainly on their being in the impressible state so often mentioned, in which suggestion and command act like magic on the patient. If this were generally known, and systematically attended to, much good might be effected. And where the patient is susceptible or impressible, there is good reason to hope that direct mesmerisation may produce the best results. Mesmerism is not less powerful on the insane than on the sane; nay, it is more so; probably because, in many cases, the essence of insanity is some disturbance of the natural distribution of odyle in the system. The effect of the moon, whose rays are strongly charged with positive odyle, on the insane, corroborates this view.

There is, however, another reason why the physician ought to study mesmerism, in its relation to insanity. It is this: many insane persons appear, when we study the symptoms as they are described by writers on the subject, to be, in fact, only in a peculiar mesmeric state. I mean they have a consciousness distinct from their ordinary consciousness, just as happens in the mesmeric sleep. Let us suppose a patient to fall, spontaneously, into a continued mesmeric sleep, in which, while his eyes are open, he has no recollection, or only an imperfect recollection of his ordinary state. He is perhaps lucid, and lives in a world of his own, entirely consistent with itself, but absolutely incomprehensible to all around. His perceptions are, to him, and indeed absolutely, real; but to others they appear mere dreams. He sees absent or dead friends; nay, he holds long conversations with them; he sees also objects, which really exist, but at a distance. He is partially or wholly dead to the objects which surround him, and is absorbed, and perhaps supremely happy, in the contemplation of the persons, places, and things seen by his lucid vision. Lastly, he becomes ecstatic, and sees, and converses with, denizens of the spiritual world.

Now every word uttered by such a person is, to those around him, positive proof of his insanity. He is shut up; and the nature of the case not being even suspected, he becomes still more firmly rooted in his new state, in which, perhaps, he dies.

But it may be asked, Is that man insane? I answer, in one sense, yes; for he is unfit, so long as he continues in that state, for intercourse with the world. But in another sense, I say, no; for his mental powers are unimpaired, and he is only in a very vivid dream, so to speak, but a dream of realities, visible and audible to him by reason of exalted odylic sensitiveness.

In such a case, admitting, for the present, the possibility of its occurrence, it would seem reasonable to expect that he might be cured, that is, simply restored to ordinary consciousness, by mesmeric treatment. The chief symptom is so intense a degree of odylic sensitiveness, that the impressions made on the sensorium by those odylic emanations of which I have spoken, and of which I shall speak more fully in the next chapter, and which are at all times acting on us, though overlooked,—that these impressions, contrary to what occurs in the normal state, are so vivid as to overpower those derived from the external senses. May not this excessive sensitiveness be removed by appropriate mesmeric treatment? This, however, cannot be done, unless the nature of the case be understood, and mesmeric treatment practically studied.

Now, it is not a mere fancy of mine, that cases are viewed as cases of insanity, and the patients shut up accordingly, without appropriate treatment. I have been informed of a case, in which a lady, confined as insane, who (in consequence of accidental circumstances, which, appearing to have a favourable effect, were very judiciously made use of,) so far recovered as to be sent home, and was regarded as cured. During her illness, her conversation was not only rational but highly intelligent, except, of course, in regard to certain delusions, probably of the nature I have described above. When sent home, she retained complete consciousness and memory of all that had passed during her confinement; and by and bye was engaged to a gentleman with whom she had become acquainted since her illness. She now had a severe illness, of a febrile character, and on recovering from this, she had lost all trace of recollection of her insane state, of her confinement, and of the person to whom she was engaged, who was received by her, to his no small surprise, as an utter stranger. She was now really cured, and just as she had been before her insanity.

Now, I cannot help thinking, that she was, during her confinement, in a mesmeric state such as I have described, and that her first recovery was no true cure, inasmuch as she retained her new consciousness, which was for ever lost when she really recovered, after an illness which strongly affected the nervous system. Might not a similar change have been effected by mesmeric treatment, when she was first confined?

I know a young gentleman, singularly susceptible to odylic and mesmeric influences, who some years ago was still more so, and very narrowly escaped being confined as insane. Being fortunately under the charge of humane and enlightened persons, he had been treated by mesmerism with very great benefit. His susceptibility has diminished so far as no longer to cause uneasiness to his friends. Had he been confined, and had not mesmerism been employed, he would probably now be in an asylum, while, with the exception of the excessive sensitiveness to odylic impressions, he never was, in any shape or degree, insane.

I am acquainted with another gentleman, who, at times, suffers acutely from odylic impressions of every kind, so that almost every person or object that he approaches is a source of the most painful and distressing sensations. He is so far from being insane, that his intellect is of a very high class, and he is quite aware of the cause of his sufferings. He has told me, that, but for this knowledge, he would almost have, at times, believed himself to be insane.

I observe in the accounts of the insane, the delusion of seeing and conversing with absent persons, or with spirits, given as an infallible sign, where it occurs, of insanity. It may be so, in some cases; but it is self-evident, that as it may depend merely on spontaneous extasis, more permanent than usual, while the mind is entirely unimpaired, just as happens in artificial mesmeric extasis, some cases, treated as insanity, may have been of this latter kind.

It is therefore much to be desired, that physicians should learn the characters of every stage of mesmerism.

I would say, therefore, and especially to medical men, use mesmerism, were it only to become acquainted with it, and in the course of your experiments, you will be sure to find some one unexpectedly benefited by it. You will then use it for the cure of disease, and although we cannot hope to enjoy its full benefits until it has been fully studied, still, so safe and so powerful a remedy should be employed, especially when the usual means have failed. The more it is used, the better shall we become acquainted with it, the more effectually shall we be able to employ it.

---

3. In regard to other useful applications of mesmerism, much cannot as yet be said. I can very well imagine it to be used for the purpose of searching more deeply than by other means we can, into the nature of the relation between the mind and the body; into the laws of thought; and even into the nature of the mind itself. I mean, that whether we regard thought, as some do, as merely the necessary result of the action of the brain, or whether we assume, as is usually done, the existence of a separate being, the soul, mind, or thinking principle, which uses the brain as its

instrument,—in either case, a careful study of the various mental phenomena observed in mesmerism must throw light on the laws of thought, perception, sensation, &c. There is, for example, a class of patients, who, in the mesmeric sleep, can accurately point out the precise part of the brain which acts in every manifestation, whether of thought, of sensation, of muscular motion and muscular sense, of memory, in short, of every act in which the brain is concerned. Some can even see, and describe consistently, the actual physical changes in the brain which accompany these acts, and it can hardly be doubted that much valuable information may thus be acquired. I have already pointed out, that the questions of the essential nature of mind, and even of matter, are beyond the reach of the human faculties; but the laws of their mutual relations are attainable. While, therefore, we confess that we know nothing, and probably never shall know anything, of the essence of mind, let us diligently use all the means in our power to acquire a knowledge of the laws of its action.

4. Again, it is quite easy to see how sympathy and clairvoyance may be turned to useful purposes. They may enable us to obtain information about absent friends or relations; nay, they are actually often used for that purpose. They may be used to discover missing or stolen goods and documents, and for this purpose also they are daily employed. In a former chapter will be found some instances of such applications of lucidity. I have already said that I think it far from improbable that this power may be so used as to throw light on obscure historical points, and to discover documentary evidence in regard to these. Moreover, I have already described the principle, which certainly has nothing impossible in it, of the use of animal sympathy in the projected Snail Telegraph, or, as it is called by the inventors, the Pasilalinic Telegraph; and I have also mentioned the application of lucidity to the inspection of the living frame, healthy or diseased, for anatomical and physiological as well as medical purposes.

All these applications are yet in their infancy. But as the number of observers increases, so will increase, not only the number of cases of lucidity, but also that of useful applications, which, if not made now, will sooner or later be discovered.

5. There is, indeed, one useful application of our knowledge of mesmerism, which has already been made, and will yet be made to a greater extent; I mean that of explaining many things, which, to the ignorant, appear supernatural, and which, the knowledge of their real nature having, in earlier ages, and even down to a period not very remote, been considered as a secret, or mystery, and confined to certain classes of men, such as priests,

adepts, magicians, sorcerers, and perhaps astrologers and physicians, some of whom may have really believed that they possessed supernatural power, acquire the names of magic, sorcery, witchcraft, and the black art.

All sorts of necromancy, divination, and oracles, may be ranked in the same category; and it may safely be said, that the more intimately we become acquainted with mesmerism, the more do we find, that every notion which has prevailed among men in regard to these matters may be referred to natural causes, connected with mesmerism in some of its innumerable developments.

In the heathen temples, the magic cure of disease was often associated with the oracle, and the belief in both was universal. If we grant the possibility of clairvoyant prevision, and the evidence seems to me to lean in favour of it, we may readily understand how the priesthood, trained in the sacred mysteries, knew how to produce the mesmeric state, including lucidity, especially in females, who are more readily mesmerised. The lucid priestess, rendered so by means partly known and partly unknown, but certainly with the aid of music and fumigations, probably also by gazing and passes, sat on her tripod, which was perhaps a mesmeric apparatus, and by means of her lucidity, described the diseases of her clients, or predicted future events. When true lucidity was not to be found, imposture was resorted to, but it is hardly possible to believe that there was no foundation whatever for the universal belief.

There can be no doubt that the priests of India, Egypt, Greece, Rome, and other Pagan countries, had secrets or mysteries, which were carefully veiled from the public eye. It is equally certain that those of Egypt had extensive knowledge of natural truths, both in astronomy and physics, as well as in medicine, and that the sages of Greece resorted to their temples for information. Such acute observers as the Egyptians could not fail to discover the leading facts of mesmerism, which indeed occur spontaneously every day. A spontaneous somnambulist, if lucid, and capable of truly describing absent persons and objects, perhaps also of predicting the inevitable consequences of what he saw, would be regarded by the people as inspired by the gods; while the priests, who studied the matter, and made it a crime for any but their own class, to do so, would soon find out how to produce the inspiration at pleasure, especially if, as is now found to be the case in India, according to the unimpeachable testimony of Dr. Esdaile, every man proved to be susceptible to mesmerism.

This is a most tempting subject of investigation. I had already, to a considerable extent, collected materials for a brief history of mesmerism, and its various developments, from the earliest ages down to its rediscovery by Mesmer; and I had intended here to show, that all the magic, sorcery, divination, and witchcraft, of the early and middle ages was, in so far as founded on natural truth, only the result of a knowledge of mesmerism, a knowledge jealously guarded by those who possessed it, and probably at their suggestion, regarded by the people as the direct gift of infernal powers. This has been often proved, and many authors, especially in Germany, have treated of the subject, although there is no English work to which the reader might be referred.

Such, therefore, was my intention; namely, to have given a brief abstract of what had been ascertained, by the laborious investigations of many learned men, in regard to the history of mesmerism, and its relation to magic, witchcraft, divination, &c.

But while occupied in this research, for which my other avocations left but little time, at uncertain intervals, I was informed and I rejoiced to find the information correct, that the proposed work was in far better hands, in those, namely, of the veteran defender of modern mesmerism in this country, Mr. Colquhoun, whose *Isis Revelata* many years ago, contributed in a great degree to create that general interest in the subject which has of late so greatly increased. Those who know that gentleman and his works, are aware that his learning is profound, his research unwearied, and his intellect clear and comprehensive. His proposed work on the History of Mesmerism is, I rejoice to say, in a state of great forwardness; and I gladly refer you to it, and refrain from here entering minutely upon the questions above alluded to.

I shall content myself with pointing out, that a great proportion of those things which are called magic, witchcraft, divination, &c., obviously depend on those principles of mesmerism which I have endeavoured systematically to present to you.

6. Thus, not only the ancient oracles, and the magical cures of the pagan temples, but also the divination which even yet survives in Egypt, and of which Miss Martineau and other writers have given us an account, are clearly to be referred to lucidity or clairvoyance, as, in regard to the latter mode of divination, I shall endeavour to show, when treating of the theory or explanation of mesmerism. The wonders of the magic mirror, and of the magic crystal, will also be found, I think, to depend essentially on the same cause. The researches of M. Dupotet have led him, it is said, to the discovery of the secret of much of the magic of the middle ages, including the magic mirror, and the world looks anxiously for the completion of his investigations. It is well known, also, that the magic crystal has of late attracted much attention, and that several ancient crystals are extant, the properties of which, it is to be hoped, with the help of Reichenbach's discoveries, will admit of scientific ascertainment.

7. The belief in witchcraft, in the power possessed by certain persons of rendering themselves invisible, which feat is now daily exhibited by such operators as Dr. Darling or Mr. Lewis, who become invisible to their subjects *in the conscious state;* in that of

assuming the form of any man or animal, which is quite as easy, by means of suggestion, as the former; in the power of travelling through the air to a distant place, and seeing what there occurs; finally, the belief in intercourse with spirits, good or evil, which occurs daily, whether as a dream or otherwise, in mesmeric or spontaneous extasis; all these are explainable by what has been said on sympathy, suggestion, lucidity, trance, and extasis.

It is notorious that, while many persons suffered for witchcraft, who died denying every allegation against them, many confessed to all the above offences, as they themselves regarded them. The truth and actual occurrence of the facts, or visions, or delusions, which are still facts, is the most satisfactory explanation of such confessions.

Mesmerism will finally dissipate all the obscurity and all the superstition which have attached to this matter; and when we see that the facts, delusions or visions, however astounding, depend on natural causes, superstition will have lost her firmest hold on the human mind.

8. Another obscure subject has been, in part, cleared up by the discoveries of mesmerism. I mean, spectral illusions, or Apparitions. There can now be no doubt, that some apparitions are visions occasioned by lucidity, that is, by clairvoyance, occurring spontaneously. In these, the absent person is really seen, and his occupation at the moment is also perceived. Or the image of a dead person is recalled by suggestion, and becomes so vivid as to be taken for reality.

There is, as we have seen above, a third species of spectral appearances, depending on the odylic light from graves, &c. This, according to Reichenbach, is described generally without specific form, a mass of faint white light, often as high as a man. The accounts I have had of it, as seen by persons in this country, are similar. It is easy to imagine, that an excited and terrified imagination may give a human form to this light, and such, in the opinion of Reichenbach, is the origin of ghosts, which are generally white. But it is remarkable, that in the very first case mentioned by Reichenbach, that of the light seen by the amanuensis of the blind poet Pfeffel, in the garden, over a spot where, on digging, there were found the remains of a human body, imbedded in a mass of lime, doubtless buried there many years before, during a pestilence, the young man saw it in the form of a female figure floating over the spot, with one arm on her breast, the other hanging down. I am not prepared to say, that in this case, in which the observer was repeatedly questioned in presence of the object, the learned Baron has not been rather hasty in regarding the form as altogether fanciful. Further investigation only can show whether the light from a body may not have the general form of the body, as asserted by the seer in this case.

But when we thus refer certain classes of apparitions to natural causes, we are not yet able with certainty to account, in this way, for all. Cases are recorded, and I shall give one or two further on, in which it is stated, and on good testimony, that apparitions, generally of persons at that moment dying or dead, have been seen, not as a faint light, but in their natural aspect and colour; not by one terrified peasant, but by two or more self-posssessed and educated men. Nay, some of these are said to have spoken, and to have done so for a purpose. Every one knows the story of Lord Lyttleton, and of the mark left on his hand, as a sign of the reality of the apparition. But, granting that to have been a dream, we cannot apply the same theory to the case of the apparition which I shall mention further on, which was seen by a whole party of officers at table, and by one among them who had never seen the person whose figure appeared. The explanation of such facts can only be looked for, when we shall have more deeply studied every branch of the subject. I am inclined to think. that mesmerism will supply us with a key for the explanation even of this class of Apparitions on natural principles. I need hardly mention, what is well known, that another class of spectral appearances, not connected with our subject, is that which depends on optical illusions, or disordered internal perception, as it has been illustrated and described by Sir W. Scott, Sir D. Brewster, and many other writers.

9. There can, I think, be little doubt, that the second sight is also a phenomenon depending on mesmerism, that is, on spontaneous lucidity. The objects of the seer's vision are commonly said to have been such as were at a distance, such as an approaching traveller, or enemy. If we suppose the seer to have become, by concentrating his thoughts, lucid, and yet conscious, or at all events, if in the sleep, yet capable of speaking, and conversing with those around him, he may have seen, by clairvoyance, the distant traveller, toiling along a mountain path, which he recognises as distant, one, two, or more day's journey, and he may thus have predicted his arrival and described his appearance. I know that, in certain persons, extreme voluntary concentration, or reverie, may alike produce the conscious lucid state, and that persons unknown to the seer may thus be seen. I have good reason to believe, that is, the testimony is such as I have no reason to doubt, save the strangeness of the fact, that I have myself been seen, in that state, by a gentleman in whom it frequently occurs. I mean that this took place before we ever met, and when he had only heard of me from another gentleman, who had corresponded with me on mesmerism, but had never then seen me. That this kind of distant vision often occurs in the mesmeric sleep, we have already seen. It is its occurrence in the conscious state which is so remarkable. Probably, to judge from the accounts we have of them, the Highland seers were sometimes conscious, that is, while lucid, still retained their ordinary con-

sciousness, and sometimes in the mesmeric sleep, or in a dreamy state closely allied to it, in which their consciousness might be more or less distinct and separated from their waking consciousness.

10. I have not, in treating of lucidity, mentioned, so fully as I am now enabled to do, the fact that Major Buckley generally produces the clairvoyance I have described, in which the clairvoyants read mottoes, &c., enclosed in boxes or nuts, and known to no one present, without causing mesmeric sleep, or affecting consciousness. He has now produced this remarkable state and degree of lucidity in 89 persons, most of whom belong to the upper, educated class. Major Buckley has most kindly furnished me with some details, which will be found further on. I shall speak of his method of operating, when I come to the attempt to explain these phenomena.

11. To return to the second sight. It is also said to extend to future events, and I am not prepared to deny the possibility of this. I have elsewhere referred to the prediction of a seer, which had become a universally believed tradition in the Highlands, that the male line of Mackenzie, Lord Seaforth, should be extinguished in the person of a "Caberfae," or head of the name, as in that clan he is called, who should be deaf and dumb. The last Lord Seaforth, whom I have seen in his unfortunate condition, deaf, unable to speak articulately, and suffering from paralysis, was in his youth a man of uncommon ability, and free from any such defects. He had a large family, and several sons, all of whom died before him, so that the title is now extinct. The family is represented by the Hon. Mrs. Stewart Mackenzie of Seaforth. The prediction was current long before the events took place, how long I know not; but the period of its fulfilment was indicated by the seer as to be marked by the contemporaneous occurrence of certain physical peculiarities in one or two chiefs of other clans. These also occurred, as has often been affirmed to me; and it is mentioned in Lockhart's Life of Scott, that Mr. Morrit of Rokeby, being on a visit to Brahan Castle, the seat of the family of Seaforth, heard the tradition in all its details, while Lord Seaforth had yet several sons alive, and in perfect health.

12. A still more remarkable instance of prevision is that of M. de Cazotte, who, some years before the breaking out of the French Revolution, predicted, with great minuteness, its bloody character, and the precise fate of many noblemen, literary men and ladies, and even that of the King and Queen. This prediction, which was uttered at a time when all in Paris looked forward with confidence to the peaceful march of Reform, excited great sensation. It was communicated, by persons who were present, to friends in England, and became the topic of conversation in the highest circles, while no one believed in its fulfilment as probable or even possible. Persons are or lately were alive, who knew of it at that time. It has often been printed, and I shall give some account of

it further on. Here I shall only observe, that we have an account not only of the vision or tradition, as in the former case, but of the seer. And it is most important to remark a fact which, although often recorded, is omitted, as not essential to it, from many editions of the story, that M. de Cazotte was frequently in the habit of uttering predictions; and that, previous to doing so, he invariably fell into a state, which is described as a kind of sleep or dreaming, but not ordinary sleep. It was, doubtless, either the mesmeric sleep, or a state of deep and dreamy abstraction, favourable to lucidity. I confess that this circumstance is, to me, the strongest evidence, if such were wanting, of the truth of the story.

13. This leads me to allude, briefly, to the subject of popular traditionary prophecies, to which, especially as regards certain predictions long current in Germany, I ventured to draw public attention in an article in *Blackwood's Magazine* for May, 1850. These predictions enter into very curious and often minute details exactly resembling such as would be given of a distinct vision. They are generally described as having been seen by the prophets or seers, some of whom appear to have been what are called in Westphalia, Spoikenkiker, that is, ghost-seers, in other words, highly sensitive to the mesmeric influence. At all events, these ghost-seers, are frequent in the country, and among the people, where the prophets also have appeared.

The predictions above alluded to, refer in general to events to happen in Germany about this time, that is soon after the introduction of railroads, and especially to a dreadful general war, in which the final conqueror, or great monarch, is to be a young prince who rises up unexpectedly. The war is also to break out unexpectedly, and suddenly, after a period of disturbance and revolutions, while all the world is crying, "Peace, peace." I need not here enter into more minute details, for which I refer to the article above mentioned. I shall only add, that the state of Europe, and the events which have occurred since that paper appeared, are much in favour of the general accuracy of the predictions, whatever their origin. Time alone can show, how far they are to be fulfilled. But their existence, as authentic and generally received traditions, is at all events a remarkable circumstance.

It appears to me, that while some predictions, which are said to have been fulfilled, may have been nothing more than the deductions made by a superior intellect, contemplating actual facts, and their most probable consequences, this explanation cannot apply to such as that concerning the Seaforth family, nor to that of M. de Cazotte; nor, should they be fulfilled, to many of those of the Westphalian and Rhenish seers, who are ignorant, illiterate peasants.

Men, such as Napoleon, have hazarded predictions as to the

inevitable occurrence of a great war, the end of which, according to Napoleon, would decide whether Europe was to be Cossack or Republican. But such guesses, dictated by profound reflection, or extensive knowledge of men and of politics, as well as of history, are always vague and general; whereas the predictions above mentioned are not only precise and minute, in many cases, but are described as being seen, in a trance, sleep, dream, or reverie, in short, as actual visions, not deductions. That such visions, even if true previsions, may be mixed up with mere dreams, and exaggerated or otherwise disfigured by preconceived or suggested ideas, is very probable. But the cases above referred to, with their minute particularity, cannot be disposed of in this way. While, therefore, I doubt not that many spurious predictions have existed, I think we are entitled to admit, that genuine prevision or second sight is a possible occurrence as a spontaneous fact, and that it is to be explained on the principles of mesmerism.

In my next chapter, I shall endeavour to show that the explanation of the phenomena of mesmerism is not so hopeless a matter as may at first sight be supposed. I have already pointed out, that we are unable to explain any natural fact, even the most familiar, in the sense of ascertaining its ultimate cause. All that we can do, is to reduce or refer facts to certain natural laws, which, like that of gravitation, are nothing more than collective facts enabling us to see the law or rule, according to which the facts occur, but not throwing any light on the ultimate question, *Why* or *how* they occur? We know that the Sun and the Earth attract, or tend towards, each other with a force, which has been shown to vary, according to a certain rule, in proportion to the masses of matter which act, and to their distance from each other. But we know no more. We cannot tell why they should tend towards or attract each other; nor what is the nature of the force; nor how it operates; but only that it does operate; or rather, we know only that a fact occurs, and we ascribe it to a force, which we conceive to be the cause of it. Our explanations, in any branch of science, reach no farther than this; and I propose to show, that we may, if we only study it, learn as much, in process of time, about the nature or cause of mesmerism, as we know about Gravitation, Chemical Attraction, or Electricity.

Chapter X.

AN EXPLANATION OF THE PHENOMENA ATTEMPTED OR SUGGESTED—A FORCE (ODYLE) UNIVERSALLY DIFFUSED, CERTAINLY EXISTS, AND IS PROBABLY THE MEDIUM OF SYMPATHY AND LUCID VISION—ITS CHARACTERS—DIFFICULTIES OF THE SUBJECT—EFFECTS OF ODYLE—SOMNAMBULISM—SUGGESTION, SYMPATHY — THOUGHT-READING — LUCID VISION — ODYLIC EMANATIONS — ODYLIC TRACES FOLLOWED UP BY LUCID SUBJECTS—MAGIC AND WITCHCRAFT—THT MAGIC CRYSTAL, AND MIRROR, ETC., INDUCE WAKING CLAIRVOYANCE—UNIVERSAL SYMPATHY—LUCID PERCEPTION OF THE FUTURE.

LET us now try whether we can, by comparing the facts which have been ascertained, throw any light on the cause or causes of the phenomena of mesmerism. It is well known that Mesmer ascribed the facts which he observed to a peculiar fluid, which he called the magnetic fluid. This he supposed to exist in the human body, as well as in the magnet, and he does not seem to have distinguished accurately between vital magnetism, and what has recently been called ferro-magnetism, that power, namely, by which the magnet attracts iron-filings, and the magnetic needle, when freely suspended, points North and South.

As it was soon and easily perceived, that the human hand does not attract iron-filings, and cannot give to the needle the property of pointing to the magnetic poles, it was rashly concluded that Mesmer's idea of a magnetic fluid, existing in the body, was altogether without foundation; and his facts were then rejected and denied. There can be little doubt that Mesmer and his followers were in a great measure to blame for this result. They shrouded their operations in mystery, and spoke with undue confidence on points of theory which had not been sufficiently investigated; nor is the memory of Mesmer quite free from the reproach of a certain amount of charlatanism, and of having preferred his own personal interests to those of science. A not unnatural prejudice was thus created against what he taught, and the progress of truth was retarded. The reports on mesmerism by the various commissions of men of science appointed, in France, to examine it, although in some points favourable to the existence of facts well worthy of investigation, yet had on the whole an unfavourable effect, as the commissioners were not successful in obtaining good evidence of the higher phenomena, and evidently leaned to the opinion that, in some cases, deceit was practised. Yet these reports were not decisive. Bertrand and Colquhoun have shown their imperfections, and they are now seldom referred to by those who know anything of the subject practically.

On the other hand, the Marquis de Puysegur, who devoted his life to the practice of mesmerism, and who was far above suspicion, continued to obtain, by the method of Mesmer, very remarkable results, especially in lucidity and the cure of disease. The experiments, also, of Dr. Petetin, of Lyons, established the fact of the transference of several senses to the epigastrium as a spontaneous occurrence; and in all countries, men continued quietly to study the subject, chiefly, however, with a view to its use as a remedy. No one undertook a regular inductive experimental investigation, but many facts were empirically observed.

At last Baron von Reichenbach, about 1843 or 1844, was accidentally led to study the effects of magnets on susceptible persons, at first, indeed, on such as were suffering from disease of the nervous system. This inquirer was peculiarly fitted for the task. To a thorough scientific training, and the reputation of an accurate observer and skilful experimenter, amply justified by his many scientific memoirs, and his success in improving the manufacture of iron, he added the advantages of an acute and logical intellect, of habits of patient and persevering thought, and caution in drawing conclusions, as well as extreme conscientiousness in reporting the facts observed. It was fortunate for science that his attention was directed to the subject. But I must here mention, that his experiments, so far as I know, were not made on persons in the mesmeric sleep, but were rather confined to the influence exerted by magnets, crystals, the human hand, &c., &c., on persons in the natural waking condition. His object was, to begin at the beginning, and to lay a truly scientific and lasting foundation for more extended inquiries.

He began his investigations with a strong prejudice against the views of Mesmer, a prejudice universally diffused among the scientific men of Germany. But he was soon compelled, by the force of facts, to admit some of the very statements which had been most repugnant to him. I may specify, the influence exerted by the magnet on the human frame; the fact that water may be magnetised, so as to be known by the patient from ordinary water; the action of the human hand; its power of magnetising water as well as the magnet does; and the appearance of light from the fingers of the operator.

He now entered on a regular series of researches, continued during nearly five years, and on upwards of 100 persons, in which he made the important discovery, that light, visible in the dark to sensitives in the waking state, is emitted not only by the hand, and by the magnet, but by crystals, and, in fact, by all bodies more or less. He further observed, that his sensitives, when in a cataleptic state, or in that of spontaneous somnambulism, became far more sensitive than before. He found, that heat, light, electricity, galvanism, chemical action, friction, animal life, and vegetation, all caused emanations of the same light seen in the magnet, in crystals, and in the human hand. And he finally ascertained, that sensitiveness is not a morbid state, but is found in healthy persons, and that one person out of three is more or less sensitive, not indeed always to the light, but to the influence of magnets, &c., as proved by their sensations.

By these laborious researches, Reichenbach was finally compelled to adopt the hypothesis of a peculiar influence, or force, to which he has given the name of Odyle, and which he regards as the cause of all these phenomena. And he was also forced to admit that odyle is identical with the mesmeric fluid of Mesmer, that is, in so far as the latter differs from ordinary, or ferro-magnetism, from which Mesmer failed to distinguish it. In magnets, odyle is associated with ferro-magnetism; in light, with light; in heat, with heat; in electricity and galvanism, with the electric influence; in crystals, &c., it is found alone, and, while analogous to all these imponderables, forces, influences, or fluids, as some of them are often called, it is distinct from all.

The name given to this influence is a secondary matter. I have called it, in the preceding pages, Mesmerism, as being a name established and known, but if it is to have a new name, that of Odyle, which conveys no theoretical opinion, is preferable, and indeed unexceptionable. In regard to its nature, we know no more than we do of heat, light, electricity, galvanism, ferro-magnetism, chemical action, cohesion, gravitation, &c. We may call it a force, as we do chemical action, gravitation, and cohesion; or a fluid, as we speak of the electric, galvanic, or ferro-magnetic fluids; or an imponderable, as we call heat, light, electricity, and magnetism. It may be essentially a motion in the particles of matter, as heat and chemical action are supposed to be; or a motion in the particles of some subtle ether, as light is now considered; perhaps even of the same ether, if such exist. All this may, or may not be. What we know is only that certain facts occur, and we cannot from the constitution of our minds, avoid referring these to a force or influence, which, when the phenomena are carefully studied, is found not to be identical with any of the other influences or forces above mentioned, and must therefore have a name to itself.

It is quite possible, and even probable, that many, or all of these influences, may, in the progress of our knowledge, be referred to one and the same common cause or principle. But till such a common principle be discovered, the attempt to classify the phenomena of Odyle under any other imponderable, such as heat or light, electricity or ferro-magnetism, would only lead to a hopeless confusion. And even after such a common principle shall have been discovered, it will be necessary to classify these different phenomena, as different modifications or phases of it.

For these reasons, I shall use the terms odyle and odylic in discussing the subject theoretically; premising, that I use odyle as synonymous with what has been called mesmerism in the foregoing chapters.

The odylic influence, like that of heat, light, and electricity, is universally diffused. In regard to Ferro-magnetism, that force was long supposed to be confined to certain bodies, such as the loadstone, and two or three metals. But even then, the earth was necessarily regarded as a vast magnet. The beautiful discoveries of Faraday, however, have shown, that in a modified form, called by him diamagnetism, because bodies under its influence place themselves, when freely suspended, at right angles to the magnetic meridian, ordinary magnetism is possessed by all bodies. And his most recent discovery, that oxygen gas is attracted by the magnet, is a pregnant proof that our knowledge of this force, and of its effects on our earth and atmosphere, admits of indefinite extension. I need hardly point out, that the discovery of diamagnetism, harmonises well with the earlier discovery of Reichenbach, that all bodies act on sensitive persons, and give out luminous emanations, as the magnet does.

The universal diffusion of the odylic influence naturally leads to the anticipation, that, like heat, light, &c., it must exercise some action on the human body. That it does so, has been experimentally demonstrated, and may at any time be proved by the action of crystals on the sensitive.

Like heat and light, odyle is transmitted through, or traverses space, by what may be called radiation, and is also, like heat and electricity, conducted through bodies. It appears to travel less rapidly than light; but it is conducted through matter much more rapidly than heat. It passes readily through all known substances, but with somewhat less facility through fibrous or interrupted than continuous structures. Thus paper or wood are not traversed so easily by it as some other bodies, but cannot long arrest it. Heat passes very slowly through most bodies excepting metals; and electricity is arrested by most non-metallic bodies, and is indeed only well conducted by metals, charcoal, and certain liquids.

Odyle may be, to a certain extent, accumulated in a substance, but is slowly dissipated again. The body charged with it retains the charge longer than does one charged with electricity.

Like electricity and magnetism, odyle has a strong tendency to a polar distribution. Thus, in magnets, in crystals, and in the human body, it is polar, and the opposite poles exhibit distinct modifications. In bodies confusedly crystallised, and in amorphous bodies, the polar arrangement cannot be observed.

Like heat, light, electricity, and magnetism, odyle tends to a state of equilibrium, and its external manifestations seem to depend chiefly on the disturbance of this equilibrium. Just as a hot body, radiating more heat than it receives, soon comes to an equilibrium of temperature with the surrounding bodies, so does a body, odylically excited or charged, tend to, and ultimately attain, an equilibrium of odylic force with the surrounding bodies. And just as ferro-magnetism is found polar and active in a magnet, which is in a certain ferro-magnetic state, a state we may call a peculiar disturbance of equilibrium in the ferro-magnetic force; so, in a body odylically polar, and odylically active, such as a magnet, a crystal, or the human body, we may suppose a like peculiar arrangement or distribution of the odylic force, the results only of which, and not its precise nature, are known to us.

Such is a very brief and popular sketch of the nature of that influence to which I consider we must, in the present state of our knowledge, refer the phenomena of mesmerism. But while I have endeavoured to show that odyle is to be ranked with the known imponderable agents, I am quite aware that we know comparatively little of the laws by which it is regulated. The observations made on persons artificially mesmerised, have hitherto been almost entirely empirical, and have not been guided by a plan of research. Those on natural somnambulists have been accidental and imperfect; and the little that we do know on the subject, we owe entirely to the reseaches of Reichenbach, made on persons in the ordinary state. These researches have laid a firm and lasting foundation for future investigations, but it must be remembered that the subject has peculiar difficulties.

First, there is the fact, that hitherto the observer has not been able to examine the most important facts by the aid of his own senses, but has had to trust to the sensations of others. In my Preface to the translation of Reichenbach, I have endeavoured to prove, that this is no argument against the observations, if made with due care, and on a sufficient number of persons. There are many facts, admitted and reasoned upon in all sciences, by those who have not personally seen them, and know them only by the reports of those who have. Nor is there any reason why this should be less practicable in odylic inquiries than in others, for example, in geology, geography, meteorology, and above all, in medicine. Many facts in medicine rest exclusively on the testimony of the patient, whose descriptions the physician cannot test, save by comparing them with those of other patients, or with those of the same patient at another time. Everything, in medicine as well as in researches on odyle, depends on the sagacity, knowledge, experience, and honesty of the observer or physician, who, if he cannot distinguish false statements from true, after a little experience, is not well qualified for such observations. An observer, possessing the necessary qualifications, will not find this difficulty insuperable ; but it is a difficulty, and should stimulate us to increased ardour and perseverance in the pursuit of truths, which are at once more interesting and more difficult of attainment than many others. The difficulty, however, becomes daily less by practice, because we learn the necessary precautions against illusion ; and our increasing knowledge of the phenomena enables us to detect fallacies and to avoid sources of error.

Secondly, we do not yet possess any means of collecting, accumulating, and concentrating the odylic force, as we can do in the case of magnetism and electricity. The odyle, which, in the magnet, accompanies ferro-magnetism, is, indeed, more powerful in proportion to the force of the magnet, but, even in the most powerful magnet yet tried, it has not reached so high a degree of intensity as would be required to enable us to investigate its properties with ease and complete success. One great desideratum is an odylic battery, just as we have a galvanic battery. Since chemical action gives rise to odyle, it is probably in some form of chemical action that we shall find the means of constructing such a battery, and when we shall have thus obtained the power of odylically affecting every man, a vast step will be gained.

Thirdly, we have not yet obtained, as in heat, light, electricity, galvanism, or magnetism, a convenient and accurate means of measuring the quantity or intensity of odylic force. We do not possess any substance, which is so affected by odyle, that its consequent changes may be rigidly measured and referred to a standard. But if we reflect, that odyle has only just been discovered as a physical agent, and that the galvanometer and photometer are but recent inventions, while daily improvements are made in all our instruments for measuring electricity, light, and the other analogous forces, the natural conclusion is, that a diligent study of all the properties of this new force, will ere long yield us an odylometer. It is not improbable that it may be of an animal nature, since the most marked property of odyle is its peculiar action on the nervous system. But whatever its nature, such a discovery would at once do the work of a century in promoting the progress of odylic science. Let us hope that it will not be long delayed.

In the meantime, let us examine whether we cannot turn to some account the knowledge which we already possess of odyle and of its manifestations, in accounting for the phenomena of mesmerism in a natural way.

1. If the human body be a perpetual fountain of odylic force, in virtue of the chemical changes at all times going on within it, in the processes of respiration, digestion, assimilation, excretion, secretion, muscular and nervous action, &c., and if any body, containing odyle, radiates it to all other bodies, it is easy to see how the vicinity of a vigorous healthy person may powerfully affect one sensitive to odylic influence.

2. If the odyle in the human body be polar, and if, as is the case, according to Reichenbach, the hands be the chief or primary poles (there are, as in crystals, secondary axes and poles, but in man the transverse axis is the principal one), then we can easily understand how the hand should be so powerful in producing mesmeric, that is, odylic effects, as we find it to be. This is confirmed by the fact that light is seen to issue from the points of the fingers, not only by sensitives in the ordinary waking state, but still more vividly by persons in the mesmeric sleep.

3. These persons, of both classes, also see light proceeding from the eyes and mouth, which are also subordinate foci of odyle, the eyes being, as the hands, and indeed the halves of the body are, oppositely polar. Hence the efficacy of gazing, or fascination, and of breathing on the head or over the heart of the patient, with the mouth close to these parts, the breath being highly charged with odyle.

4. Supposing a current of odyle, like that of galvanic electricity in an open circuit, to tend to pass from the positive or + pole, through any interposed body, to the negative or — pole, then, as the left hand is odylo-negative, and the right hand is odylic positive, the odylic current must pass from the right hand, through any substance held in the hands, or touched by them, to the left hand, thence up the left arm, through the chest, and so down the right arm, till the circuit is completed. When the circuit is not closed, as we close it by joining the hands, or by holding a conductor with them, the odylic current does not take place, just as in an open galvanic circuit; the odylic force is in a state of tension, and polar, that is, strongest at the poles.

If now the operator, in whom a similar state of tension exists as in the patient, takes hold of his hands, right in left, and left in right, which is the natural or normal way, the current from his right hand, the patient acting as a conductor, and closing the circuit, will flow into the patient's left hand, up his left arm, through the chest, down the right arm, thence to the operator's left hand, up his left arm, through his chest, and down his right arm, thus completing the circuit.

This current is very strongly felt by the sensitive, and as it harmonises with, and is added to their own, the sensations are agreeable, although they often become too violent, and if continued would cause, in some cases, fainting or convulsions. This general fact, observed by Reichenbach in waking sensitives, I have often confirmed, as to the great power and agreeable nature of the sensation, which is often compared, by the patient, to something flowing exactly in the direction described.

5. But if the operator's hands be crossed, before he takes hold of the patient's, then the two currents are opposed, instead of being united. The result is, that a species of contest occurs, accompanied by sensations so horrible, that very sensitive persons

cannot endure the experiment for a minute, and can hardly ever be persuaded to allow it to be repeated. This remarkable fact, observed by Reichenbach, I can confirm to the fullest extent; having been fortunate enough to meet with a highly sensitive patient, who, when I tried the experiment, without saying one word to her, tore away her hands after a few seconds, and declared the sensation to be so intolerable, that, had it continued only a few seconds longer, she must have fainted and been convulsed. No entreaties or bribes could induce her to let me repeat the trial; indeed her expressions were almost verbatim those of one of Reichenbach's most sensitive patients, and this poor woman had never heard either of him or of his book, which at that time, early in 1846, was not yet known in this country. In a less striking degree, I have seen the same fact in many other cases.

6. It is obvious that the action of magnets and crystals, both of which are polar, on the patient, admits of the same explanation. Both, in fact, cause currents, differing according to the pole held in the hand, or to the hand which holds the pole. The pole, which causes a cold sensation in the right hand, produces a warm one in the left, and *vice versâ*. This I have verified more than a hundred times.

Non-polar bodies are altogether, according to their nature, cold or warm to the patient. It is odylo-negative bodies which are generally cold, such as oxygen, acids, &c.; and odylo-positive bodies which are warm, such as hydrogen, alkalies, &c.

7. With regard to what is called the mesmeric state, whether that extend to mesmeric sleep or not, we may attempt to explain it in the following manner. Ordinary sleep has been proved by Reichenbach to be connected with a change in the distribution of the odylic influence in the body. During sleep, the head, generally, is less odylically charged than in the waking hours. For details, I refer to the work of Reichenbach. Now, by the action of the operator, who, whether by passes, gazing, or contact, throws some of his odyle into the system of the patient, a change is produced in the relative amounts of odyle in different parts of the body or of the head, or, in other words, the distribution as well as the absolute quantity of odyle in the patient is changed. The precise nature of the change is not known; but we can readily conceive how, if different from the normal nightly change, as it undoubtedly is, it may produce a peculiar kind of sleep, in which the intellect remains awake, while the external senses are drowned in slumber. Such is the general view I would propose to take of the production of the mesmeric state, and of the sleep. The essential character of it I take to be this, that while most or all of the external senses are cut off from action on the sensorium, the internal senses are, perhaps in consequence of this, more alive than usual to odylic impressions of all kinds. It is certain, that persons in the mesmeric sleep, or somnambulists, spontaneous or

artificial, are always highly sensitive to odylic impressions, such as the light, the sensations of heat and cold, &c. Such persons as are moderately sensitive in the waking state, often become, in the state of somnambulism, sensitives of the highest order.

8. I would regard the spontaneous occurrence of somnambulism as nothing more than the spontaneous occurrence of that peculiar distribution of odyle which is caused in the mesmeric processes, but without any addition to its quantity. And the same view will apply to the impressible state produced by Dr. Darling's process, as well as to the mesmeric sleep of Mr. Braid's method, in neither of which is odyle added from without.

9. The power of the operator over the volition, sensations, perceptions, memory, and imagination of his subject, when the latter is in the mesmeric or odylic sleep, and without any suggestion, may be supposed to depend on the odylic force of the operator being superior to that of the patient, while, from the conductibility and ubiquity of odyle the operator continues in communication with that portion which has passed into the system of the patient. If odyle be the nervous force, or vital force, and it is at least as likely to be so as electricity, then it may be the odyle of the operator, overpowering that of the patient, which moves the muscles and determines the sensations, &c., of the latter.

10. The attraction of the patient towards the operator, both mental and physical, may be explained on the same principle. This supposition, as well as the preceding ones, receives considerable support from the fact, that the sleeper constantly speaks of a light round his mesmeriser, or of a luminous vapour, which extends to and embraces himself.

All the facts ascertained in regard to odyle point to an external influence, passing from one body to another, and here we have a visible something, which is seen to pass, not only from the points of the fingers, but from the whole person of the mesmeriser, to that of his patient.

11. The disagreeable and injurious effect of cross-mesmerism arises, or rather may be supposed to arise, partly from the conflict of different odylic influences, and partly from the accidental reversal of currents, which must often occur, when others take hold of the patient's hands, or touch him in various ways. The facts are notorious to all who have looked at the subject.

12. The antipathy of the sleeper to certain objects or persons, probably depends on the strong and disagreeable odylic sensations they cause, according to whether they are negative or positive in reference to him.

13. Sympathy most probably depends on the harmonious action of the odylic force of the operator, or of any other person, on the temporarily exalted odylic state of the patient. And we may, without much difficulty, conceive odyle to be the medium through which the impressions of sympathy are conveyed.

The existence of sympathy cannot be doubted. It is constantly seen in spontaneous manifestation, as I have already mentioned. Nay, it is often used, by those who are unwilling to admit the existence of direct clairvoyance, as furnishing an explanation of that phenomenon less repugnant to their preconceived opinions. Such persons, rather than admit that the clairvoyant possesses the power of vision without the use of the external eye, at once adopt or even suggest the hypothesis of such a degree of sympathy with the operator, as enables the subject to read all his thoughts with perfect accuracy. They do not stop to inquire whether this be in reality less wonderful, or less unaccountable on ordinary principles, than the notion of direct clairvoyance; nor do they consider that such sympathy is as truly a new sense as anything can possibly be. But all men know, that sympathy, to a very remarkable degree, is a daily recurring fact, and, although quite unable to explain it, having never perhaps thought on the subject, they embrace it at once as a refuge from the obnoxious idea of direct clairvoyance.

I need hardly remind you here, that, even if we admit, as I unequivocally do, sympathy and thought-reading as sufficient to explain if they be admitted, many instances of clairvoyance, namely, such as I have called sympathetic, mediate, or indirect, clairvoyance, yet there are many facts which this explanation will not reach.

I wish, in this place, only to point out, that, admitting sympathy, odyle is most probably the agent or medium. The odylic atmosphere of the operator, and that of the subject, interpenetrating each other, and the former predominating over the latter, the subject becomes for the time, a partaker in the thoughts and feelings of the operator; and thus, although the details of the process are shrouded, at present, from our sight, thought-reading is brought about. I have no doubt that such sympathy does occur, and I consider it highly probable, that the explanation here suggested, as far as it goes, is the true one. Of course, I understand it to apply to spontaneous, as well as to artificially excited sympathy. It is obvious, that to an influence like odyle, as to light, distance may be a matter of no importance. If odyle exist, it traverses space, as light does; only, as appears from the observations of Reichenbach, not quite so rapidly.

14. We now come to the explanation of direct or immediate, or true clairvoyance, which I have already given my reasons for admitting, as an ascertained fact, whether we can explain it or not.

This phenomenon is the great stumbling-block. Some boldly declare that they never *will* believe it, forgetting that belief is involuntary, and that, on sufficient evidence, they must, if they attend to that evidence, believe it. In my opinion, after reading, I will not say the whole recorded evidence, but as much of it as I could procure, the recorded evidence of the fact is sufficient. But I have not expressed that belief, until after I had seen it myself. Now I have observed two things among those highly estimable persons who do not yet admit the fact of direct clairvoyance. The first is, that they are, in general, quite unacquainted with the recorded evidence. The second is, that their objections, when sifted, always assume ultimately the shape of an assertion,—that it is impossible; or that, as it cannot be accounted for, it must be rejected. I might add a third, namely, that these persons have rarely, if ever, investigated the matter for themselves. Now I do not quarrel with the philosophic caution which declines to adopt an entirely new and startling fact, unless on unexceptionable evidence, or on ocular demonstration. But when the witnesses are numerous their character unimpeached, and the fact not physically or mathematically impossible, caution is not entitled to go further than to say, "I am not satisfied; I must inquire into these things." The most cautious philosopher has no right absolutely to reject facts thus attested, because he cannot see their explanation; and, above all, he has no right to brand the witnesses with the charge of deceit or imposture, *without full and careful inquiry*. If he will not, or cannot, investigate, let him, in decency, be silent. I do not invent; I speak of what happens every day; and I say, that those men of science who, declining to investigate, have nevertheless fulminated denial and accusations of falsehood against those who have investigated, have not acted on the golden rule, "Do unto others as ye would that they should do unto you," and their conduct is as illogical and irrational as it is unjust and impolitic.

But while I protest against this conduct, on the part of men of science, who ought to know better, I make every allowance for those not trained to scientific pursuits, many of whom, unwittingly confounding belief and understanding, really have a difficulty in admitting anything for which a plausible explanation cannot be found. I have more than once pointed out, that if we were to reject all that we cannot explain, little indeed, if anything, would be left; but that our explanations are, at best, but attempts to classify phenomena under natural laws, of which we know no more, than that they exist or appear to us to do so. Yet such explanations are of great use in facilitating the apprehension of scientific truths, and therefore I shall now endeavour to give such an imperfect explanation of clairvoyance as occurs to me, in the present state of our knowledge, as being admissible. It is only an attempt, however, and is not to be regarded as truth, but only as an allowable hypothesis. It is again to odyle that I refer, as the cause or rather medium of the manifestations of this kind of vision.

First, let me remind you, that the first observation of many lucid persons is, that they see, with closed eyes, the operator's hand, as well as his person, and other objects, and that all are luminous; indeed, they are often described as surrounded by a luminous vapour or atmosphere, which, as I have already mentioned, embraces the subject, and mixes with his own atmosphere. Now I think this is clearly an odylic phenomenon. The objects seen are seen in odylic light, to which lucid persons are invariably highly sensitive.

Secondly; the eyes are not used, but the objects, if not clearly seen, are placed on the head or forehead, commonly on the anterior coronal surface. I have seen a clairvoyante, who, in trying to write, *looks at* the paper with the top of her head close to it, her eyes closed and useless from their position. I have a specimen of writing thus executed by her. She is the same, who, when she wishes to place an object, such as a letter or lock of hair, "right before her eyes," places it in contact with the anterior coronal region of her head. If odyle or odylic light be here the agent, the cranium is no barrier to its passage to the brain, for odyle traverses all solid bodies that have been tried.

Thirdly; when distant objects are seen, the clairvoyant, if asked how he sees them, often speaks of a luminous cloud or fog, extending from them towards him, and joining a similar cloud from himself; in this combined cloud he then sees the object, at first dim and grey, afterwards plain and in its natural colours. This description tallies well with our hypothesis of the universal diffusion of odyle as the medium for lucid perception.

All this may be learned from the accounts given of their own sensations, given by intelligent lucid subjects, and it corresponds closely to the statements, on many points, of the sensitives of Reichenbach, who were in the waking condition.

15. Now, if we would proceed a little further, and endeavour to discover how lucid perception is obtained, I must again remind you of an opinion, which is not new, that every influence emitted by any body acts, so to speak, on all other bodies. The heat, light, electricity, and sound, emitted by any bodies, fall upon all other bodies, and consequently on our organs of sense, but so weakened as to be utterly overlooked among the stronger impressions caused by nearer objects of sense.

16. Now, let us suppose that the odylic emanations, which appear certainly to be emitted by all bodies, fall on our inner sense, they also are entirely overlooked, in persons of ordinary sensitiveness, because they are very feeble, when compared with those of sight, hearing, smell, taste, and touch. The sensitives, however, perceive them, when their attention is directed to them, and under favourable circumstances; and the lucid are always very highly sensitive.

17. Next, let us see what happens in the mesmeric sleep. In that state, the two most marked characters are, the closing of some one or more of the external senses, especially of sight, and of

hearing (for all sounds save the operator's voice), the two which are constantly receiving impressions from without. The consequence is, that the inner sense, no longer distracted by the coarse impressions of these senses, becomes alive to the finer odylic emanations (which do not require the usual modes of access, as we have seen), and may even preceive the faint pulses or reverberations of the distant sights, sounds, &c., alluded to in the last paragraph but one, the odylic atmosphere aiding perhaps to convey them by their new route. If the subject be highly sensitive, and the external senses closed, he is in the very best condition for lucid perception; but the impressions he notices are not new! they were formerly overlooked because of their faintness; they are now attended to because of their intensity; for they are the strongest of all that now reach the sensorium.

18. One powerful argument in favour of this view is derived from the fact that the lucid state occurs spontaneously, and is then always preceded by abstraction, concentration of thought, reverie, sleep, or somnambulism, all of which states render us more or less dead to the impressions of the external senses, and by consequence, alive to odylic impressions.

19. I may here allude to the state of conscious lucidity, which, as I have mentioned, some individuals can produce in themselves; and this is always done by concentration or abstraction; and which Major Buckley has been so successful in producing in others. I presume some part of his process implies concentration of thought; but it is truly remarkable, that the chief part of it, after lucidity has appeared, consists in his making passes over his own face, and over or towards the object to be deciphered. Both manipulations are said by the clairvoyants, to shed a flood of blue light (negative odylic light) over the object. Too many passes render the blue too deep, and reverse passes clear it up again.

20. Clairvoyants who see the intimate structure of their own bodies, or those of others, often describe the frame as bathed in beautiful light, and entirely transparent to them. This accords, in a remarkable manner, with the observation of Reichenbach's sensitives, to whom a thick bar of iron, shining in odylic glow, appeared transparent like glass.

21. When a clairvoyant takes into his hand a lock of hair, or a letter, it would appear that the odylic emanations adhering to these bodies, according to the account given by lucid subjects, enable them to trace and discover the person to whom they belong. Can it be that the hair and the writing are never totally disconnected from him, and continue in odylic communication with him? Certain it is, not only that he is thus discovered, but that much confusion and difficulty occur when the objects have been handled by various persons. The clairvoyant sometimes sees the last person who touched them, but recognises, by an instinctive feeling, at least in many cases, that that is not the right person. He

often requests the operator, by certain manipulations, to banish the intruding images, and never hesitates about the right one when found.

22. I have been informed by a gentleman, who is able to bring himself into a conscious lucid state, and in that state to see persons quite unknown to him, at a distance, that he can only compare the process by which he finds the person asked for, to that by which a dog, liberated from confinement traces his master. He first finds the (odylic) trace in the inquirer's mind, then follows it back to the point where the person asked for last parted with the inquirer. From this point he pursues only the trace of the former, and soon finds him. I regard this observation, on which I have every reason to place entire reliance, as extremely interesting and important, in reference to the theoretical inquiry. And when similar observations shall have been multiplied, we may hope to advance much further.

23. When a clairvoyant sees past events we may suppose that he follows their (odylic) traces upward, instead of downward, from a given point. Here our theory agrees with that old one, which maintains that every event leaves an indelible trace, which continues to exist as long as the world to which the event belongs. In point of fact, in mesmeric experiments with hair and with writing, past events are described every day as vividly as present ones, whatever may ultimately prove to be the medium of communication; and this, too, where the action of sympathy with the operator or inquirer is excluded.

24. We may apply the same principles to the explanation of witchcraft, magic, and sorcery, as practised in various countries from the earliest periods. I have already alluded to this, when considering the application of mesmerism, and I have there given my reasons for not entering fully into the interesting historical questions connected with it, which are in far better hands, in those of the author of *Isis Revelata*.

It is evident that the priests of India, Egypt, and Greece, were well acquainted with mesmerism, and that they had probably various methods of producing artificial clairvoyance. The Egyptians, as is proved by paintings, used processes like our modern ones, and it is well known that soft music, a dim light, and fumigations, were used by magicians in all countries. Heathen tradition and mythology contain many traces of mesmerism. Thus the transformations of the gods into men and animals, are founded on the power possessed by the mesmeriser over the mind of his subject, even in the conscious state. I have seen this power, in regard to the perceptions of the patient, who was all the time quite conscious, so complete, that he was made to believe the operator to be any other man, or any animal, or any inanimate object; and also to believe the same of any third person.

Now, our present knowledge of these matters is quite in its

infancy, and there would be nothing marvellous or incredible in the news that some one had discovered, or rediscovered, the means of bringing any number of persons, at pleasure, into the impressible state in which suggestion acts so powerfully, and of acting on all at one time. By such means, jailers might be eluded, and he who had the secret might escape from a room full of people, eager to catch him, by simply deceiving or deranging their perceptions by such subjective metamorphoses, which they could not but firmly believe in. The ring of tradition, which conferred invisibility, is easily matched nowadays; for nothing is so easy as for the mesmeriser to render himself, and any other person or object, invisible to those under his control. The power of acquiring, silently, this control over several at once is, as I have said, now a desideratum; but that desideratum may soon be discovered if sought for. I saw, two days ago, a gentleman made, by Dr. Darling, without any preparation whatever, but just as he entered the room, to believe a watch to be a turnip, a friend to be a lamp-post, and a huge balloon to ascend majestically from the floor. From this, it is hardly a step to seeing the witch ascend on her broomstick, or the devil flying through the air to the Brocken, on the back of a goat.

We can readily understand how any one, possessed of this knowledge in an age of ignorance, might acquire the reputation of being a bought slave to Satan, from whom, at the cost of his soul, he had obtained his powers. Nay, we can even imagine, the story of the Devil and Dr. Faustus, to be merely an allegorical warning against the risk, to him who too curiously pries into the mysteries of nature, of losing his hold on religion and his hopes of a future existence, an opinion not yet extinct.

25. Possession, which was universally believed in, was obviously founded on the occurrence, spontaneously, perhaps as a symptom of a nervous disease, of a high degree of mesmeric susceptibility. The patient saw and felt what no one else did. He believed himself, as others believed him, to be under the influence of an evil demon. If subject to extasis, he saw in his visions the spirit that possessed him.

26. Clairvoyance was doubtless used, or abused, to inspire confidence in the sorcerer. When he could truly describe absent and unknown persons, unknown places, the events there occurring, nay, the very thoughts of the inquirer's mind, and the desires of his heart, he might easily pass for a prophet, even without prevision. An adroit use of the present, known perhaps by clairvoyance to him alone, might convince the world of his power to read the future.

The magic of the modern Egyptians, as Miss Martineau has concluded, depends on clairvoyance. The boys who see are chosen, because the young are most susceptible. An operation is performed, including fumigations, which, when tried on Miss

Martineau, produced to a certain degree the well-known sensation of being mesmerised. The boys are then told to gaze at a surface of ink on the palm of the hand. Now, we know that the gazing thus, without the ink, as in Dr. Darling's method, produces a mesmeric state, and even the sleep; while Major Buckley produces clairvoyance, and that too in the conscious state, by a process as easy. Probably, some passes are also made, and the boys, becoming lucid, see and describe, often by thought-reading, the persons asked for. That they sometimes fail and blunder is what happens in all such experiments, if genuine, both because the subjects are of unequal power, and because even a good subject varies in his power. But it is not possible to doubt that they often succeed, although, no doubt, there are sham magicians and spurious subjects occasionally. The ink serves chiefly as a mirror, but may also act in virtue of its odylic influence in aiding the induction of the mesmeric state.

27. The magic crystal belongs to the same category. The high odylic virtue of rock crystal seems to have been known to the adepts of the middle ages; and crystals were cut into a round or oval shape, that they might act as mirrors. Several of such crystals, said to be magic ones, exist in this country. It is now pretty generally known that one of these, said, on what authority I do not know, to have belonged to the magicians Dee and Kelly, who certainly had one, came into the possession of a noble lady, distinguished in the literary world, who has died since that time. She was told it was a magic crystal, but could not discover any of its powers. At the sale of her effects, it was purchased by a gentleman who knew its history. One day, on entering the room, he found a group of children round it, who declared that the crystal was alive; and it appeared that they saw in it the images of absent persons, it is said, even of such as they had never seen, and of some that were dead. I cannot vouch for the details of this story, however, and I shall only say that I am not prepared to reject the statements made concerning the crystal, although they may have been distorted and exaggerated.

I conceive, that when the children gazed at it, its odylic influence, added to the effect of gazing, may have thrown them into a conscious mesmeric or lucid state; and that these visions appeared *in* the crystal because they were looking into it. The subject requires to be rigorously investigated, and in the meantime we must suspend our judgment.

I have been informed of two other magic crystals, both of which are said, in the same way, to act on children. A fourth is now in my possession, and I hope to obtain its history. I am trying its effects on children, and will give the results further on. But I have tried it on a young gentleman in the mesmeric lucid sleep, who, without knowing what it was, as I put it into his left hand after he was asleep, saw in it, the first time, a light, which was so bright as to be painful, and was accompanied by a very strong odylic sensation up his arm, so that he disliked it. Another time he found the light not so painful (I rather think the crystal was in his right hand), and expressed great delight at its beauty. With shut eyes, he described it as full of bright bands, confusedly crossing each other, of the most splendid rainbow colours. On further inquiry, or rather, on leaving him to himself, he began to speak of lucid visions, which, although caused by the crystal, and quite distinct from his usual ones, he did not see *in* the crystal, but felt as if transported to the places he described. He began, and without my asking him any questions whatever, to speak of a very remarkable man, whom he saw, and whom he connected, in some way, with the crystal. I shall give further on some curious details about this man, whom my clairvoyant repeatedly visited, till, in consequence of his illness, I could make no further experiments with him. I found the same crystal to be strongly felt by other mesmerised persons, who also saw light from it, with their eyes shut.

It has lately been said, by various observers, that susceptible persons, looking into a glass of mesmerised water, see clairvoyant visions. While I write this, I have been informed of some experiments of a very satisfactory nature with the crystal above referred to as having been for a few days in my possession. These I shall give further on, along with any other facts that I can obtain in reference to the action of mirrors, crystals, or mesmerised water.

28. I have been informed on good authority, that round or oval masses of glass are made in England, and sold at a high price to the ignorant, for the purposes of divination. The persons who sell them perform a certain process, which they say is necessary to their virtue. It is probably a process of mesmerisation, as water is mesmerised. The purchaser is then directed to gaze into the crystal, concentrating her thoughts, for it is generally females who resort to them, on the person she wishes to see. She then sees her lover, or any other person in whom she is interested. Now, I believe, that by the gazing and concentration of the thoughts, aided by the odylic influence of the glass, she may be rendered more or less lucid, and thus see or dream of the absent person. So that the dealers in these crystals are not mere impostors, but, as I suppose, trade on a natural truth, imperfectly known to themselves.

29. The magic mirror is to be explained on the same principle as the magic crystal. It was a contrivance, probably a very effectual one, and depending on portions of knowledge now lost or dormant, for causing the conscious lucid state. It would appear that the researches of M. Dupotet have led him to the rediscovery, among other portions of the magic of the middle ages, of the mode of manufacture of these mirrors, which are not, I understand, of glass, but of a black substance with a smooth

surface, on which the visions are well seen. Now I should anticipate that the mass of the mirror will prove to be formed of some substance, either by nature highly odylic, or at least capable of being, by some means, strongly charged with odyle, and of retaining the charge.

The inquirer, in a darkened room, surrounded by all those objects which act powerfully on a lively fancy, in perfect silence, except for the strains of a solemn music from time to time, and steeped in balsamic and narcotic odours, is shown the mirror, on which he is told to look earnestly, and he will see the absent friend or lover, and how they are occupied. He does so, at first; after a time, sees a cloud on the mirror, which clears up, and exhibits the image on which the thoughts are bent.

Now, every circumstance favourable to lucidity is here present. The inquirer is deeply interested, the accessories are all such as promote tranquillity and concentration of thought, while they deeply affect the imagination, and thus produce, with the aid, no doubt, of the fumigations, and of the odylic influence of the mirror, the impressible state, open to suggestion. The magician, when he judges, perhaps by signs well known to him, that this state has been induced, directs attention to the mirror, and perhaps artfully suggests the nature of the vision. Or he commands the visitor, trembling with excitement, and now under his control as to his perceptions, to see the vision. Or, finally, the lucidity produced may be such as to yield a true clairvoyant vision. The appearance of the cloud, mentioned in all accounts of the crystal, of the mirror, and of the Egyptian magic, is also a circumstance almost invariably attendant on clairvoyant vision, in ordinary mesmerism.

Such are the notions which I would form of the different kinds of magic described. I give them only as suggestions, and to show that we may at least conceive these things to depend on natural causes, and the magic of former days, in this point, as it has been shown to be in some others, to be Natural Magic.

30. You will easily perceive, that the preceding attempt to give an explanation of the phenomena of mesmerism, and especially of clairvoyance, while it excludes, in regard to lucidity, the theory of sympathy *with the operator*, as applicable to all cases (for to some facts, and to some cases it undoubtedly does apply), yet admits and requires another kind of Universal Sympathy, existing at all times, at all distances, and among all things, the conducting medium and essential condition of which is Odyle. I have the less hesitation in offering my views, as I regard the existence of that influence, whatever name be given to it, as established, by the researches of Reichenbach, beyond all doubt.

I have endeavoured to show a remarkable agreement on many points, between the statements of lucid somnambulists and those of Reichenbach's sensitives, who are in the ordinary waking state.

This agreement is, I think, sufficient to justify the attempt I have made to devise what may, by courtesy, be called an explanation of these facts.

But I am sure you will do me the justice to believe, that I know the rules of scientific research too well, to attach any value to these theoretical considerations. If they shall be found justified, as suggestions merely, by the present state of our knowledge, and if they shall, in any degree, contribute to enable any inquirer better to understand and appreciate the facts, the true explanation of which will not, I fear, soon be attained, they will have served the purpose for which they were written. My own conviction is, that nothing but an infinitely more extended investigation of the phenomena of the mesmeric sleep than has ever yet been made, can enable us to make any approach to a true theory of these phenomena; and if I have here ventured, for a moment, out of the region of fact, into that of hypothesis, it has been solely from the desire to convince others that the pursuit is not a hopeless one; and to stimulate as many well-qualified observers as possible, to add to the store of well-observed facts, waiting patiently, till, out of the fulness of experience, the true theory shall be developed, as was Newton's theory of gravitation 200 years ago, and as have been, in our own day, the undulation theory of light, and the atomic theory of chemistry.

31. You will observe, that I have not attempted to explain lucid prevision. It would be mere fancy to say, that future events, as I have already hinted in the form of a query, "cast their shadows before," in the same way as past events leave their traces behind them. Can we imagine that which has not yet existed to cast a shadow, or any thing else, before it? It is true that we may again resort to a very old opinion, namely, that which holds that present events and existences contain within them, to be developed in necessary sequence, the germs of all that is future, and that the gifted eye can see, in a succession of pictures, all the steps or some of the steps of the process, as regards the persons or objects in whom the seer is interested. We come here on ground, which quakes and gives way beneath our feet; and this is the case also, in any attempt to explain the phenomena of extasis, or what may be called the spiritual chapter of the subject. While I freely admit my inability, in either case, to throw any light on the matter, I hold fast to the facts, which I regard as too well attested to be sneered down, and too interesting to be longer neglected. I cannot waver in the conviction that patient and persevering research, carried on in the sincere love of truth, will, in the end, enable us better to understand even these obscure phenomena.

Chapter II.

INTEREST FELT IN MESMERISM BY MEN OF SCIENCE—DUE LIMITS OF SCIENTIFIC CAUTION—PRACTICAL HINTS—CONDITIONS OF SUCCESS IN EXPERIMENTS—CAUSE OF FAILURE—MESMERISM A SERIOUS THING—CAUTIONS TO THE STUDENT—OPPOSITION TO BE EXPECTED.

I HAVE partly completed the task which I originally proposed to myself, when asked to record my experience and my views in regard to mesmerism. I have endeavoured, all along, to treat the matter as I conceive all obscure subjects of research ought to be treated; that is, I have dwelt on and described the facts to be studied; and urged, on all men, the duty of inquiring into them before pronouncing an opinion. I have only attempted a theory or explanation, for the reasons already given, and because I find that almost all persons, even those who admit the facts, when I converse with them on the subject, unite in declaring, that a theory of some kind, even if it be only a temporary scaffolding, destined to be swept away before the completion of the building, is necessary to enable them to reach and grasp the facts.

It would be ridiculous to deny that I have written these chapters with a view to publication. In fact, I have often been urged to publish on the subject, but many causes have hitherto prevented me from doing so, and but for your request, might still have continued to prevent me.

It has happened, opportunely, that about the time when these chapters were projected, the subject of mesmerism occupied the attention of many of my friends, especially since the beginning of the winter 1849-1850; and more recently, the visits of Mr. Lewis and of Dr. Darling to Edinburgh, have caused a much larger number of persons to feel an interest in the matter.

Under these circumstances, as I have for many years attended to the subject, I have had opportunities of discussing it with most of those who, in Edinburgh, have become interested in it, and I have also corresponded with many in all parts of the country.

The result has been a conviction, that from want of personal familiarity with the subject, and perhaps from the want of any work, in the English language, in which all the branches of the subject are briefly but systematically treated, a great amount of misconception prevails, even among such as have seen enough to become satisfied of the truth of the facts; and that, among those who reject, or but lately rejected them, a still greater amount of misconception, and, consequently, a great deal of prejudice, are to be found.

Acting on this conviction, I have ventured to hope, that, as having, for so long a time, studied the subject, and also, as having some practical knowledge of it, I might be able to produce a work, which would teach the inquirer what has been observed, what to look for, and how to look for it. The recent publication of the researches of Reichenbach, and their very favourable reception, have greatly encouraged and stimulated me to attempt to produce such a work; but no one can be more sensible of its imperfections than I am. It is not with the view of claiming indulgence for these, but merely to explain why I have not undertaken, and cannot help to undertake, such an investigation of mesmerism as that subject deserves, that I remind you that other, and imperative, duties claim the greater part of my time and attention. My most earnest wish is, to see the subject taken up by some one fully qualified to do it justice; and if these chapters should be so fortunate as to stir up any such observer to the laborious task, they will amply repay me for the pleasant labour of writing them.

I rejoice to know, as I have already mentioned, that several distinguished men of science have recently become convinced of the truth and of the importance of some of the facts I have adduced above. There never could be a doubt that this must sooner or later take place; and now, in reference to other, and still more remarkable facts, which several of these gentlemen, not having *yet* seen them, continue to reject, no one who is familiar with the subject can for a moment doubt, that these also, in their turn, will be admitted. Nay, I venture the prediction, that those who have recently seen, and admitted, the beautiful and interesting facts of suggestion, as exhibited by Mr. Lewis and Dr. Darling, and who now perceive that these facts are not entirely new, nor contrary to what is known, will, ere long, if they examine for themselves, as I have good reason to believe they will, not only meet with the mesmeric sleep, with the irresistible evidence of an external influence capable of producing it, with divided consciousness, the power of silent will, sympathy, clairvoyance, and possibly also trance, extasis, and prevision, but will also then discover, that they have hitherto clothed these things in an imaginary and repulsive dress, from which, and not from the real facts, they have recoiled; and that not one of these facts is truly new, any more than that of the power of suggestion; although the knowledge that we may produce them at pleasure, may, in both cases, be new *to them*. This I predict with confidence, provided these gentlemen study for themselves; because I have never known any one who did so, who did not come to that conclusion, as far as his experience went. We cannot insist on any man's going further;

and I would only venture to recommend to these, and to all inquirers, pending their inquiries, to refrain from expressing any decided opinion, above all, any opinion unfavourable to the character of preceding investigators.

It must be admitted, moreover, that while every man is entitled, if he choose, to withhold his assent to alleged facts of a startling character, until he shall have practically convinced himself of their truth, there is yet nothing illogical or irrational in admitting, nor is it always a proof of weak credulity to admit, even startling facts, if properly attested, without a personal repetition of the necessary experiments.

For example, when the power of the vapour of ether to produce insensibility to pain, when inhaled, was first announced, on the testimony of gentlemen certainly respectable, but altogether unknown to most of us, the fact was at once received, and is to this day received, by many persons who have neither tried the experiment, nor seen it tried by others. In this, I see no proof of absurd credulity, but only a proper respect for the testimony of persons unknown to us, and whose characters are therefore unimpeached, and to be presumed respectable.

Yet, when a far larger amount of even better evidence, inasmuch as the witnesses were often known to us, had been produced in favour of the same fact as produced by mesmeric passes, the fact was scornfully rejected. I maintain that it is absolutely impossible to point out any difference between the cases, *as far as concerns those who tried neither mode*, which can justify the opposite reception of the two truths, for such they are. Nay, at this moment, many still deny the mesmeric fact, and that, too, without appealing to experiment. Even the admirable work of Dr. Esdaile is often tossed aside with contempt, apparently for no other reason than that he, in his researches, having met with many other mesmeric facts, has also recorded these. Did those who now admit, that insensibility to pain may be produced by suggestion, admit or not Dr. Esdaile's results before seeing Dr. Darling's experiments? Nay, do they now admit them, as proving that the same thing may be done by passes, without suggestion? If not, why not? Surely the evidence of several hundred painless operations is sufficient. A hundredfold less number, without a trial of their own, sufficed to convince millions of the power of ether.

Now I wish to point out, that the treatment which such works as the truly admirable one of Dr. Esdaile, distinguished as it is by care, caution, and good logic; or as that of Reichenbach, a model of cautious scientific research into a most obscure and difficult subject; as well as many other records of well-observed facts in mesmerism, have met with from many scientific men, and others, and indeed from some of those to whom I have referred as being now satisfied of such mesmeric facts as they have seen, goes far beyond what is dictated by pure scientific caution. Caution would have dealt with Morton and Jackson as with Esdaile, had such been her dictates. And he who is not convinced by Dr. Esdaile's work, of the truth of the facts therein recorded, must labour, I say it with all respect, under some lurking prejudice, possibly unknown to himself. Truly, such a one would not believe, though one rose from the dead.

The drift of all this is to urge, on the class of inquirers alluded to, the extreme probability that Dr. Esdaile, Reichenbach, and others, whom they have now discovered to be right in certain points, in which they agree with other writers on mesmerism, will also prove to be right in those points on which our inquirers have not yet experimented. I wish them all success in their search after truth; but I wish also to see them get rid of that peculiar kind of incredulity, if it should yet lurk in their minds, which led many men, to my certain personal knowledge, without a single experiment in either case, to adopt at once the conclusions of Jackson and Morton, and to reject those of Esdaile and Reichenbach, because, to use their own words, they were "impossible."

I wish to see every respectably alleged fact in mesmerism treated as was the discovery of the power of ether; that is, either accepted on the faith of the observer (an every-day occurrence in all other sciences), or tested by a reference to observation and experiment. I shall never cease to protest, in the name of truth and science, against the system, already too long prevalent, of rejecting such alleged facts, and branding the observers with the charge of deceit, without a searching investigation, nay, without any investigation at all.

I would conclude by offering some practical considerations to such as may resolve to investigate mesmerism for themselves.

1. The first is, that, next to an ardent and sincere desire to ascertain the truth, which is to be presumed of every observer, he should be armed with patience and perseverance, without which nothing can be done in these investigations. He must not allow himself to be deterred by a few failures or apparent contradictions; but must remember, that it is only through failures and blunders that we can hope to attain to a knowledge of phenomena so little understood, and consequently so liable to the action of unknown causes of disturbance. The conviction of the necessity of patience must be a living motive, not merely a belief. In my own case, I had long believed it, as strongly laid down by all the writers on the subject; but I had failed to act on it, and not meeting with marked success in my first independent trials, I rashly concluded that I did not possess the necessary mesmeric power, and contented myself with taking every opportunity of observing the experiments of others, whom I held to be more powerful. Afterwards, however, observing that many of those who were successful did not exhibit externally, any peculiar indications of unusual

powerlor vigour, while they all agreed that, with patience, success was a most certain, I took courage, tried again, and, although not at once successful, with patience soon became so. Indeed, I am not inclined to believe, that any one, whether lady or gentleman, of average health and vigour, possesses enough of the mesmeric or odylic influence, which indeed is generated in the human body by chemical action, to be able to mesmerise any person of average susceptibility, provided patience and perseverance be practised. Nay, I would go further, and say, that I am most convinced, that every one, if the process be continued and repeated for a certain time, may be mesmerised, especially if the operator be powerful. But it is only the more susceptible who can be mesmerised in a few minutes, or at the first trial.

I have no reason to suppose that I possess more than average mesmeric power. Yet, on lately trying, with a view to this question, three young men, who had never been mesmerised at all, I was successful with all three. One of them was very slightly affected at the first trial, and at the second and third was much more strongly acted on. The effect continued, after this, gradually to increase, but at the end of eight trials, although many marked effects had appeared, sleep, that is, full mesmeric sleep, had not been attained. I regard it as certain, that in this case, a few more operations will produce complete mesmeric sleep. In the second case, some effect was also produced in the first trial, and on the third, deep and sound sleep took place. In the third subject, sleep occurred in one of the early trials, but for some time was not deep, but easily interrupted by speaking to him, and did not become at all deep till about the ninth sitting, after which there was no difficulty.

2. The second remark I would make is, that we are not to consider it a failure, if the sleep be not produced at all. Many effects, and especially many of the curative effects, may be fully brought out independently of the sleep, or of divided consciousness. It is even possible, as we have seen, to produce clairvoyance in the ordinary conscious state.

3. The variety in the minuter details of the phenomena is so great, that no two cases are exactly alike. This is just what might be expected from the fact, that no two human beings are exactly alike, either in person or in mind. If, then, in any given case, we do not obtain the same results as were obtained, perhaps, in that case, the seeing of which induced us to take up the subject, we are not to imagine, either that the former case was spurious, or that we have failed. Our duty is, to study our own case, as it is presented to us by nature, and we shall certainly find, if we do so, with variations in many details, a very great agreement in the leading and essential facts.

4. Even in the same case, the phenomena vary prodigiously at different times, more especially in degree. The subject who, to-

day, is highly lucid or sympathetic, may, to-morrow, be dull and irritable, and may fail in every trial. In such cases, we should at once desist; for the patient, if urged, only becomes less lucid, and perhaps, resorts to guessing, to satisfy his questioner. If he guess wrong, this is set down as a proof that all clairvoyance is imposture.

5. Mesmeric experiments ought to be conducted with entire privacy, no one being present but the subject, the operator, and one or two assistants or witnesses, if experiments are intended, on which conclusions are to be founded. Some experimenters, anxious to demonstrate the facts, make of every experiment a kind of exhibition, to which all their friends are summoned; but we should always bear in mind, that every additional person is possibly an additional cause of disturbance to a susceptible subject. When we have, in private, satisfied ourselves of any fact, we may then try to satisfy others, but we should take but a few at a time, and endeavour to diminish their influence on the subject by keeping them at a certain distance.

6. This precaution is more especially required, when we wish, as many do now, to repeat Reichenbach's experiments on the light from magnets, and crystals, or from the human body. Indeed, so many precautions are necessary, that unless the experiments be directed by some one who has practical experience, failure is far more probable than success.

In order to have the odylic light seen and described to us, we must strictly attend to the following conditions, and if we neglect any of them, we must not hope to succeed. 1st, We must have a truly sensitive subject, one, for example, who in the darkness of night has observed light from objects or persons. It is not enough that the subject be nervous, or hysterical, or subject to spasmodic attacks, although these are things usually favourable to sensitiveness. He should feel the magnet strongly; but after all, we must try him with the light, before we can pronounce him sensitive to it. 2nd, The darkness must be absolute. In any ordinary room, and during the day, this condition is not attainable; but with care, it may be secured at night. 3rd, The subject shall remain an hour, an hour and a half, or even two hours, uninterruptedly, in this total darkness, that the eye may acquire its full sensibility, and the pupil be enlarged to the utmost, before any trial be made. The time varies in different cases. 4th, Not a ray, nor even the faintest glimmer, of daylight or candle-light, must be admitted after the subject enters the dark chamber. All arrangements must be previously made, and no one must come in, or go out, during the whole time. For the light admitted by opening the door, is sufficient, even if feeble, to dazzle the subject's vision, so as to render him blind to odylic light for half an hour or longer. 5th, The magnet should be powerful. A permanent steel horse-shoe, carrying 60 or 80 lbs.

will suffice for most experiments, and it is easy to have an electro-magnet much more powerful. Highly sensititive persons will see, in a totally dark room, the light even from a pocket horse-shoe magnet, if of great intensity, but it is of course a light of small size. 6th, No one should hold the magnet in his hand, or on his knee, or touch it at all, while the subject looks at it. When the light is seen, the close approach of the operator, or of any one else, to the magnet extinguishes it, because its odylic influence neutralises that of the magnet, and tends to cause a reversal of its odylic polarity. A straight bar-magnet, indeed, if held in the right hand by its northward or negative end, or *vice versâ*, will exhibit a larger flame than before at the farther end, the two influences being now combined; but in the case of horse-shoe magnets, they should be set upright on a table, and the operator, after disarming them, should retire to a distance. 7th, No one should sit or stand near or close to the subject; for if they do, their influence destroys the sensitiveness more or less. When they retire, the subject often sees the light for the first time. 8th, The subject, to see distinctly the odylic flame, must be at a certain distance from the magnet; for, at a greater or less distance, the light may be visible, or only seen as a faint general luminousness. Now this distance is different in every subject. Some can see at nearly 40 inches from the magnet, others not till within 2 or 3 inches of it; others again, at intermediate distances, different for each. Few see the flame at a greater distance than four feet. In each case, the specific distance must be ascertained, and, ever after, strictly attended to. Short-sighted persons find their vision, as regards odylic light, improved by the glasses they commonly use. This condition of distance is absolutely essential, so that, even if all other conditions be fulfilled the neglect of this one will cause failure. 9th, The subject should be placed, sitting, with the body in the plane of the magnetic meridian, and the back towards the north, the feet tending towards the south, the head in the opposite direction, looking, however, to the south.

There is not one of these nine conditions, the neglect of which may not cause failure in an ordinary sensitive in the conscious state. With highly sensitive subjects, some of them exert only a secondary and modifying influence. Persons in the mesmeric sleep are, as a general rule, so intensely sensitive to odylic light that they see it in daylight.

7. Mesmeric experiments, or those of clairvoyance, should never be tried, *for the first time*, in the presence of a crowd of eager spectators, whose involuntary action on the subject confuses him, and who are sure to touch and speak to him, each striving to clear up some doubt of his own, and never once reflecting how delicate is the susceptibility of the subject. Failure is the almost unavoidable result; and, ignorant of how much of that failure is to be attributed to their own proceedings, the spectators, who had

been rashly summoned to a trial, as if it were an exhibition certain to succeed, go away under the impression that the alleged facts are not true, whereas, in truth, the trial has not been fairly made. When a good case has been found, and proves not too susceptible to accidental causes of disturbance, the facts may be exhibited, even to a considerable number, if kept at a proper distance, and under strict discipline, by the operator; but, as a general rule, two or three spectators are as many as can satisfactorily see the facts at one time. Such spectators should be near enough to see and to watch every turn of the countenance, and to hear every inflection of the voice in the subject, for these things furnish by far the most convincing evidence of his sincerity, and may often enable us to judge whether he sees directly or by sympathy. At any considerable distance, this is impossible. To avoid the disturbance caused by a near approach, the spectator should be placed by the operator *en rapport* with the subject, who will then become accustomed to the new influence, and will no longer be confused by it. This, however, can only be done to one at a time.

8. It is sometimes quite distressing to see the measures pursued by some sceptical inquirers, in cases where the very idea of deceit is not only absurd, but insulting. They will inflict severe injuries, twist and pinch the arms, and suspend heavy weights to rigid limbs, as if rigidity and insensibility to pain implied invulnerability. They will cause the subject, when excited, by mesmeric attraction, or by suggestion, to muscular effort, to contend against the whole force of three or four men, each stronger than himself; as if, to prove muscular effort, it were necessary to strain it beyond the subject's powers. They will feel his pulse, two at a time, and then be astonished if he faint, or fall into convulsions, which, every tyro in mesmerism knows, are often the results of interference or cross-mesmerism. All these things, and a great many more, among which I may mention, only to protest against them, stuffing concentrated ammonia up the nostril, without regard to its subsequent action, or applying corrosive acids to the hand, or thrusting pins through the skin, are the results, partly of ignorance of the subject, partly of want of consideration in regard to the mode of testing. It is very easy to ascertain the presence of rigidity or of muscular effort, or the absence of sensibility to pain, without doing one thing that can have a painful or disagreeable effect after the patient is awakened. No good or skilful experimenter ever resorts to such cruelties, because they are to him quite unnecessary, and prove no more than can be equally demonstrated without them.

9. It is hardly necessary to say, that leading questions, and all other forms of suggestion, ought to be carefully avoided, except in showing the power of suggestion. But, obvious as this rule is, many experiments are violated and rendered quite valueless, from

neglect of it. The somnambulist should be allowed or induced to tell his own story, and no questions should be asked, unless they are indispensable to understanding what he says, till he has told it fully.

10. The observer should carefully abstain from doing or saying anything tending to imply a belief that the subject is dishonest. It is hardly possible to convey in words the suffering which is often in this way recklessly, and indeed unknowingly, inflicted on persons whose sense of truth is as pure as their sensibility to reproach is intense. Not only have we no right to inflict this suffering, but we shall find, if we do, that it sadly mars all our efforts, and often causes very simple and easy experiments to fail. Many subjects refuse, and properly, to answer questions put in this thoroughly unscientific spirit, while they will readily and satisfactorily answer the same questions, if put by one who has confidence in their truth. One of the most striking facts in mesmerism, is the ease with which many subjects instantly detect the opposite states of mind referred to, even when they are not proclaimed in any way; as well as the sympathy and attraction they feel for those who are candid, the approach of whom is a source of pleasure and of increased power; and their decided antipathy to the uncandid, domineering sceptic, whose mere approach distresses them, and diminishes their lucidity.

11. I would once more repeat, that mesmerism is not a plaything, not a toy, not the amusement for an idle hour, not a means of gratifying a morbid craving for novelty, or for the marvellous. All such uses of it I abjure, and protest against, as abuses. Neither is it a thing to be exhibited to gaping crowds for money; to crowds who stare and laugh, and go away, thinking it very strange or very funny. It is a serious subject, well worthy of the most earnest and devout attention we can give to it. It is painful to see it abused to raise a laugh; and so strongly do I feel this, that, for my own share, I object to all *exhibitions* of it, public or private, and only admit an exception, when the exhibitor is in earnest, and his object is to convince those who feel an equally sincere desire for the truth, of the reality of the facts investigated. All exhibitions, in which mere amusement is the object, tend to degrade science, and to retard its progress.

12. In another sense, mesmerism is no plaything. I mean that rash experiments, by persons ignorant of the practical details of the subject, may probably lead to very unpleasant results. I have already discussed this matter fully; and I would only here repeat, that while such disagreeable consequences may arise from ignorant and rash meddling, I do not know of any instance of such results in the hands of a judicious operator, well acquainted with the practice of mesmerism; unless, indeed, the patient have been rashly interfered with by another. No beginner ought ever to try such experiments, without the presence of an experienced mes-

merist; but I would have every one know how to mesmerise, as this knowledge may often prove extremely useful. Of course, this knowledge ought to be acquired, not by groping in the dark, but from some experienced operator, by seeing his methods, and learning his rules.

If the above practical hints be attended to, every one who makes the trial will obtain satisfactory results. I do not say that all will see the same phenomena manifested in the same forms, for not only do subjects differ *ad infinitum*, but the operators also differ, so that one, perhaps, may never see, in his own practice, clairvoyance; or another may never meet with trance, or extasis. But few, if any, will altogether fail, and above all, most healthy persons will succeed in producing the curative effects, and will often be able, by means of mesmerism, to assuage pain and to dissipate disease.

There can, I think, be no doubt that mesmerism will now be studied like any other science, and with the same satisfactory results. Let us all do our best to promote this consummation, so devoutly to be wished.

Scientific men and learned bodies have neglected it long enough, to be quite secure against the charge of weak credulity in now directing their attention to it. Like all new truths, it has met with opposition here, and neglect there; but this is not to be wondered at, and hardly to be regretted. The tendency to oppose and reject new ideas is natural to man, although it may be pushed too far, and is designed, no doubt, to serve a good end, probably to ensure the thorough sifting of the new ideas, and the rejection, ultimately, of such as are false.

Those who cultivate mesmerism, therefore, if they know anything of human nature, should be prepared, not only to meet with opposition and prejudice, but to make allowance for these. There is no occasion for them to get angry about it, for anger never promotes the progress of truth, but, on the contrary, by exciting additional prejudice, greatly retards it.

It is true that human nature finds it difficult to remain patient and cool, when not only assailed by bad logic, and met by abuse instead of argument, but also accused of fraud and falsehood, though entirely innocent of such offences. But has not this been the fate of discoverers and innovators, of the advocates of new truths, in every age? Do we mend the matter by returning abuse for abuse, and by retorting on those who accuse us of deceit, with the charge of want of candour? For my part, I think not. I believe the opponents of new ideas to be sincere, though mistaken; and I do not so much object to their caution and incredulity in reference to strange facts, as I am amazed at their boundless credulity in regard to fraud, which, without hesitation, and without inquiry they ascribe to thousands of respectable men.

But I should regard any man's conviction of truth, as existing in mesmerism or elsewhere, to fall far short of the deep earnest feeling which it ought to be, if it did not enable him to look calmly down, while his science and himself are passing through the fiery ordeal provided for all truth, on the errors and prejudices of those who decide, as too many do, without inquiry.

There has been more than enough of angry and personal controversy on this subject. Let us try whether we cannot now discuss it calmly and rationally, as becomes men of science. And let us welcome every inquirer, even were he but yesterday an opponent, who is willing to aid in the investigation of a subject, at once so interesting, so extensive, and so obscure

The PSYCHOLOGICAL PRESS ASSOCIATION beg respectfully to announce that they now offer for publication by Subscription,

"PRESENT DAY PROBLEMS,"
By JOHN S. FARMER,
AUTHOR OF
"*A New Basis of Belief in Immortality;*" "*How to Investigate Spiritualism;*" "*Hints on Mesmerism, Practical and Theoretical;*" "*Ex Oriente Lux,*" &c.

This work, first announced a year ago, has been unavoidably delayed, owing to the Author's numerous engagements. It is now, however, ready for press, **as soon as a sufficient number of copies have been subscribed for.** The plan of the work has been considerably enlarged; its scope may be gleaned from the following draft synopsis of the sections into which it is divided.

    I.—Introductory: Giving brief résumé of ground to be traversed, and present position of Psychological Science, embracing—(*a*) What is known based on personal observation; (*b*) What is believed on reasonable grounds; (*c*) What is speculation only; (*d*) The Tendency of Material Science towards the Realm of Spirit.
    II.—Methods and modes of investigation, with suggestions.
    III.—General difficulties experienced by investigators (*a*) on Scientific grounds, (*b*) on Religious grounds.
    IV.—The Present Day Problems and their general bearing on Modern Thought.
    V.—Mesmerism. Its Rise, Progress, and Present Position. Recent Investigations, Comparison and Analysis of Results, &c.
    VI.—Thought Transference.   VII.—Clairvoyance.
    VIII.—Reichenbach's Researches and the Luminosity of the Magnetic Field.
    IX.—Apparitions, Hauntings, &c.   X.—Spiritual Phenomena.
    XI.—Summary.

This book is intended to present to the student of Psychological Science a succinct and bird's-eye view of the subjects enumerated, in each case narrating and discussing the results of recent research, and attempting to shew how each new development of science is bringing us nearer, step by step, to the Unseen Realm of Spirit. It advocates the existence of the Counterparts of Natural Laws in the Spiritual world, and proves by scientific methods that the Spiritual is not the projection upwards of the Natural; but that the Natural is the projection downwards of the Spiritual,—in short, that the Unseen World is the world of Causes, and this is the world of Effects. The Author also endeavours to trace out some of the laws which appear to govern the abnormal phenomena with which he is concerned in this volume.

To Subscribers only:—*Single Copies,* **7s. 6d.**, *or Three Copies for* **£1**.
THE BOOK WILL BE PUBLISHED AT **10s. 6d.**

ORDERS AND REMITTANCES TO BE ADDRESSED TO
THE PSYCHOLOGICAL PRESS ASSOCIATION.

Printed on superior paper, cloth, bevelled edges, with portrait of author; price 3s., post free.

A NEW BASIS OF BELIEF IN IMMORTALITY.

This book was specially mentioned by Canon B. Wilberforce at the Church Congress. He said:—"The exact position claimed at this moment by the warmest advocates of Spiritualism is set forth ably and eloquently in a work by Mr. J. S. Farmer, published by E. W. Allen, and called 'A New Basis of Belief,' which, without necessarily endorsing, I commend to the perusal of my brethren."

Mr. S. C. Hall, F.S.A., and Editor of the *Art Journal*, says:—"Your book is both useful and interesting; a very serviceable addition to the literature of Spiritualism."

"One of the calmest and weightiest arguments, from the Spiritualist's side, ever issued. . . . Those desirous of knowing what can be said on this present-day question, by one of its ablest advocates, cannot do better than procure this volume."—*Christian World.*

"This is an exceedingly thoughtful book; temperate, earnest, and bright with vivid and intelligent love of truth. Mr. Farmer is no fanatic, if we may judge of him by his book, but a brave seeker after the truth. . . . We commend this book to the attention of all who are prepared to give serious attention to a very serious subject."—*Truthseeker.*

"Mr. Farmer writes clearly and forcibly."—*Literary World.*

"M.A. (Oxon.'s)" new work, 291 pp.; price 10s. 6d., post free.

SPIRIT TEACHINGS.

The work consists of a large number of messages communicated by automatic writing, and dealing with a variety of Religious, Ethical, and Social subjects of general interest. Among the subjects thus treated may be mentioned Mediumship and Spirit Control—Spheres and States of Spiritual Existence—The Spirit Creed: God, Heaven, Hell, Faith, Belief, Inspiration, Revelation—Orthodox Theology and Spirit Teaching—The Old Faith and the New—Spiritualized Christianity—Suicide and its Consequences—The Final Judgment of Souls—Capital Punishment—The Treatment of the Insane—The True Philanthropist, &c., &c., &c.

The volume contains many cases of proof of the identity of communicating Spirits. The writer has connected the message by an autobiographical narrative, giving many details of personal experience.

Just published; crown 8vo.; paper, 1s.; cloth boards, 2s.; postage 1d.

THE USE OF SPIRITUALISM.

By S. C. HALL, F.S.A., BARRISTER-AT-LAW, ETC., Editor (during 42 years) of the *Art Journal*. Author of "Retrospect of a Long Life"—1883; &c., &c.

Being a Letter addressed to Clergymen and others containing a reply to the oft repeated question—"What is the Use of Spiritualism?"

In demy 8vo., sewed, price 3s. 6d., postage 2d.

EXALTED STATES OF THE NERVOUS SYSTEM.

In explanation of the Mysteries of Modern Spiritualism, Dreams, Trance, Somnambulism, Vital Photography, Faith, Will, Origin of Life, Anæsthesia, and Nervous Congestion. By Robert H. Collyer, M.D., Discoverer of Anæsthesia.

Price 1s., Postage 1½d.

THOUGHT-READING; or, Modern Mysteries Explained.

Being Chapters on Thought-Reading, Occultism, Mesmerism, &c., forming a Key to the Psychological Puzzles of the Day. By Douglas Blackburn.

New and Cheaper Edition. Price 5s.

THE PIONEERS OF THE SPIRITUAL REFORMATION.

Life and Works of Dr. Justinus Kerner (adapted from the German). William Howitt and his Work for Spiritualism. Biographical Sketches. By Anna Mary Howitt Watts. The two Pioneers of a new Science, whose lives and labours in the direction of Psychology orm the subject matter of this volume, will be found to bear a strong similarity to each ther in other directions than the one which now links their names, lives and labours.

Cloth, demy 8vo., price 3s.

PSYCHOGRAPHY.

Second edition, with a new introductory chapter and other additional matter. Revised and brought down to date. Illustrated with diagrams. A collection of evidence of the reality of the phenomenon of writing without human agency, in a closed slate or other space, access to which by ordinary means is precluded.

172 pp., cloth, price 3s. 6d.

THE OCCULT WORLD.

By A. P. Sinnett. A New Edition. Contents: Introduction—Occultism and its Adepts—The Theosophical Society—Recent Occult Phenomena—Teachings of Occult Philosophy.

Price 2s. 6d.

THE HIGHER ASPECTS OF SPIRITUALISM.

By "M.A. (Oxon)." A Statement of the Moral and Religious Teachings of Spiritualism; and a Comparison of the present Epoch with its Spiritual Interventions with the Age immediately preceding the Birth of Christ.

Chapter III.

PHENOMENA OBSERVED IN THE CONSCIOUS OR WAKING STATE—EFFECTS OF SUGGESTION ON PERSONS IN AN IMPRESSIBLE STATE—MR. LEWIS'S EXPERIMENTS WITH AND WITHOUT SUGGESTION—CASES—DR. DARLING'S EXPERIMENTS — CASES—CONSCIOUS OR WAKING CLAIRVOYANCE, PRODUCED BY PASSES, OR BY CONCENTRATION—MAJOR BUCCKLEY'S METHOD—CASES—THE MAGIC CRYSTAL INDUCES WAKING LUCIDITY, WHEN GAZED AT — CASES — MAGIC MIRROR—MESMERISED WATER—EGYPTIAN MAGIC.

I NOW proceed to give you some details concerning the various phenomena which have been considered in the preceding chapters. These details would have seriously interrupted the course of the general description, if introduced in the midst of it; and as many of the cases are more than once referred to, it seemed best to collect them separately. Moreover, many of them are derived from the observations of others, and a few have been already published, although I have referred to these last only in case of necessity.

In general, however, the difficulty is not to obtain facts, but to select from the number that are accessible. I shall endeavour to lay before you only as many as may serve to explain and illustrate what has been said; but there is hardly one fact in the whole series, of which, did time and space permit, many additional instances might not easily be given.

I propose to arrange the facts and cases to be adduced, in the following order. First: Those which illustrate the phenomena observed in the ordinary conscious state; that is, in a state, differing from our ordinary waking state, inasmuch as it exhibits a high degree of impressibility; but in which our consciousness is not altered, and is continuous with our ordinary consciousness. This section will include, of course, the phenomena of suggestion, whether produced by the usual mesmeric process, or by that of Dr. Darling, so far as these are seen in the conscious state.

Secondly: The phenomena observed in the true mesmeric sleep, with divided consciousness. This will include the production of the sleep; the effects of suggestion in the sleep; the phenomena of sympathy; those of direct clairvoyance; those of trance; and those of extasis.

Thirdly: The foregoing phenomena, as occurring spontaneously. And,

Fourthly: I shall add a few general remarks on the therapeutic agency of mesmerism.

First, then, we shall attend to the phenomena seen while the subject is conscious, that is, while he retains his ordinary consciousness, and is, therefore, not only sensible of all that passes, and capable of conversing and reasoning about it, but able to recollect all his sensations, when no longer in the impressible or impressed condition. I need not here dwell on the mere sensations perceived in the very first stage. These are often slight, and always varied, so that hardly any two subjects feel the same. The most common sensations are heaviness, drowsiness, tingling, numbness, pricking, creeping, and quasi-electric or galvanic sensations; a warm or cold aura, or a strong feeling of heat or cold without the aura. The eyes, especially when the subject is made to gaze fixedly on a point, are also affected. A dark veil often seems to come before them, and the object gazed at appears dark or black, or becomes multiplied, or vanishes. Some of these sensations may be observed in every susceptible case; and where they are strongly marked, we are pretty sure of producing the sleep if we persevere. But if we wish to study the phenomena of the conscious state, we stop short of the sleep and try the effect of suggestion. A case or two will illustrate this.

A. Effects of Suggestion in the Conscious State.

Case 1.—In a large party at my house, Mr. Lewis acted on the company *en masse*, standing at one end of the room, while all present were requested to gaze at him, or at any fixed point in the same direction, and to keep themselves in as passive a state as possible. Mr. Lewis gazed on the company, beginning at one end of the circle of fifty persons, and slowly carrying his gaze round, with the most intense concentration I have ever seen, as expressed in his face, attitude, and gesture. In much less than five minutes, although the necessary silence was but partially observed, several persons were distinctly affected. Among these, Mr. D., a student of medicine, very soon appeared to be the most susceptible. Mr. Lewis, observing this, directed his attention more particularly to him, and made a few distant passes, gradually approaching Mr. D. The latter bent forward with fixed insensible eyes and heaving respiration, and seemed to be attracted towards the operator. It soon appeared, however, that he was so rigid as not to be able to move forward, although he evidently tried to do so. Mr. Lewis then came near, and, by a pass or two, stopped the laborious respirations, and removed the general rigidity, when the eyes became natural. Mr. D. was then made to close his eyes, and on being told he could not open them, he found it impossible to do so. His mouth being closed, he was then told he could not open it, nor speak, and this also he found impossible. His right arm being raised, Mr. Lewis, who had not touched him, told him that he could not lower it, which proved to be the case. It very soon became hard, rigid, and immovable, and was held out horizontally for a long time. In fact, a pass or two, over any limb, rendered it instantly rigid. Mr. Lewis then desired Mr. D. to gaze at him for a second or two, he gazing in return; when the eyes at once became fixed, the pupil dilated, and utterly insensible, so that no contraction ensued when a candle was passed close across the eye, or held close before it. The pulse being 76, Mr. Lewis pointed with one hand over the heart, while a medical man felt the pulse. It rapidly rose to 150, and became so feeble as hardly to be felt, while the patient became pale, and would certainly have fainted, had this experiment been continued a minute longer. Mr. Lewis then caused both the arms and legs of the patient successively to move, in spite of all the efforts of the patient, according as he, Mr. L., chose to direct them. They first moved to a certain extent, and then became rigid, and all this without contact. When his hand was laid on that of Mr. L., and he was defied to remove it, he found it quite impossible to do so.

Mr. Lewis, having thus shown his control over the muscles, both voluntary and involuntary, next showed his power of controlling sensation. A penknife being placed in Mr. D.'s hand, he was told that it would soon become so hot that he could not hold it. Within about two minutes he began to shift it from one part of the hand to another, and soon threw it away as if it had been red hot. The knife was again placed in his hand, and he was told that it would become so heavy as to force his hand down to the floor. He very soon began to make efforts to keep it up, but in about three or four minutes, in spite of the most violent resistance, which caused him to be bathed in perspiration, and to be out of breath, his hand was forced down to the floor.

Mr. Lewis next caused Mr. D. to forget his own name, and the perplexity of his countenance, while seeking for it in vain, was very striking. In this, as in all the other experiments, the effect was instantaneously dissipated by a snap of Mr. L.'s fingers, or by the words "All right."

This was the first case in which I saw this peculiar form of experiment in the conscious state; for Mr. D. was throughout perfectly conscious, and explained his sensations, except when, from cataleptic rigidity of the muscles of speech, he could not articulate. Mr. Lewis had never seen him before, nor had Mr. D. ever seen any mesmeric experiments. Indeed, he came to my house utterly sceptical on the subject. I should add, that when first acted on, and when his respiration was so much affected, he felt smart shocks, like those of a galvanic battery, in his arms. I had reason afterwards to think, that this depended on his being closely surrounded by many gentlemen at that time, whose influence, crossing that of Mr. Lewis, produced disturbance; for one evening, at the house of Dr. Simpson, when Mr. Lewis had affected him, and had requested two medical gentlemen, one at each side, to feel the pulse, the result was, an appearance of great suffering, which was not fully removed for at least a quarter of an hour. Mr. D. then told us, that he had not really suffered as much as we supposed; but that he had felt, after the two gentlemen felt his pulse, a succession of severe shocks, stronger than those of Mr. Kemp's battery in full action, which continued to return at intervals. It was impossible to doubt that the crossing of the influences of three powerful men, acting on a delicate and highly susceptible frame, was the cause of this singular disturbance. When I mesmerised Mr. D. in private, no unpleasant sensations ever occurred. Mr. D. was, at the time, in a very delicate state of health, caused, as I afterwards discovered, in great part, by excessive study. He was, in fact, in the first stage of a severe illness, affecting chiefly the chest, by which he was soon after confined to bed for some weeks. There can be no doubt that his extreme susceptibility was morbid, for, in the interval between the above experiments and his illness, I produced the mesmeric sleep with the greatest ease, whereas, after his recovery, he was, although still capable of being mesmerised, far less susceptible in every way, and exhibited, when mesmerised, a train of phenomena quite distinct from those I had at first obtained.

On various other occasions, Mr. Lewis operated on Mr. D., and produced the same results, as well as others, including various forms of control over his sensations, perceptions, and memory. Those above described, however, sufficiently illustrate the conscious phenomena, as they appeared in this case. I found, as above mentioned, that Mr. D. was very easily thrown into the mesmeric sleep, and I shall describe the phenomena observed in him in that state in their proper place.

Cases 2 and 3.—Two lads, who were sent with some message to Mr. Lewis, one evening when I was with him and several other gentlemen, were tried in the same way, and found highly susceptible. Both were stout and healthy, and about 16 or 17 years of age. They exhibited the whole train of phenomena connected with the muscular motions, and were rendered by Mr. Lewis's expressed will or suggestion, quite unable to perform any motion no matter what efforts they made, as for example, to pick up anything, or to drop anything, to raise the hand to the head, or to take it down when laid on the head. They were so strongly attracted by him, in spite of their strongest efforts at first, that very soon these efforts of resistance changed into efforts to follow him, powerful enough to overcome those of persons who tried to hold them back, while all the time they were urged to resist, and did their utmost to resist, the tendency to move towards him. When the point of the middle finger of one was laid against the point of the middle finger of the other, so as just to touch it, and Mr. Lewis made a rapid pass over both, they could not, with their utmost exertion, separate the fingers. Nay one, being stronger than the other, dragged him across the room by no other hold, he resisting with all his might.

Their sensations and perceptions were entirely under control. When they drank water, and were told that it was milk, coffee, rum, whisky, or wormwood, they tasted it as such. Nay, after drinking it as whisky, they were told that they were drunk, and in a minute or two became, in every particular, very drunk indeed. The expression of the face was perfectly that of intoxication, and they could not walk a step without staggering or falling. They were easily made, by suggestion, to fancy themselves any other persons, and acted in character. They shot, fished, swam, lectured, and exhibited every feeling suggested to them. They were as easily made to suppose a stick to be a gun, a rod, a sword, nay, a serpent; or a chair to be a tiger or a bear. From these animals they fled with extreme terror. They were made to see, hear, and feel a dreadful storm, and to creep for shelter under a table or a chair, supposed by them to be a house. From this they were soon expelled by the serpent, or by the flood rising, when they swam lustily for their lives. This was the first time that either of them had been tried; and the control exercised by Mr. Lewis over their sensations, perceptions, and emotions was perfect, although their consciousness was entire. They knew the suggested impressions to be false, but could not resist them. It was most interesting to watch closely their countenances, when an object, for example, a handkerchief, was placed in the hand, and after they felt quite sure of what it was, they were told it was a rat, &c. The gradual change to doubt, from doubt to certainty, and from that to disgust or anger, was inimitable, and conveyed at once, to those near enough to see it, complete conviction of their sincerity.

Case 4.—Mr. F., acted on by Mr. Lewis in the presence of ten or twelve persons, of whom I was one, exhibited several of the phenomena. He was sceptical at first, but soon found that his perceptions were under control. For example, an apple was given to him, and he was then told it was an orange. At first he denied this, but by degrees he began to feel doubtful. At last he said, "It is certainly very yellow" (it was dark brown). He then took a sly glance round the company, each of whom had an apple, but found them all yellow too. He next cut out a piece with his finger, looked at the inside, smelt and tasted it, and concluded with, "Well, it *is* an orange, but yet I know I took an apple into my hand."

I could give at least twenty similar cases, in which I saw

Mr. Lewis produce numerous effects of suggestion in the conscious state; but the above are sufficient to illustrate the effects, and it would be tedious to give a repetition of the same experiments. I shall add, however, one more case, in which Mr. Lewis produced, in my presence, very striking results.

Case 5.—Mr. J. H., a young and healthy man, could be rendered instantly and completely cataleptic by a glance, or a single pass. He could be fixed in any position, however inconvenient, and would remain ten or fifteen minutes in such a posture, that no man in a natural state could have endured it for half a minute. Thus he stood for about ten minutes, fixed and rigid, the eyes insensible to light, on one foot, the body lying horizontally, the head forward, the other foot stretched out behind. He was made to place his feet without shoes, the toes pointing in opposite directions along the wall, and the feet resting on a narrow foot-board, about two inches wide, while his back was placed flat on the wall, and the arms stretched out like a cross. In this awkward position, he was rigified and fixed by a pass or two, and stuck there at least five minutes. Mr. Lewis then demesmerised the upper parts down to the knees, when Mr. H. felt in great danger of falling off, but the feet adhered so firmly to the foot-board, that I could not move them. When the feet were demesmerised, he instantly fell down in a heap in our arms. He was made to go down on all fours, and in this position rendered rigid, so that, with all his efforts, he could not lie down on the floor. Mr. Lewis fixed him in a standing posture, and left the room, accompanied by another gentleman. In their absence, I saw Mr. H. move his arms up and down, and when they returned, the other witness told me that Mr. Lewis had made corresponding motions, willing the subject to repeat them. This he had done, but evidently to a less extent, that is, where Mr. L.'s hand moved two or three feet, that of Mr. H. moved perhaps only one foot or six inches, but in the same direction. At my request, while Mr. H. stood opposite the door, at ten or twelve feet from it, Mr. Lewis slipped out, while the company stood round Mr. H., talking with him, and tried to attract him to the door and into the lobby, of course without a word being said. In half a minute, Mr. H. began to look fixedly at the door; he then made a step or two towards it; but becoming rigid, he bent forward, resting on one foot, while his arms were stretched towards the door, and the other leg was raised behind him. In this position, with an expression of fixed and earnest desire to reach the door, he was finally fixed, and rigid from head to foot; and by this time he had got into a position in which, had the limbs been flexible, he must have fallen. As it was, when the balance became untenable, he made a hop forward, and did not fall, although every one thought he must. It was evident, that, but for becoming rigid, he would have gone straight to where Mr. Lewis was. This same patient was made to sleep and to wake by a word, and when asleep, to fancy himself Shakespeare or Campbell, &c., and to recite long passages with great earnestness and feeling. He was also made, by the silent will of Mr. Lewis, while talking to others, to move towards Mr. L., and follow him, till stopped by the rigidity which never failed to appear. When Mr. L. stood on a chair, and tried to draw Mr. H., without contact, from the ground, he gradually rose on tiptoe, making the most violent efforts to rise, till he was fixed by cataleptic rigidity. Mr. Lewis said, that had he been still more elevated above Mr. H., he could have raised him from the floor without contact, and held him thus suspended for a short time, while some spectator should pass his hand under the feet. Although this was not done in my presence, yet the attraction upwards was so strong, that I see no reason to doubt the statement made to me by Mr. Lewis and by others who saw it, that this experiment has been successfully performed. Whatever be the influence which acts, it would seem capable, when very intense, of overpowering gravity. But of course I cannot speak with certainty on this point. I saw, however, this subject kept by Mr. Lewis's influence for some time in a position leaning backwards, in which he could not have remained for a second without falling, in his usual state, and in which he instantly fell, when Mr. Lewis's influence was removed.

I have given the above cases, as instances of the effects which I saw produced by Mr. Lewis on persons in the conscious state. I could easily multiply these instances, but my space is too limited, and what I have given will suffice to illustrate the principle. With all the subjects except Mr. H. the experiments were often repeated on different occasions, when Sir David Brewster and many other scientific gentlemen were present, all of whom were satisfied as to the genuineness of the facts, as far as they saw them. It will be observed that most of these effects were the result of suggestion, acting on persons in a peculiarly impressible state. But they were not all the effects of suggestion; for the effects produced on the pulse and on the eye, which were also shown in many other cases, besides that of Mr. D., were produced without any suggestion. Moreover, the impressible state itself was produced without direct suggestion, and in such a way as to prove, in my opinion, the existence of an influence proceeding from Mr. Lewis. This influence was further shown in those experiments in which he acted on Mr. H. from the lobby, or from another room. I shall now proceed to give some instances of what I saw done, in the way of suggestion, on persons in the conscious state, by Dr. Darling. I have already described his method of producing the impressible state, for which he does not employ his own mesmeric influence, but causes the subject to gaze at a small coin in the hand.

Case 6.—Mr. W., an officer, met Mr. Darling at my house.

Col. Gore Browne had ascertained, some weeks before, that Mr. W. was susceptible, but had made no further experiments, and Dr. D. had never before spoken to him. He was found, in about two minutes, quite susceptible or impressible. His muscular motions were controlled in every possible way. He was rendered unable to raise his hands, or to let them fall; he was made unable to move one, while he could move the other; unable to sit down or to rise up; or to take hold of or let go an object. One arm was deprived of sensation, or both arms, or the whole frame. He was made to feel a knife burning hot, and the chair on which he sat equally so. When he started up, he was made to feel the floor so hot that he was compelled to hop about, and wished to pull off his boots, which burnt him. He was made to feel the room intolerably warm, and actually perspired with the heat; after which he was made to feel it so cold, that in a minute or two he buttoned his coat, and walked about rubbing his hands. In about five minutes, his hand was really chilled, as I found, like that of a person exposed to frost. He was made to forget his own name, as well as that of Col. Gore Browne, who was present, and to imagine Col. B. a total stranger. He was compelled, for a time, to give a false answer to every question asked; and then was forced to give true answers to every question, in spite of any effort he might make to do otherwise. He was told he was on duty, at drill; and began to give the word of command, as if in the barrack-yard. He was compelled to sing and whistle, in spite of himself; to laugh immoderately, and then to feel sad, and even to weep, all in spite of his own will. He was told that a stick was a gun, and with it he shot and bagged a grouse, which he was made to see before him. He was told the pianoforte was a horse, and after feeling and closely examining it, he specified its points and defects, and appraised its value. He tasted water precisely as was suggested to him, as lemonade, tea, or wormwood. He was told that Dr. D.'s hand was a mirror, and in it he saw himself with a black face, as Dr. D. told him to do. He was made to look at his watch, and then convinced that it pointed to a different hour from the true one. He was then made to believe the watch to be a daguerreotype of Col. Browne, and again of a lady. Dr. D.'s empty hand became a snuff-box, from which he took a pinch, which made him sneeze violently, and this passed into a most severe cough, as if he had inhaled snuff, which sensation was not removed for about half an hour. He was made to go to sleep in one minute, and in his sleep to be deaf to the loudest sounds. He was made to see, in Dr. D.'s empty hand, a bank note for £10, to read its number, to fold it up, and put it in his pocket. And when afterwards asked, he declared he had done so, and was surprised not to find it there. He was rendered quite unable to jump over a handkerchief laid on the floor; and was compelled, according to Dr. D.'s command, and in spite of every effort, either to come down on it, or on one or other side of it, or straddling across it. In every one of these experiments, Mr. W. was quite aware that the suggested idea was false, but found it impossible to resist the impression. About fifty persons were present, including Sir David Brewster, and other men of science. On another occasion, Mr. W. exhibited many of the same as well as other proofs of impressibility, without any preliminary process whatever. Dr. D. made him take a gentleman for a lamp-post; his watch for a turnip, the chain for a string; he told him that a gentleman was insulting him, when he demanded an apology. He caused him to see the great Nassau balloon ascend from the floor of my drawing-room, &c., &c. On both occasions, the suggested idea was always instantly dissipated by the words "All's right"; and Mr. W.'s countenance then expressed confusion and shame at what he had just done or said.

Case 7.—Mr. B. was discovered by Dr. Darling to be susceptible, at the house of a well-known and popular authoress. He was so obliging as to meet Dr. D. and a large party at my house There he exhibited many of the effects above described, chiefly, however, the control of Dr. D. over his movements, sensations, perceptions, and memory. His movements were controlled in many ways, which it is unnecessary to repeat. But what rendered the case peculiarly interesting, was, that he described his feelings, and reasoned on every experiment as it was made, and told us that, in spite of perfect consciousness, he found it impossible, by any efforts, to resist the suggestions of the operator. He was made to forget his own name, or that of any other person; to be unable to recognise persons whom he knew quite well; to forget and be unable to name a single letter of the alphabet. It seemed to him as if he saw the letters in motion, but could not lay hold of one of them. These experiments were very painful to him, and he informed me, that when thus compelled to forget his own name, not only was the sensation most unpleasant, but he felt ill for a day or two in consequence. This rarely happens, but there can be no doubt that such violent and false impressions may do harm, and that such experiments should be made with great reserve and caution, or not at all, at least for the mere gratification of curiosity. Mr. B. was rendered insensible to pain in one arm only, while the other arm and the rest of the body retained their sensibility. Dr. D. wished to render the arm insensible even to touch, but at first a slight degree of that sensibility remained. Even this was removed by a second suggestion, and the arm became utterly insensible. The hand was well pinched, and pricked with pins, but Mr. B. was not even aware of this, except when he looked and saw it done. He wished, for his own satisfaction, to be cut or burned to the bone. I declined this, however, and contented myself with forcing a blunt pin through a thick fold of skin on the back of the hand, which is a very painful operation. To this he was utterly insensible, and indeed would not believe that he had been fairly tried, till, after the insensibility was removed, I applied the pin gently, when he quickly withdrew his hand, and declared himself satisfied. Mr. B. was also made to sleep in one minute, and rendered deaf to the loudest noises, till the magic words "All right" awoke him.

Case 8.—Mr. H. B. was found susceptible, at least, in so far as concerned the control of his movements by Dr. D. The case was remarkable, from the very violent efforts made by Mr. H. B. to act in opposition to the suggestions of Dr. D. Thus, when he was told he could not rise up from his seat, he made the most desperate exertion of all the muscles of his body, but could not combine them so as to rise in the usual way. Once, he projected himself into the air by a violent effort against the sofa, but instantly fell down. Another time, he again succeeded in forcing himself up by a jerk from an ottoman, but fell down on his back on the seat. He made such efforts as I have never seen made, but always failed to perform the movement which he was told he could not make, while any other motion was quite easy. I am not certain whether the experiment was tried with Mr. H. B., but I think it was, and it certainly was tried with the preceding subjects and with several others, of defying them to hit Dr. D. a blow on the face with the closed fist. In every case, the blow fell to one side or the other, and never touched the face. In most of the cases I saw, the subjects were also rendered unable to pick up a handkerchief from the floor, although they could touch it; and when they had it in the hand, they were rendered equally unable to drop it. I could adduce ten or a dozen similar cases, in which I saw Dr. Darling operate; but the above are sufficient, and I have selected them, without in any instance detailing all the experiments, which would be tedious, as fair examples, all of which occurred in my own drawing-room, in the presence of large parties, including many scientific and medical gentlemen, all of whom were perfectly satisfied of the facts.

In these experiments of Dr. Darling, as I have formerly explained, the impressible state was not produced by his mesmeric influence, but by the subjects gazing at a small coin, placed in the left hand. If Dr. D.'s influence was at all exerted, it was to a very limited extent, as he occasionally made the subject gaze for a moment at him, while he gazed in return, perhaps holding one hand, or laying his finger on the middle of the lower part of the subject's forehead.

It is also to be observed, that many persons have found it quite easy to produce the same effects, in susceptible cases. The Earl of Eglinton and Col. Gore Browne, have been quite as successful as Dr. Darling, and I know of many others who have also succeeded. I have myself found no difficulty, by his method, in producing similar results, but have merely satisfied myself that it could be done, without pushing the experiments very far.

These experiments demonstrate only the power of suggestion on persons in the impressible state, which state, as we have seen, may be produced, either in the method of Dr. Darling, or in that of Mr. Lewis, in which it is done by the direct influence of the operator. This also I have tried with success; but I find that, as a general rule, Dr. Darling's method is the easier of the two. We must bear in mind, however, that the other method enables us, not only to obtain the phenomena of suggestion in the conscious impressible state, but also, if carried further, the mesmeric sleep, with its peculiar phenomena and its divided consciousness.

It is self-evident, that the power of suggestion, thus acquired, over persons in the conscious state, may be usefully applied in medicine. Thus, Dr. Simpson, who has made many experiments with success, has been enabled to cause patients, by a command given in that state, to sleep for a certain number of hours at a subsequent period, in cases where sleeplessness had long prevailed.

I am informed by the Earl of Eglinton, that he was enabled, while travelling by railway with a party, one of whom, a lady, was very deaf, to restore her hearing to such an extent, that she heard whispers inaudible to him. This proved at once that the deafness was not hopeless, and it was found, that the improvement lasted for some hours after that single trial. It is probable that a course of such treatment might permanently restore the hearing, and at all events it is worth while to make the attempt.

B. CLAIRVOYANCE IN THE CONSCIOUS STATE, BY PASSES, OR BY SIMPLE CONCENTRATION.

It is generally supposed that clairvoyance belongs only to the higher stages of the mesmeric sleep; but it now appears, that it may, in certain cases, be produced without the sleep, and when the subject is in a state of ordinary consciousness. Indeed, if we are to regard clairvoyance, as I am disposed to do, as simply the power of noticing or observing certain very fine or subtle impressions, conveyed from all objects to the sensorium, by the medium of a very subtle agent, influence, fluid, or imponderable, which we may call vital mesmerism, or, with Reichenbach, Odyle, the impressions caused by which are usually overpowered by the coarser impressions conveyed to the sensorium through the external organs of the senses, it is evident, that the essential condition of clairvoyance is not the sleep, but the shutting out of the impressions of senses. This occurs, no doubt, in the sleep, but it also occurs in the state of reverie and abstraction, and may, in some cases, be effected at pleasure by voluntary concentration. I have not myself had many opportunities of seeing this phenomenon, but as it is extremely interesting, I shall give a brief account of the experiments of Major Buckley, which appear to have been made in a manner perfectly satisfactory, and on a large number of persons. It would certainly appear that Major B. has a rare and very remarkable power of producing conscious clairvoyance in his subjects.

Before describing Major Buckley's method, I should mention that he had been for some time in the habit of producing mesmeric sleep, and clairvoyance in the sleep, before he discovered that, in his subjects, the sleep might often be dispensed with. The following details are abridged from letters with which Major B. has very kindly favoured me..

Major Buckley first ascertains whether his subjects are susceptible, by making with his hands, passes above and below their hands, from the wrists downwards. If certain sensations, such as tingling, numbness, &c., are strongly felt, he knows that he will be able to produce the mesmeric sleep. But to ascertain whether he can obtain conscious clairvoyance, he makes slow passes from his own forehead to his own chest. If this produce a blue light in his face, strongly visible, the subject will probably acquire conscious clairvoyance. If not, or if the light be pale, the subject must first be rendered clairvoyant in the sleep. Taking those subjects who see a very deep blue light, he continues to make passes over his own face, and also over the object, a box, or a nut, for example, in which written or printed words are enclosed, which the clairvoyant is to read. Some subjects require only a pass or two to be made, others require many. They describe the blue light as rendering the box or nut transparent, so that they can read what is inside. (This reminds us of the curious fact mentioned by Reichenbach, that bars of iron or steel, seen by conscious sensitives, without any passes, shining in the dark with the odylic glow, appeared to them transparent, like glass.) If too many passes be made by Major B., the blue light becomes so deep that they cannot read, and some reverse passes must be made, to render the light less deep. Major Buckley has thus produced conscious clairvoyance in 89 persons, of whom 44 have been able to read mottoes contained in nut-shells, purchased by other parties for the experiment. The longest motto thus read, contained 98 words. Many subjects will read motto after motto without one mistake. In this way, the mottoes contained in 4,860 nut-shells have been read, some of them, indeed, by persons in the mesmeric sleep, but most of them by persons in the conscious state, many of whom have never been put to sleep. In boxes, upwards of 36,000 words have been read; in one paper, 371 words. Including those who have read words contained in boxes when in the sleep, 148 persons have thus read. It is to be observed that, in a few cases, the words may have been read by thought-reading, as the persons who put them in the boxes were present; but in most cases, no one who knew the words has been present, and they must therefore have been read by direct clairvoyance. Every precaution has been taken. The nuts, enclosing mottoes, for example, have been purchased of 40 different confectioners, and have been sealed up until read. It may be added, that of the 44 persons who have read mottoes in nuts by waking or conscious clairvoyance, 42 belong to the higher class of society; and the experiments have been made in the presence of many other persons. These experiments appear to be admirably contrived, and I can perceive no reason whatever to doubt the entire accuracy of the facts. It would of course be tedious to enumerate so many experiments, all of the same kind; but I shall select one or two of the most striking as examples.

Case 9.—Sir T. Willshire took home with him a nest of boxes belonging to Major Buckley, and placed in the inner box a slip of paper, on which he had written a word. Some days later he brought back the boxes, sealed up in paper, and asked one of Major Buckley's clairvoyantes to read the word. Major B. made passes over the boxes, when she said she saw the word "Concert." Sir T. Willshire declared that she was right as to the first and last letters, but that the word was different. She persisted, when he told her that the word was "Correct." But on opening the boxes, the word proved to be "Concert." This case is very remarkable; for, had the clairvoyante read the word by thought-reading, she would have read it according to the belief of Sir T. Willshire, who had either intended to write "correct," or in the interval, forgot that he had written "concert," but certainly believed the former to be the word.

Case 10.—A lady, one of Major Buckley's waking clairvoyantes, read 103 mottoes, contained in nuts, in one day, without a pass being made on that occasion. In this, and in many other cases, the power of reading in nuts, boxes, and envelopes, remained, when once induced, for about a month, and then disappeared. The same lady, after three months, could no longer read without passes, and it took five trials fully to restore the power. This may be done, however, immediately, by inducing the mesmeric sleep and clairvoyance in that state, when the subjects, in the hands of Major Buckley, soon acquire the power of waking clairvoyance.

Case 11.—The words, "Can you see inside?" were written on a narrow slip of paper, which was then laid on a quarter sheet, and folded over 11 times. The folded paper was placed in a thick envelope, and sealed with three seals, in such a way that it could not be opened undetected. It was then sent to a clairvoyante, who returned it with the seals uninjured, having read the contents in waking clairvoyance. Mr. Chandler has published an account of a precisely similar experiment, in which the sealed envelope, after the contents had thus been read, was shown to many persons, all of whom were quite satisfied that it could not have been opened. I have in my possession one of the envelopes thus read, which has since been opened, and I am convinced that the precautions taken, precluded any other than lucid vision.

I regard the experiments of Major Buckley as not only well devised, but of very great value, as proving the existence of waking or conscious clairvoyance, which, as I have mentioned further on, undoubtedly occurs as a spontaneous natural fact. Indeed, it is to the occurrence of this state that we must refer many well-attested instances of persons acquiring intuitively a knowledge of distant events, many of which are recorded.

The case of the lady, mentioned further on, is far from being the only case in modern times, of spontaneous and waking clairvoyance. Mr. Atkinson, to whom I owe that case, in the work recently published by himself and Miss Martineau, gives an instance of its occurrence in his own person. I have also mentioned further on, that Mr. Lewis, by concentrating his thoughts, can produce, in himself, the state of waking clairvoyance, and has frequently thus seen persons, myself among the number, at the time quite unknown to him. I shall here give an instance of Mr. Lewis's power in this way, on the testimony of the gentleman who made the experiment, who is well known to me, and on whose accuracy the utmost reliance may be placed.

Case 12.—Mr. B., the gentleman alluded to, arranged with another gentleman, that, at a certain hour, the latter should be present in the drawing-room of his aunt, Miss C., with that lady and another, these three being all the inhabitants of the house, except servants. At the same hour, Mr. B., in his own house, at a distance of 14 miles, requested Mr. Lewis, who had never seen Miss C. nor any of her family, nor her house, to try to mesmerise that lady. Mr. Lewis proceeded to concentrate his thoughts on the subject, and soon saw and described the house, its situation, the house-dog, the drawing-room, and the persons in it, all quite accurately. But he saw in the room not only the three inhabitants of the house, but two other persons. While this was going on, two visitors had previously come into the room in Miss C.'s house, and Miss C., who was susceptible to mesmerism, just at the time when Mr. Lewis saw her, described her, and endeavoured to mesmerise her, was so strongly affected, that she declared her nephew must be mesmerising her, and begged him not to do so.

I mention this case, as a proof of the power possessed by Mr. Lewis in a certain state of concentration, not only of seeing, but of acting on persons whom he has never seen. I have already, further on, described a case in which he acted on a lady in my house, whom, however, he had before seen, from a distance of 500 or 600 yards. In that case also, he saw the lady when acting on her from a distance, and subsequently pointed out accurately where she sat, and the direction in which he saw her move away from the pianoforte, when compelled to leave it and to lie down.

On the whole, it must, I think, be admitted, and the experiments of Major Buckley are alone sufficient to decide this point, that conscious or waking clairvoyance may be produced artificially; and this being the case, it is in the highest degree probable that it may also occur spontaneously, even if its spontaneous occurrence had never been observed. I consider it highly probable, that many dreams, concerning events passing at the time, or just before, which prove correct, depend on the spontaneous occurrence of this state during sleep or reverie. It can hardly be doubted, that if the subject be duly investigated, very interesting results must be obtained. But we must not expect to find that all, or many mesmerisers possess the power of inducing conscious clairvoyance in the same way, or in the same degree, as Major Buckley. On the contrary, that gentleman appears to have a very peculiar influence on his subjects. Yet it is not probable that he is the only person possessed of this power.

There is another method of producing conscious clairvoyance, or, at least, phenomena which appear to me to be the same; namely, gazing at what is called the Magic Crystal, or at the Magic Mirror, or other objects. It will be observed, that as the mesmeric sleep may be caused either by passes or by the patient gazing at an object, so it is quite conceivable that gazing at an object, as well as passes by the operator, may produce conscious or waking clairvoyance. I shall now, therefore, adduce a few facts connected with the production of

C. Conscious clairvoyance by gazing at an object.

Many persons, especially the young, who are more susceptible, when they are made to gaze steadily at an object, pass, without going into sleep, into a state in which they see persons or things not present.

1. *The Magic Crystal.*—This is generally a round or oval-shaped piece of clear glass. Several exist, and one is now in my hands, which were made long ago, and used for the purpose of divination, as in the case of the crystal of Dr. Dee. It is said that Dr. Dee's crystal is still extant, and, according to some, it was a polished mass of jet; but it does not appear that the nature of the substance is of much importance, or rather, it would appear that Dr. Dee had a globe of glass or of rock crystal, and also a magic mirror, probably the piece of jet alluded to. The essential point is, that persons who gaze earnestly on the crystal, often see the figures of absent persons, nay, as in ordinary clairvoyance, of such as are unknown to them. The crystal of which I speak, is of the size and shape of a large turkey's egg, and was sold some years since, by a dealer in curiosities, as an old magic crystal, with a paper containing certain mystical and magical rules for its use. In the few experiments I shall mention, it was used by simply desiring the person to gaze earnestly at it.

Case 13.—A boy quite ignorant of what was expected, after gazing at the crystal for about half an hour very steadily, saw a dark cloud appear in it, which soon cleared up, and he then saw his mother in her room. By-and-by, his father appeared. I then asked him to look for a lady, whom he saw walking in the street in which she lived, and accurately described her walking dress, which he had never seen, although he may have seen the lady for a moment in the evening. I then asked for a boy and a servant who I was sure he had never seen. He saw and described most accurately the persons and dress of both. I asked for another servant, whom he saw opening the street-door to admit the lady. I marked the time, and found that this lady had been walking in the dress described, and had entered her house at the time when the boy had seen her.

In all this, I could see nothing but conscious or waking clairvoyance, produced by long gazing. I conclude, that the figures appeared in the crystal, because the boy was looking there, and I see no reason to doubt, that by intense gazing on some other objects, he might have been made to see the same. I made several similar experiments, both with that crystal, and with others, two of which I knew to have been recently made, one several years ago, the other only a week or two previously. The experiments were also made with two other boys, and the general result was, that when they gazed long and steadily, they generally saw figures of some sort, sometimes of a father, mother, or brother, but sometimes also of persons quite unknown to them, without such persons being asked for, and of course, in such cases, I could not tell who the persons seen were. But when, as often happened, their attention wandered, they saw nothing. I could not observe, in any one of these boys, the slightest tendency to deceive me. On the contrary, I was surprised as well as pleased at the patience with which they submitted to these tedious experiments, and at their reserve in declaring that they saw anything. It often happened, that they saw nothing during the whole time; but when they did see anything, they were very precise in stating how much or how little they saw. I generally asked no questions, but encouraged them to tell their own story.

The impression made on my mind by these trials was, that the gazing produced an impressible state (as I ascertained several times by trying some of Dr. Darling's experiments on them), and that when they gazed very steadily, conscious clairvoyance was developed, to a greater or less extent. I resolved to investigate the matter more fully, but as the means of doing so have only very recently been in my possession, I must wait until I shall have time to pursue the investigation. In the meantime, I consider it as certain, not from these experiments alone, but from many others of which I have been informed on good authority, that conscious clairvoyance may be thus produced.

I shall here mention some other instances of visions seen in crystals.

Case 14.—A globular mass of crystal, rather larger than an orange, was lying on a table, when a little girl entered the room, and accidentally looked at it. She exclaimed, "There is a ship in it, with its cloths (sails) all in rags. Now it tumbles down, and a woman is looking at it and leans her head on her hand." Her mother afterwards came into the room, and without having heard what the child had seen, immediately saw the ship and the woman. This accidental observation was communicated to Earl Stanhope by the person in whose room it happened, and by his lordship to me.

Earl Stanhope informs me that he has made experiments with three crystals, in one or other of which visions have been seen by fifteen children of both sexes and of different ages, and by seven adult females, one of them upwards of sixty years of age. In regard to these visions, his lordship observes that "In many cases it is very remarkable, that they could not have been presented by memory to the imagination; as, for instance, visions of a dog wearing a crown; of a bed with a black counterpane; of a house with 126 windows and 33 doors," &c., &c. All this corresponds with the strange visions of ordinary clairvoyance.

"The objects seen in succession were often, as in dreams, unconnected with each other, and while they were exhibited, no other objects in the room were visible to the seers." This proves that the power of seeing them was, as in ordinary clairvoyance, connected with abstraction, or reverie, the result of long and concentrated gazing. Earl Stanhope adds, that very often those who had previously seen visions saw nothing, and that none of the persons he tried showed any disposition to deceive. This agrees perfectly with my own experience, in the few trials I have made.

Case 15.—We have seen that Mr. Lewis possesses, at times, the power of conscious clairvoyance, by simple concentration of thought. He finds that gazing into a crystal produces the state of waking clairvoyance in him much sooner and more easily. On one occasion, being in a house in Edinburgh with a party, he looked into a crystal, and saw in it the inhabitants of another house, at a considerable distance. Along with them, he saw two gentlemen, entire strangers to him. These he described to the company. He then proceeded to the other house, and there found the two gentlemen whom he had described.

Case 16.—On another occasion, he was asked to see a house and family, quite unknown to him, in Sloane-street, Chelsea, he being in a house in Edinburgh with a party. He saw in the crystal, the family in London, described the house, and also an old gentleman very ill or dying, and wearing a peculiar cap. All was found to be correct, and the cap was one which had lately been sent to the old gentleman. On the same occasion, Mr. Lewis told a gentleman present, that he had lost or mislaid a key, of a very particular shape, which he, Mr. L., saw in the crystal. This was confirmed by the gentleman, a total stranger to Mr. Lewis.

Mr. Lewis is distinctly of opinion, that the crystal is only a means of producing conscious clairvoyance by gazing at it; and from what I have seen, such is my own opinion. But it is quite possible, that, besides the gazing, the mesmeric or odylic influence of the crystal, or rather glass, may assist in producing the effect. Mr. Lewis has frequently been so kind as to look into crystals for me, and although this has chiefly been done in reference to persons and things at a distance, and in cases in which what he saw cannot yet be verified, I am convinced that he saw what he described to me. Whether the things he saw, in these cases, were only dreams, or whether his visions were of actual facts, is another point, which, after a time, I may be enabled to ascertain. But I may here state, that a very large crystal globe, belonging to myself, had, in a short time, so strong an effect on him, as nearly to throw him into mesmeric sleep, while a much smaller one had no such effect. This seems to indicate that the odylic influence of the crystal may assist in producing the effect.

2. The Magic Mirror. Of this, I have no experience; but I conceive its action to be the same as that of the crystal. The mirror may be of jet, as Dr. Dee's is said to have been, or of metal, or even a simple black surface, blackened by charcoal. It is at all events an object which must be gazed at for some time, before visions appear. Now we know, that both metals and charcoal act strongly on susceptible persons. M. Dupotet has found, that many persons, on gazing for a while at a surface of charcoal, see visions of a most exciting kind, the nature of which they are generally most unwilling to disclose. But sometimes they do mention what they see. In one case, a lady saw a ship in a storm, and described it, in the presence of my informant, who is a lady of very high rank, and of the highest character. When these experiments of M. Dupotet with his mirrors are long continued, the subjects not only become much excited by what they see, but are frequently rendered quite unconscious of what is passing around them. The experiments are indeed very remarkable, but must be tried with great caution, in consequence of the violent effects produced. I am disposed to agree with M. Dupotet, in thinking that he has, in this discovery of the powers of the mirror, rediscovered a part of the magic of the middle ages, which like all magic, is founded on natural facts. The whole subject requires a thorough investigation.

3. Water. It is found that susceptible persons may be made to see visions, by gazing into a glass of water, especially if the water be mesmerised, in which case we know that it acts on the susceptible.

Case 17.—Major Buckley caused a lady to look into a bottle of mesmerised water, who had been found to be rendered consciously clairvoyante by looking into a crystal. She saw an alligator in the water.

Case 18.—A lady of rank caused a clairvoyante to look into a bottle of mesmerised water, when she let the bottle fall from fright declaring that she saw a serpent in it.

All the facts above mentioned tend to prove, that conscious clairvoyance, or visions, or dreams, may be produced by gazing at a variety of objects, and probably most easily by gazing at crystals, metallic or carbonaceous surfaces, and mesmerised water. But it will probably be found, on trial, that many other substances will produce a similar effect.

Many persons, on reading the accounts that have been published of the visions seen in the crystals, &c., are disposed to reject the whole as sheer imposture. But it appears to me, that we cannot thus get rid of the subject. It is quite conceivable that some seers may have endeavoured to deceive; but it is not conceivable that all should have done so. If some of the statements which have appeared seem very absurd, it must be remembered, that the subject has not yet been scientifically investigated, and that while most of the seers are children, often very young and ignorant, the operators have also frequently been unaccustomed to experiment, and may have vitiated true phenomena by suggesting their own ideas. It appears certain, that many children and adults in different places have seen visions in crystals; many of them have been not only trustworthy, but have been much alarmed and agitated by what they saw; the visions have very often been exactly such as are seen in ordinary clairvoyance; and on the whole, it appears that there are very interesting facts, whatever be their true nature, which require and deserve the most careful investigation.

I have not alluded to those still more wonderful visions said to have been seen in crystals, &c., of persons long dead, of good and evil spirits, and of answers to questions exhibited in written or printed characters. I have had, as yet, no opportunity of investigating these matters, and I can easily see many sources of fallacy. But even here, I do not feel myself entitled to reject, summarily, and without investigation, all that has been asserted. Believing, as I do, that the state of clairvoyance may be induced by gazing at crystals, &c., I think it quite possible that higher states, such as that of extasis, may also be so produced. Now, as clairvoyance and extasis, are states as yet hardly studied, and certainly not fully investigated, I cannot affirm the impossibility of things far more strange than any I have yet seen. But I dare not venture to bring forward such things as facts, until I shall have been enabled to investigate them, which I hope to be able to do.

In concluding what I have to say on the subject of conscious clairvoyance, as produced by gazing, it is unnecessary to do more than to advert to the method employed by the Arabian sorcerers in Egypt at the present day, as that has been fully detailed by Miss Martineau, and other authors, who have seen it. Every one knows, that a boy is made to gaze on a large drop of ink (a liquid mirror) in his hand, while fumigations and magnetic manipulations are employed. There appears to be no good reason to doubt that, in this way, which is merely a variation of the crystal, boys have seen and accurately described absent persons quite unknown to them. In many cases, I conceive thought-reading to have been the medium, as when Shakespeare was seen, as he is generally represented. But I have already shown that thought-reading, or the highest degree of sympathy, is, in truth, not less wonderful than direct clairvoyance. Both sympathy and direct clairvoyance occur in the mesmeric sleep, and both may also occur, as clairvoyance does, in the conscious state, under favourable circumstances.

Chapter XIII.

PRODUCTION OF THE MESMERIC SLEEP—CASES — EIGHT OUT OF NINE PERSONS RECENTLY TRIED BY THE AUTHOR THROWN INTO MESMERIC SLEEP—SLEEP PRODUCED WITHOUT THE KNOWLEDGE OF THE SUBJECT—SUGGESTION IN THE SLEEP—PHRENO-MESMERISM IN THE SLEEP—SYMPATHETIC CLAIRVOYANCE IN THE SLEEP—CASES—PERCEPTION OF TIME—CASES; SIR J. FRANKLIN; MAJOR BUCKLEY'S CASE OF RETROVISION.

I NOW proceed, secondly, to give some illustrations of the mesmeric sleep and its phenomena.

A. Production of the Sleep.

The method which I have generally employed consists in sitting opposite and close to the subject, a little higher than he is, pressing gently his thumbs with mine, and gazing steadily in his eyes, or in one of his eyes. As soon as some effect is produced, which may be, when it happens, in a few minutes, or after a quarter of an hour, or half an hour, I make passes with both hands downwards over the forehead and face to the chest. Sometimes I begin with passes, and frequently I alternate gazing and passes.

Case 19.—Mr. D., a student of medicine, nineteen years of age, in delicate health, had been found by Mr. Lewis very susceptible in the conscious state, as already described, Case 1. I wished to try whether, as he was susceptible, I could produce in him the mesmeric sleep. In the first trial, after 25 or 30 minutes of gazing, as above described, alternated with passes, he slept, but not deeply, and was easily roused when spoken to. I persevered with similar results for nine successive trials, during which the sleep was gradually produced in a shorter time, and on the eighth trial, in 15 minutes. Finding that he was still disturbed by noises in the street, I silently willed that he should not hear them, and thus succeeded in producing a deeper sleep, so that in the ninth trial he slept in twelve minutes, and was, for the first time, perfectly unconscious, on waking, of what passed during his sleep. He spoke now, which he had not done before, in a voice quite different from his usual voice, and, in answer to questions, said he would sleep deeper every time, and would be able to see; but at present a thick mist prevented him from doing so. I desired him to sleep exactly half an hour, which he promised to do, and when the time came, which happened to be five p.m., he woke suddenly while the clocks were striking. I commanded him, before he woke, to sleep next day in five minutes, which he promised to do, and did. From this time I found it quite easy to produce the sleep, and by the thirteenth trial I had got him to sleep in one minute. In this case the full, true mesmeric sleep was not obtained till the ninth day, after which it was easily produced, and gradually became deeper, presenting many interesting phenomena, to be mentioned in their proper place.

Case 20.—Mr. T., a student of chemistry, aged 21. Having found that this gentleman, who enjoys perfect health, was easily rendered impressible, in the conscious state, by about seven minutes' gazing at a coin in his hand, as is Dr. Darling's process, I tried next day to produce the sleep in him, and in this first trial, by alternate gazing and passes, succeeded in putting him to sleep in about twenty minutes. The sleep was not very profound. I proceeded with him as with Mr. D., and soon reduced the time necessary to cause sleep from twenty minutes to two minutes. In this case also I was able to fix the length of the sleep. He was quite unconscious, after waking, of all that passed during the sleep, but I found that, by telling him to remember anything, I could cause him to do so.

Case 21.—Mr. H. W., a German, blind, having lost his sight from a complication of diseases. The left eye is entirely destroyed, having been operated on unsuccessfully for cataract. The right eye is said to be affected with opacity in the posterior part of the capsule, and also with amaurosis. During the last two years, he has only been able to distinguish, very feebly, between daylight and darkness, and he is to all practical purposes stone blind. On looking at the eye, it appeared to me that the opacity in the capsule, which had been expected to form into cataract, had not yet proceeded to any great extent; and as the iris was to a certain extent moveable, although generally much dilated, I had some hopes that the retina might not be quite insensible to light. Not being familiar with diseases of the eye, and not thinking at the time of trying to restore the sight, the state of the eye was not particularly examined till nearly two months after I first mesmerised Mr. W. At that period, as I shall afterwards mention, a considerable improvement had taken place, and Sir D. Brewster, who kindly examined the eye, thought he saw some degree of separation of the layers of the lens, caused by deficient moisture. By that time, the iris contracted very decidedly when the eye was suddenly exposed to strong daylight, after having been shaded by the hand. My chief reason for trying to mesmerise Mr. W. was, that I thought I might improve thereby his general health, which was delicate; and as he was most anxious to make the trial, I also felt very desirous to ascertain the effects that might be produced on a blind patient.

I operated on him exactly as on the two gentlemen above mentioned, and soon produced decided effects and strong sensations, without, however, at first causing sleep. It was not till the twelfth sitting, that I succeeded in inducing the true mesmeric sleep, after which I found it quite easy to do so; and in this also, the time required was soon reduced to two or three minutes. I beg here particularly to point out, that as far as I could see, the blindness of Mr. W. was no obstacle to his being affected. On the contrary, the sensations produced were from the first much stronger than in the other cases, and indeed it was a very peculiar and strong quasi-galvanic sensation in the region of the lumbar vertebræ which kept him from sleeping so long, in spite of great drowsiness. When I knew this, I removed it by a few passes, and thus obtained the sleep. I may here add, briefly, that from the first trial, Mr. W.'s health rapidly improved, and that by the fifth operation, long before sleep was produced, not only had the eyes, previously dry, red, and unhealthy-looking, become of a natural colour, with a due proportion of moisture, but also a very copious, tough, and extremely offensive discharge from the nose, from which he had suffered more or less for years (but which he had not mentioned to me) having been very severe when I first mesmerised him, had disappeared entirely, and been replaced by a perfectly natural secretion. Nor has this distressing affection once returned since, although Mr. W. has had frequent slight catarrhs, which, till I mesmerised him, invariably aggravated the offensive discharge. I regard that affection as permanently cured, and it must be borne in mind, that I was not even aware of its existence, while Mr. W. had never thought that it was to be removed by mesmerism. He declares that it had rendered his life burdensome to him, and no words can express the relief he has experienced. Such was the effect produced, by five operations, on the state of the mucous membrane of the eyes and nose. But this was not all. His general health and spirits, from the time of the first trial, rapidly improved, so that the change was visible to every one, while he himself was so sensible of it, that he daily longed for the hour at which he was to be mesmerised. When this had been done fifteen times, he found that his sight had begun to improve, so that he saw the full moon, which for two years he had been unable to do, and about the same time began to perceive the gaslights in large shops, so as sometimes to be able to count them. It was only when he mentioned this, that I ventured to hope that by perseverance his sight might be permanently improved; and in hopes of this, I have continued to mesmerise him, till, when I write, he has been mesmerised 40 times, at first daily, more recently every other day. A slow improvement in the sight continues to appear, but in such a case, to obtain a satisfactory result, if that be possible, the treatment must be persevered in for a long period. I have mentioned in this place the effects of a short course of mesmerism on the health of Mr. W., because, from his being blind, it was necessary to describe the case. When I come to make some remarks on the therapeutic agency of mesmerism, I shall, instead of repeating what has here been said, simply refer to it.

Case 22.—Mr. C. M., a student of Natural Philosophy. This young gentleman was acted on in the same way, and in the first sitting had convulsive twitches of the eyelids and of the arm, but no sleep. In the second, after I had gazed at him, holding his hands, for fifteen minutes, he slept for five minutes. After this I got him to sleep in three minutes, but was only able to do so four times, as he left town. I could, in this, as in the preceding cases, fix the duration of the sleep.

None of these four gentlemen had ever been put to sleep by mesmerism before I tried them, except Mr. D., Case 17, whom Mr. Lewis, at my request, tried one night to put to sleep at my house, as I wished to ascertain whether he could be put to sleep before beginning my experiments. Mr. Lewis soon produced a sleep, which was, however imperfect, similar to that which I also produced in the first few trials, so that, even in that case, the true, deep mesmeric sleep, or state of somnambulism, had never been produced until I succeeded in evolving it.—These cases will give a fair idea of what I experienced as to the induction of that state. They are not selected, but are simply the first four cases which I tried this season, and in which I persevered till sleep took place. In all of them, had I stopped short after one or only a few trials, I should have failed as to the sleep, and I think the results are sufficient to justify the conclusion, that most persons, if not all, may be thrown into somnambulism by perseverance and patience on the part of the operator. I possess no unusual mesmeric power, probably less than the average, and in most of the cases in which I had previously tried to cause the sleep in persons not yet mesmerised by others, I had, just as in these, failed to do so on the first or second trial, and had therefore supposed that I had not the requisite power.

I may here mention that I have this season tried to produce the sleep in five other persons, not previously mesmerised, and have succeeded in all but one. Thus, out of nine persons, not before mesmerised, I have put to sleep eight. The ninth was only tried three times, which is not nearly enough to show that I should have ultimately failed with him. Among those with whom I succeeded was another blind man. But as I was unable to make further trials, except with the four cases first enumerated, I refrain from detailing the others. The method was the same, and the results exhibited nothing peculiar.

I need not dwell longer on this subject, nor need I quote the published experience of others on this point. Enough has been said to show that the peculiar state called the mesmeric sleep may be easily induced, in a large proportion of persons, with the aid of patience and perseverance. I have seen Mr. Lewis produce the sleep in numerous instances, generally by gazing alone, without even holding the hands. But, as I have already said, his power of concentration is very remarkable, and from what I have seen, I believe that there are very few persons whom Mr. Lewis could not put to sleep, if not at the first trial, as very often happens with him, at all events after a few trials. It is quite obvious that such experiments succeed infinitely better in private than in public, or in a large party, because in the latter cases, the excitement of the patient, or his alarm, and the proximity of other persons, very much interfere with the result.

It may be proper here to state, that in the case of Mr. D., who was frequently put to sleep in half a minute, or even a quarter of a minute, I found that I could produce the sleep not only without contact, but without his knowing my intention. On one occasion, while he was intently engaged in conversation, and looking another way, I gazed steadily at him, from the side, at a distance of five or six feet, with the intention of putting him to sleep; in about 25 seconds his eyes closed, and he was found in a sleep as deep as I have ever seen. I desired him to sleep an hour, which he did, and on waking, his first words were to complain that I had not told him what I intended to do. Finding him thus susceptible, I intended to make a series of experiments in regard to the power of causing sleep at a distance, but unfortunately, Mr. D. was about this time seized with a severe illness, which had been impending over him for some time, in consequence of too severe study. It was chiefly an affection of the chest, which confined him to bed for some weeks, and after his recovery, his extraordinary susceptibility was gone, having evidently depended on a morbid state. It was therefore in my power only to make two experiments; one, similar to that just described, and with the same result; the other, in which, at a distance of about half a mile, I attempted to put him to sleep by concentrating my thoughts on him with that intention. I was interrupted before I had done so for more than two minutes; but when I saw him, he spontaneously told me that, precisely at the time at which, unknown to him, I made the trial, he felt an inclination to sleep, as well as the usual sensation when mesmerised. I very much regret that I was unable to repeat this experiment; for after his recovery, I found it far more difficult to mesmerise him than before his illness, although I could still produce full sleep.

B. Effects of Suggestion in the Sleep.

Here it is only necessary to say, that every effect above described, as being produced by suggestion in the conscious impressible state, may be even more easily produced in the sleep.

Case 23.—When Mr. D. was asleep, I could render any limb, or the whole body, rigid at pleasure. I did not multiply experiments of this kind, but only satisfied myself that suggestion acted on him as well when asleep, as it had done in the conscious state. I could also, with ease, fix the duration of the sleep by a command.

Case 24.—Mr. J.D. (a different subject), put to sleep in my presence by Mr. Lewis, exhibited, in great perfection, all the effects of suggestion. Whatever Mr. Lewis told him, he acted on it with a perfect conviction. He was thus made to fish, to shoot, to sing, to imagine himself a general or a lecturer, to take a stick for a sword or gun, a chair for a wild beast, to feel the pelting of a pitiless storm, to hear the thunder, to be drenched with rain or frozen with cold, to swim for his life in the flood, to taste water as beer, milk, lemonade, or whisky; and when he had taken a little under the last-named form, to be so utterly drunk, that he could not stand without support. Indeed he continued so perseveringly drunk, that it took Mr. Lewis a quarter of an hour to sober him.

This subject exhibited a remarkable tendency at all times to persevere in any state in which he was put, so that it was often difficult to get him under the influence of a new suggestion, or to wake him when asleep, unless by causing him to promise that he would awake when desired to do so. On one occasion, this subject was exhibited in public with great success.

Not only are most subjects strongly influenced by suggestion while asleep, but they may in general be strongly influenced, in the waking state, by a command given in the sleep.

Case 25.—Thus, I could cause either Mr. D. or Mr. T. to forget everything that passed during the sleep, or to remember a part, or the whole of it, by commanding them to do so. I could also by a command given in the sleep, of which they had no recollection, fix the time necessary to put them to sleep next day, and I made use of this power, to reduce that time from 20 or 30 gradually down to two minutes or less. I could further determine the kind of feeling they should have after waking. Having observed that Mr. D. occasionally felt languid, he being, unknown to me, in bad health, I used, in the sleep, to desire him to wake without that feeling of languor, and I found that when I did so, he always felt light and well. When I omitted it, his habitual languor prevailed. In his case, as in all the others, when the subjects were in good health (and Mr. D. was so after his severe illness), I found the effect of the sleep always to be, that the patient was refreshed and felt happier.

I have not myself made many experiments of this kind, having very soon seen enough to convince me that impressions made in the sleep are retained in the waking state, so as to influence the sensations and even the actions of the subject. Indeed, this phenomenon is one which presents itself so frequently, that it is found recorded in a large proportion of published cases. My space will not permit me to quote such cases, as I require it all for what

I have seen and done in regard to other phenomena, as well as for some interesting cases kindly communicated to me by others. I therefore pass on to another branch of the subject.

C. Phreno-Mesmeric Phenomena in the Sleep.

I have already, further on, pointed out that there is considerable variety in these phenomena. We find many cases, in which touching the head has no effect in exciting manifestations of any kind. This is the case with Mr. D., Mr. T., Mr. H. W., and Mr. C. M., at least in that state in which I have examined these gentlemen. There is another class of cases, in which the subject sympathises to such an extent with the operator, that the expressed and sometimes the silent will of the latter will produce any desired manifestation, whatever part of the head, or even of the body, be touched. Such cases, of which however I have seen little or nothing, fall under the head of sympathy, where the silent will operates, and of suggestion, where the volition is expressed. I shall not dwell on these, but shall only say, that I think there is good evidence of the existence of such cases, and that of course they can furnish no evidence whatever of the localisation of the cerebral organs.

But there is a third class of cases, in which, so far as I can perceive, no other explanation is possible but this, that touching any part of the head excites to action the corresponding part of the brain. This I have often seen. I have already stated that no such effect took place with the four gentlemen on whom I have lately operated, but I shall here give a case or two to illustrate the phenomenon in question.

Case 26.—A. F., a young man, was put to sleep by me in a few minutes. In this state, every part of the head that was tried, yielded striking manifestations of the corresponding phrenological faculty. I had no reason to think that this young man knew the position of the organs, nor any thing about phrenology; but even if he had some general notions on the subject, the effects produced appeared so rapidly that it was impossible for him to have simulated them, even had he been disposed to do so, which I am sure was not the case. Benevolence, Destructiveness, Combativeness, Secretiveness, Acquisitiveness, Self-Esteem, Love of Approbation, Veneration, Cautiousness, Adhesiveness, Philoprogenitiveness, Tune, &c., were all tried, first in rapid succession, and all yielded strong manifestations, although very often they were quite different from what I had expected, or were distinct when I had no clear idea of how they were to be manifested. Benevolence being touched, he instantly began to give away all his money to me, taking me for an object of charity, and when I continued the contact, took off his coat to give me. This is the almost universal manifestation of Benevolence, obviously because, when the feeling is excited, its most natural result is to give to those in want. Cautiousness produced the most vivid picture of terror I ever saw; he said there was a fearful abyss before him, and felt as if he was to fall into it. Tune instantly caused him to sing; Imitation, to imitate not only every sound he heard, but also, with closed eyes, the gestures made by those near him. It is impossible here to give all the details; suffice it to say, that although it all looked like first-rate acting, a close study of his countenance showed the most entire truthfulness. Besides, as I moved my hand from one organ to another, so rapidly as to confuse any one not very much in the habit of guessing what organ is touched, the effects never failed to follow. To test him further, I tried touching two organs at once, and invariably obtained combined manifestations. Thus when Benevolence and Acquisitiveness were touched, he put his hands into his pockets as before, but instead of giving me the contents, he treated me to a lecture on the heinousness of begging, and declared that he thought giving money the worst kind of charity. Veneration alone caused him to pray humbly and devoutly; Veneration and Self-Esteem combined, gave rise to a prayer, in a standing position, in which he returned thanks for having been made so superior to other men in religious knowledge. This combination was accidental, Self-Esteem having been first much excited, with very amusing results, and Veneration having been touched before the excitement of Self-Esteem had subsided, with the desire of reproducing the former humble devotion. Many similar trials yielded analogous results. I found also, that when, intending to touch one part, my hand accidentally glided to another, the manifestation was always that of the part really touched, not of that which I intended to touch. In the region of the supposed organ of Alimentiveness, I found, within a small space, three different points; the touching of one of which produced excessive desire to eat; of another, the desire to drink; of the third, sensations of smell. To obtain these results, which could not be known to the subjects, since they were not then published, nor generally known to phrenologists, although I had heard of them, it was necessary only to move the point of the finger one-fourth or one-eighth of an inch, the three points certainly lying in less than the surface of a shilling. In all these trials, it did not signify what I wished, nor what I said, only such organs were excited as I touched. I had complete evidence that the subject did not sympathise with me or with my thoughts, but that my touch excited the faculty corresponding to the part touched.

This case occurred to me in 1843, and at that time I had three other similar cases, in persons absolutely ignorant of phrenology; nay one of them, a girl of ten or eleven, of the lowest class, ignorant of everything, and very nearly imbecile. In some of these cases, certain organs could not be excited, while others were easily brought into action. In all of these, I assured myself that neither sympathy nor suggestion played a part. I shall now adduce a recent case.

Case 27.—Mr. C., a young man, had been several times mesmerised four years ago, but not since. I put him to sleep in one minute, and found him even more susceptible to the touch than A. F. The manifestations were very similar, but came out so rapidly, that it was hardly possible to be sure that a part was touched, before the effect was produced. If, while Benevolence was in action, I touched Acquisitiveness he instantaneously collared me to recover what he had given me; if Combativeness were touched, before I could remove the finger he had struck out with his fist, and assumed a very pugnacious attitude. When I combined Benevolence and Acquisitiveness, he pulled out money and offered it, but on my attempting to take it, always withdrew it, his eyes being closed, and told me he required it more himself. In short, whatever he was doing, the slightest touch, even accidental, or with the cuff of my coat, on any organ, at once arrested him, and changed his action and expression. When in the act of falling on his knees, Veneration being touched, the slightest touch on Self-Esteem sent him up like a shot, or Combativeness made him attack, in the fraction of a second, whoever happened to be before him. In short, I could play on him, exactly as on an organ, producing any expression, gesture, or action I pleased, simple or combined. There was no silent or occult sympathy with me, and my expectations or wishes had no effect in modifying the results. It was quite impossible to doubt the entire sincerity of Mr. C., who was, besides, in a deep mesmeric sleep. This case, like that of A. F., could only be explained by supposing that touching the head excited to action the subjacent parts of the brain.

But this case presented some other peculiarities. I could excite laughter by touching the organ of Gaiety or Mirthfulness. But I could also cause laughter by touching the angles of the mouth, when it often became very violent. In either case, I had only to touch the middle of the chin, in order instantly to change the laugh into the profoundest gravity. This fact was pointed out to me by Mr. Bruce, who had studied the case four years before. He also told me, that touching a certain part of the leg caused the young man to dance. I tried this, but probably did not touch the right spot, or touched it too strongly, for the result was a sudden and most violent kick, fortunately received by a table, and accompanied by a very angry pantomime. This I saw several times. When I placed my finger, for less than half a second, on his left breast, he instantly sank down, as if fainting; but observing this, I placed my hand on Self-Esteem and Firmness, when he instantly rose into a posture of defiance. I am convinced that I could have caused him to faint entirely in a few seconds; nay I think, in that state, death might be produced by keeping the hand over the heart. The effect of touching certain parts of the body, no doubt depends on their nervous connection with the brain.

Case 28.—P. G., a boy who has a bad impediment in his speech. I put him to sleep in two minutes, and found him susceptible to touch over the cerebral organs, but not so instantaneously as Mr. C. His stammer was much diminished. The only peculiarity was, that when I touched the spine, about the third or fourth cervical vertebra, all the symptoms of intoxication appeared, particularly an absolute inability to keep his balance.

These cases will suffice to illustrate this part of the subject. They appear to me especially interesting, from the evidence they afford of the existence of an external influence, which passes from the operator's hand to the subject. I may add, that in the case of Mr. C., contact is not indispensable. Pointing with the finger often brings out the effects. But even where contact is employed, the very marked and violent effects in so short a time, prove that an influence passes, to which the subjects, in a certain state, are particularly sensitive, as in the experiment of placing the finger over the heart of Mr. C.

If it be asked why these effects did not occur in Messrs. D., T., H. W., and C. M., I can only say, that I report the facts as I have found them; but that I think it probable, that the sensitiveness to touch over the cerebral organs belongs to a particular stage of the sleep, into which these four gentlemen did not come, but which occurred in the others. There is some reason to think that it is a stage not so deep as that in which Sympathy and Clairvoyance appear. But we are not yet able, at pleasure, to produce the stage we desire, and while some easily pass into clairvoyance, others only exhibit that stage in which the cerebral organs are excitable by touching the head or other parts. We now proceed to

D. SYMPATHETIC CLAIRVOYANCE IN THE MESMERIC SLEEP.

This is a phenomenon which is exhibited in various forms, such as Thought-reading, the power of perceiving the state of health of those directly or indirectly in communication, or *en rapport*, with the sleeper, and sympathetic or mediate Clairvoyance. The cases which I have myself mesmerised, have not exhibited, as yet, this power, but I have seen it in subjects mesmerised by others. It is found either alone, or combined with the power of direct clairvoyance. E., the girl mentioned before as under the care of Dr. Haddock at Bolton, possesses both powers to a high degree, and, in particular, exhibits sympathetic clairvoyance, and the intuitive perception of the state of health, in regard to those placed in communication with her, either directly, that is, by contact, or indirectly, by means of their writing, or a lock of hair. As Dr Haddock very kindly afforded me opportunities of examining this interesting case, I shall mention a few, out of many instances in which I tried her powers.

Case 29.—1. Before I had seen E., I sent to Dr. Haddock the writing of a lady, without any details, requesting merely to know what E. should say of it. I did not even say it was a lady's writing, and, indeed, as the hand is a strong bold one, Dr. H. supposed it was that of a man. E. took it in her hand, she being in the sleep, and soon said "I see a lady. She is rather below middle height, dark-complexioned, pale, and looks ill." She then proceeded to describe the house, the drawing-room in which the lady was, her dress, and the furniture, all with perfect accuracy as far as she went. She said the lady was sitting at a long table close to the wall, something like a sideboard, writing a letter; that on this table were several beautiful glasses, such as she had never seen. (In fact, this lady writes at a long sofa-table at the wall, on which then stood several Bohemian glasses.) She further detailed, with strict accuracy, all the symptoms of the lady's illness, mentioning several things, known to the lady alone. She also described the treatment which had been followed, and said, among other things, that the lady had gone over the water, to a place where she drank "morning waters" for her health; that the waters had a strange taste, but had done her good. (The lady had been at a mineral water in Germany, and had derived benefit from it. The water was always taken in the morning.) I need not enter into all the details; it is enough to state, that not only Dr. H. did not know the lady, nor even her name, but that he had had no means of knowing any one of the details specified, and indeed rather supposed E. was wrong when she spoke of a lady, until he found that she was positive on that point. I received his answer, with the above and many more details, almost by return of post, and, in short, I was perfectly satisfied that E. had seen or perceived somehow, from the handwriting, all that she said, as I knew she had done in other cases.

2. Some months later, I went with the same lady to visit E. She had never been told the lady's name, and was introduced to her and me as to two strangers. When she was put asleep, Dr. H. desired her to take the lady's hand. As soon as she did so, she said, "Oh! you are the lady I went to see." "Which lady?" said Dr. H. "Don't you remember? The lady who sat at the table with the pretty glasses." She then proceeded to say, that the lady had been lately again at a place, over the water, where she took morning waters, and where the people spoke gibberish; that she was better now, but had been worse, and that a doctor had repeatedly put something down her throat, which hurt her very much. (The throat had been cauterised with lunar caustic.) She specified exactly the present symptoms, and entered into various minute details concerning what she had formerly seen, many of which Dr. H. had forgot, but which, on referring to his notes

made nearly six months before, he found to be correct. Whenever she wished to recall anything of the former experiment, if she could not at once do so, she referred to her *book*, as she called it, that is, her right fore-arm (I think), on which, at the time, she makes notes, in the shape of imaginary signs to aid her memory. With the help of her book, she now proceeded to describe, as she had formerly seen them, the exterior of the house in Edinburgh, where she has never been; the street (Prince's Street), which she said she had looked down upon from a point where she also saw houses above her. (The Earthen Mound.) She spoke of the long street, with a garden, and trees, and a deep hollow between it and the height opposite, and spoke of it all as if she were then looking at it, having, by reference to her *book*, recalled her former vision. While engaged in this, her book recalled to her a circumstance, which she had not mentioned to Dr. H. at the time, because, as she said, it had given her a fright, and she said nothing of it. Dr H., much surprised, inquired what it was; when she told us (for she answered my questions readily), that on her way to my house, something had induced her to enter an old house on the height opposite the long street. I could not make out the precise spot, whether the upper part of the High Street, the Castle-hill, or the Castle itself. She said that she found her way into a room, in which sat a lady, richly dressed in a strange fashion; that on seeing this lady, she felt as if the lady was fond of having people's heads cut off, and thinking her own would be cut off, she got out as fast as she could, and proceeded on her journey.

3. I mention this vision, to which I have briefly alluded before, because I am inclined to think that it was not a mere ordinary dream, but that she had, somehow, got into sympathy with a past period, a phenomenon which now and then presents itself. Being curious to inquire into this, I requested permission to ask a few questions, and E. agreed, if Dr. H. would take her, to go again to this house, and tell us more about it. He then, by some trifling manipulation, brought her into the travelling stage, in which she can go, mentally, to any distance, but only hears what is spoken with the lips touching the points of her fingers, and is stone deaf to all other sounds. She now described the room she had seen. The walls were of stone, covered with loose hangings, on which she saw pictures of beasts, &c. (evidently tapestry). The lady was on a peculiar kind of sofa, and, as before, dressed in a strange but rich fashion. She wore a stiff ruff, standing up about her neck, and a cap, with a point down the middle of the forehead, and rising, curved, over the temples. This she explained, by drawing the shape of it with her finger. She was a great lady, and cried much over a baby. Her husband and she did not agree; they differed on religious matters, and the lady was very fond of priests, catholic priests. Thinks the lady was imprisoned in one of the highest houses (qu. ? in the Castle ?), at all events she was

there. Here, in answer to questions, she said that she saw the child let down in a basket from a window, and, she thinks, the lady also, or at least a lady. The lady left that place, down below, after walking a short distance, in a strange kind of carriage (from the description, a horse litter). She could see that the great lady was kept confined in another place, in a house with trees round it. Could not see beyond the trees. Saw the lady, another time, on horseback, riding very fast, to a water, which she crossed, and then gave herself up to people there. When asked, why she did so ? said, "Oh, you know, she thought they were friendly, but they were not." As some of these details led me to suppose that E. had got on the trace of Mary, Queen of Scots, I asked her to tell me what more she could see. She said that the people whom the great lady supposed to be friendly, put her in confinement. I then asked what the lady died of ? E. said she could not then see, being tired, but would be able to tell next morning. Next day, when put into the same state, Dr. H. asked the question again ; when, after looking for a short time, E. said, "Oh, dear, she died of this," drawing her hand across her neck, and added, with a smile, "I daresay, as she liked to cut people's heads off, they cut off hers, to see how she would like it herself." She has told us, on being asked, when she first saw the lady, that she was *shelled*, that is, dead; for E., like many other subjects, will never use the word death or dead. She had also told us, that the house was no longer as she saw it, but that the large room in which she saw the great lady was subdivided, by partitions, into smaller rooms, and entirely changed ; that she saw it as it had been formerly.

Now, even if we regard this vision as a mere dream, it is curious. Her whole manner, and the way in which she answered, proved to me that she was describing what she saw, and was as much surprised as I was. I conjecture, that having somehow been brought into sympathy with past events, and persons long dead, she saw confusedly, and mixed up togther different persons. It appears to me, that some part of what she saw refers to Mary, and the rest to other persons ; that she had no distant idea of dates, and confused various periods and events, but that, on the whole, she was, to a certain extent, in sympathy with the past. The question is, if this be so, how was she brought into this sympathy ? and the only conjecture I can form is, that Dr. H., who had passed through Edinburgh some little time before, having visited the Castle, and there heard a variety of traditions, more or less erroneous, about Mary, sympathy with his thoughts had led E., when in the travelling stage, in which her power of sympathy is singularly developed, to diverge from the direct route to my house, and to enter, mentally, some room in the Castle, probably that in which James VI. was born, and that she had there found traces of past events and persons. I must observe, that at this time E. could neither read nor write, and in her ordinary waking state, appeared to have no knowledge whatever of history. Even if she had, which, however, I do not believe, many of the details are not such as are found in popular histories; for example, when she spoke of the great lady as confined in a house, surrounded by trees, she could not, even after I asked her, see water near it. Now, any one who was dreaming, or pretending to dream, from a knowledge of popular history, would naturally have thought of Lochleven Castle, and the escape of Mary from it, the most popularly known of all the events of her history, not excepting the murder of Rizzio, which also was not alluded to. I conclude, therefore, that whether this be a mere dream or not, E. was not, consciously or unconsciously, thinking of Mary, or dreaming of her, from what she knew of her, but, by means of some vague sympathy with Dr. H., was brought into a confused state of perception of past events. I would repeat, that having observed her very closely, I am thoroughly satisfied of her sincerity, and that I have every reason to believe her utterly ignorant of everything relating to Scotland and Scottish history. If I were fortunate enough to possess the handwriting of Mary, or a genuine lock of her hair, or any trinket which certainly had belonged to her, I should try E. with it, and I am confident, from what I have seen, that she would then be able to perceive more clearly.

4. I gave E. a letter, which Dr. H. supposed to be written by a lady. E. did not look at it, but felt it in her hand, and laid it on her head. She began to speak of a lady, who kept coming before her, but was not the writer of the letter. On the contrary, this lady prevented her from distinguishing the writer. She requested Dr. H. to remove this influence, which he did by blowing on the letter, and passing his hand briskly over it several times. She then put it on her head, and said that it was written by a little boy, whom she described very accurately, dwelling particularly on the peculiarities of his disposition, his old-fashioned ways, as she called them, his love of reading, and various other points all more or less characteristic. His dress astonished her very much, and she described it most minutely in every part. It was the Highland dress, and she gave the colours and pattern of the tartan, as well as every other detail of the boy's dress and accoutrement. It appeared that she had never seen the Highland dress worn, and she thought it must be very cold. The boy was my own son, then in Edinburgh, and neither E. nor Dr. H. knew that I had a son, or that he wore the Highland dress. She told us that the lady she had first seen was one who was much attached to the boy, and described her accurately. This lady had charge of the boy during my absence, and his letter had been enclosed in one from her, from which it had just been taken when it was given to E. This accounted, Dr. H. told us for her seeing the lady. When E was asked whether she could see or discover the mother of the boy,

she said that she had at first supposed the lady whose figure first came before her to be the mother, but had soon discovered that she was not. She said she would try to find her out, and would, as she said, ask the boy to tell her where his mother was. After a silence, she said, "The mother left home some time since, and went over the water, but I cannot see her there now, although I see her marks in the place where she was. If Dr. H. will bring me back to Bolton, I shall be able to find her." Dr. H. then, by a few manipulations, brought her back to her original mesmeric state, and the boy's mother, who was present, having touched her hand, she exclaimed with surprise, "Why you are the mother of the little boy." She then said that she had been looking for the lady over the water, and had asked the people whom she saw to tell where she was, but that they spoke gibberish which she could not understand. She felt, however, that on returning to Bolton she would be able to find that lady. I have mentioned these experiments with E. in some detail (although I have omitted many particulars), because they gave me the opportunity of ascertaining that E. was perfectly honest and sincere, and could be put into a genuine and very deep mesmeric sleep, in which her sympathetic clairvoyance was truly remarkable, being exercised at great distances with the same facility as on the spot, provided a means of communication were provided. I also saw that Dr. Haddock operated with great care and judgment.

After I returned to Edinburgh, I had very frequently communication with Dr. H., and tried many experiments with this remarkable subject, sending specimens of writing, locks of hair, and other objects, the origin of which was perfectly unknown to Dr. H., and in every case, without exception, E. saw and described with accuracy the persons concerned. In other cases, two of which are very interesting, I sent writing or hair, belonging to persons unknown to myself, and obtained accounts of them, which I cannot yet verify. But possibly I may be able to do so, in time for a note, to be added at the end of this work. It would be entirely useless to publish her statements in those cases, until I shall have ascertained whether they are correct or not. But from what I have seen of her powers, I expect that these statements will be found correct. I shall here adduce one case, in which E. exhibited, not only sympathetic, but direct clairvoyance.

Case 30.—A nobleman of high rank, much devoted to science, found one day, among the gravel in his garden-walk, a small flint arrow-head such as was in former ages used by the Britons, and is often called a "celt." This I folded in several folds of thick white blotting-paper, enclosed it in an envelope, which was sealed, and placed this in a second envelope. I then sent it to Dr. H., requesting him to ask E. to look at it, and tell us what she could about it. When given to her, the sealed envelope was enclosed in a second, and from the way in which I had folded it up, no one, out of several whom I tried, could guess the form of the arrow-head by feeling it. E. first held it in her hand, and then laid it on her head, and very soon drew an outline of the form of the object, which she said was enclosed in several folds of blotting paper, nearly white. As it was very small, only about an inch long, and very sharp at one end, E. at first took it for the tooth of some large animal. She said its colour was yellowish white, with a few dark streaks, and pointed out where the edges were chipped. On pursuing her examination, she said that it could not be a tooth, as it was made of stone, and after (mentally) biting it, in doing which she merely approached the packet to her mouth, and appeared to be biting something, she declared without further hesitation that it was made of flint. Every detail she gave I found perfectly accurate, and as the packet was returned to me intact, I have no doubt that E. saw the object perfectly by direct clairvoyance. She could not, however, tell its use, but by sympathy, she went on to say, that a gentleman had found it in a gravel walk in a garden; that he had worn it, that is, carried it, in his waistcoat pocket (I think she said the left), for some time; that this gentleman was a very great gentleman, and, in answer to successive questions, she gave the title appropriate to his rank. She was asked to observe more about him, and then said she saw him in a palace house; she spoke in whispers out of respect, and when her attention was drawn to the point, described the nobleman's person very correctly. This was done on a subsequent occasion, as I had requested Dr. H., when I found E. had discovered the finder of the arrow-head, to ask her further questions about him. In sending the packet to Dr. H., and until I had heard all that E. had to say, I carefully avoided giving the slightest information, either as to the object or the finder.

Case 31.—Sir Walter C. Trevelyan, Bart., having received a letter from a lady in London, in which the loss of a gold watch, supposed to have been stolen, was mentioned, sent the letter to Dr. H., to see whether E. could trace the watch. She very soon saw the lady, and described her accurately. She also described minutely the house and furniture, and said she saw the *marks* of the watch (the phrase she employs for the traces left by persons or things, probably luminous to her) on a certain table. It had, she said, a gold dial-plate, gold figures, and a gold chain with square links; in the letter it was simply called a gold watch, without any description. She said it had been taken by a young woman, whom she described, not an habitual thief, who felt alarmed at what she had done, but still thought her mistress would not suspect her. She added, that she would be able to point out the writing of the thief. On this occasion, as is almost always the case with E., she spoke *to* the persons seen, as if conversing with her, and was very angry with her. Sir W. Trevelyan sent this information, and requested the writing of all the servants in the house to be sent.

In answer, the lady stated that E.'s description exactly applied to one of her two maids, but that her suspicions rested on the other. She also sent several pieces of writing, including that of both maids. E. instantly selected that of the girl she had described, became very angry, and said, "You are thinking of pretending to find the watch, and restoring it, but you took it, you know you did." Before Sir W. Trevelyan's letter, containing this information, had reached the lady, he received another letter, in which he was informed, that the girl indicated as the thief by E. had brought back the watch, saying she had found it. In this case, Sir W. Trevelyan was at a great distance from Bolton, and even had he been present, he knew nothing of the house, the watch, or the persons concerned, except the lady, so that, even had he been in Bolton, and beside the clairvoyante, thought-reading was out of the question. I have seen, in the possession of Sir Walter, all the letters which passed, and I consider the case as demonstrating the existence of sympathetic clairvoyance at a great distance. I have before alluded to various other instances, in which E. has traced, and been the means of recovering, lost or stolen property and documents, when put in communication with the proprietors of them. I shall now proceed, merely referring to these cases, some of which have been published, to describe some experiments, also alluded to before, which were made by Sir W. C. Trevelyan, to test E.'s power of sympathetic clairvoyance, with the view of ascertaining whether it was effected with or without thought-reading, and also her power of observing the hour of day at distant places, visited, mentally, by her, in the mesmeric sleep. These experiments were, I think, extremely well devised, and satisfactory in their results.

Case 32.—Sir W. T. requested the Secretary of the Geographical Society to send him the writing of several persons, unknown to him, and without their names, they being in different parts of the world. Three handwritings were sent.

1. E. soon discovered No. 1, and described his person, as well as the city in which he was, and the surrounding country. When asked the hour there, she looked, but said she could not tell. It appeared, on subsequent inquiry, that No. 1 was in Rome, and that E.'s description of him, as well as of the city, &c., was exact, As she generally finds the hour by looking at some clock or watch, it would appear that she had been puzzled by the clocks of Rome, which have 24 hours, instead of 12.

2. In the case of No. 2, E. soon discovered where he was, and gave the hour there; but it is remarkable, that she could not see the person himself. She described the country, and spoke of crops of large yellow corn then standing (late in October). The longitude, calculated from the hour she gave, corresponded to that of a part of Tuscany; and on inquiry it was found, that No. 2 resided in Florence, but was in the habit of travelling about the

country. The corn appears to have been the second crop of maize, which was then standing in Tuscany.

3. In the case of No. 3, E. found and described him, and said he was in a street which she described, in a large city: the time she gave differed from that of Bolton by 2½ or 3 minutes only, and indicated the longitude of London. On inquiry it appeared that when the writing was sent, No. 3, whose person was accurately described, was supposed to be at a much greater distance than the others; but that, before E. saw his writing, he had unexpectedly returned, and was then in London.

In these experiments, which were communicated to me by Sir W. Trevelyan, thought-reading was out of the question; for Sir W. T. did not know even the names of the persons, and if he had known all about them, he was not at Bolton, but in Edinburgh. Dr. Haddock had no knowledge whatever of the persons whose writing was sent.—I include these and the preceding cases under sympathetic clairvoyance, because in all, a communication of some kind, by writing or otherwise, was established. But it would appear as if this were only necessary, in order to put the clairvoyante on the trace of the persons seen; and that when that trace is found, it is followed up, in E.'s case, by direct clairvoyance. At least, she always speaks as if she saw before her what she describes, or rather, as if she were in the place described; and in Sir W. Trevelyan's experiments, that form of sympathetic clairvoyance which consists in thought-reading, is excluded by the circumstances. It is possible, that when E. examined the letter of my son, she might have read my thoughts, as I was present; but I do not think this was the case, for the letter was given to her by Dr. H., who supposed it to be from a lady, and during the whole conversation, although we were thinking and speaking of the same subject, I could not observe any relation between what I thought and what she said, beyond the subject of our conversation. On the contrary, her attention was constantly attracted to details of which I was not then thinking. But even if E. did exhibit the power of thought-reading, which she very probably may do at times, we must not forget, that that power is really as wonderful, and as difficult to explain, as sympathetic or direct clairvoyance.

I have mentioned before, in this work, a remarkable case, in which this same clairvoyante, with the aid of handwriting, traced the progress of a gentleman, Mr. W. Willey, then in California, as well as of another person who accompanied Mr. W., and whose writing was also shown to her. In this case, which was published in the newspapers, E. gave a multitude of details in regard to the persons, their voyage, their occupations, and various occurrences, the whole of which details were, in so far as concerned the period subsequent to their embarking at Liverpool, entirely unknown to their families, but were afterwards fully confirmed in every point by Mr. W on his return. It is worthy of remark, that E.

minutely described the country, the houses, and the mode of life of the place in which she saw these persons. It was evidently San Francisco and California; but although she spoke of their digging sand, and even, when desired to look, said she saw shining particles in the sand, and gave its value in dols or dollies, as she called them, which information she seemed to obtain by conversing with the people she saw, she never spoke of gold and wondered much why Mr. W. took the trouble to go so far to dig sand, which he might have found at home. Had she been dreaming of California from ideas suggested to her mind, she certainly would have noticed the gold. I conclude, therefore, that she merely described what she saw, and did not understand it. That her descriptions of the persons, the places they had passed through, and the events that had happened to them, both in the voyage out to Panama, in crossing the Isthmus, and in the voyage to San Francisco, as well as in that city, were exact, cannot possibly be accounted for by any suggestion, even had such been attempted, or in any other way than by the same power of sympathetic or of direct clairvoyance, which I found her to possess in the cases in which I tried her. It was in this case that E. spoke of Mr. Morgan, the companion of Mr. Willey, as having had a fever, and having also had, during his illness, a vision or dream of his wife coming to see him. She also said that he had fallen overboard. All these details, and many others, were exact, but quite unknown to any one in England at the time, and she gave them as if they were Mr. M.'s answers to her questions. I have mentioned, that E. always speaks *to* the persons she sees, and holds long conversations with them.

Case 33.—It is pretty generally known, that this clairvoyante was tried with the writing of Sir John Franklin, and a part of what she said has appeared in the newspapers. I had the opportunity of becoming acquainted with what she did really say, and, although of course the greater part of it cannot be verified until the return of Sir John, yet I am bound here to testify, although she has probably mixed up and confused many things, which we have not the means of distinguishing, that E. has said nothing concerning him which may not prove correct. It appears that some clairvoyants, of whom I knew nothing, went so far as to predict the return of Sir John during last autumn. If such predictions were made by genuine and honest clairvoyants, I conjecture that they have been of that class who are strongly affected by sympathy with the feelings and wishes of those who consult them, which feelings and wishes they, as it were, reflect. But this is not the case with E. She has made no prediction in the matter, but has simply, at various times, with the aid of Sir John's handwriting, gone, in her phrase, to see him. She was not told, and does not, I believe, even yet know, whose writing it was, but she has found the writer in one of two ships, fixed in ice, and surrounded with walls of snow. These ships she first saw in the winter of 1849—50, I believe ; I saw several of Dr. Haddock's letters about it in February and March, 1850. Since E. had been right in so many cases at a distance, it was probable that she was also right in this one. She described the dress, mode of life, food, &c., of the crews. She saw and described Sir John, and said that he still hoped to get out, but was much surprised that no vessels had come to assist him. She frequently spoke of his occupations, and when asked the time of day, found it either by looking at a timepiece in the cabin, or by consulting Sir John's watch. During the winter and spring of 1849—50, and part of the summer of 1850, she uniformly indicated the same difference of time, which I cannot at present give precisely, but which was nearly seven hours. At whatever hour she was mesmerised and sent there, she always made the same difference. Nay more, when the time there was nine or ten a.m. (four or five p.m. at Bolton)she would say that such was the hour, but that it was still dark, and lights were burning in the early part of summer. Now it is quite absurd to suppose that this totally uneducated girl has any notion of the relation of longitude to time, or of the difference between an arctic day and one in our latitude. E- also, being shown the handwriting of several of the officers of the expedition, found and described them. One was dead (shelled, as she said) when she was asked. Another, at a later period, was dangerously frost-bitten, but recovered. She said, that in one of the ships the provisions were exhausted, but that the other contained provisions. She described the fish, seals, and other animals hunted and killed for food and oil by the crews. Of, or rather to, one officer she said that he was the doctor, although not dressed like a doctor, but like the rest, in skins ; that he was a first-rate shot, and was fond of killing animals to preserve them. (This is really the case with Mr. Goodsir, whose writing she was then examining.) She added a multitude of curious details, for which I have no space, and they will no doubt be published by Dr. Haddock. But I may mention, that on a Sunday afternoon in February, 1850, she said it was about ten a.m. there, and described the captain (Sir John) as reading prayers to the crew, who knelt in a circle, with their faces upwards, looking to him, and appearing very sorrowful. She even named the chapter of St. Mark's Gospel which he read on that occasion. She also spoke, on one occasion, of Sir John as dejected, which he was not before, and said that the men tried to cheer him up. She further spoke of their burning coarse oil and fish refuse for warmth, and drinking a finer oil for the same purpose. All this time, she continued to give the same difference of time, from which the longitude might be calculated. This time, seven hours, or nearly, from Bolton, gives a west longitude of about 100° to 115', which corresponds very well with the probable position of Sir John. But at a later period, all of a sudden she gave a difference of time of somewhere between six and seven hours, indicating that the ships had moved eastward. She was not, after this, quite so uniform in the difference of time as before, and seemed not to see it so clearly; but she persisted that they had moved homeward, and if we take about 6½ hours as the later difference, this would indicate a longitude of about 97° 30' W. After this change, she also said that Sir John had been met and relieved, and has always since then seen three ships, which, for a long time past, are said by her to be frozen up together. The last observation of which I have heard, 17th February, 1851 gave a longitude of 101° 45' W. At the same time, from Captain Austin's writing, which has also been frequently tried, she gave, for him, the longitude of 95° 45' W. She does not know whose ship it is, that, according to her, has met with Franklin, but she still speaks of three ships together. I should add, that when E. has been sent there at such an hour and season that it was night in those latitudes, she has, quite spontaneously, described the aurora borealis, which she once saw, as an arch, rising as if from the ground at one end, and desending to it again at the other. From this arch, coloured steamers rose upwards, and some of these curved backwards. She was much surprised and delighted with it, and asked if that was the country the rainbow came from. She had never been told anything whatever about the aurora, and knows nothing of it.

Now, in all these details, and many others of a similar nature, there is nothing impossible, nothing which may not be found correct. Many of E.'s observations on this subject have been published, and all have been communicated to various persons by Dr. Haddock, so that, when the ships return, her accuracy, or inaccuracy may be tested. I have lately heard, that she speaks of a second officer as dead, and thinks he died about six months ago. Considering what I have myself seen E. do, and the numerous cases in which I know her to have been correct in her visions, I am disposed to think that she may possibly be found right in many points, with regard to Sir John Franklin. Certainly, it is not easy to see any greater difficulty in his case, than in that of Mr. W., whose proceedings in California were accurately described. But it is not likely that she has always been right, and it is probable that she may often have been misunderstood, and may, as I have already said, have confused different persons and times.

I might give many more instances of E.'s powers, but the instances I have given are sufficient, I think, to illustrate the form of clairvoyance, as it occurs in this very remarkable subject.

I mentioned in case 29, the singular circumstances of E. having been led, while on her way, mentally, to visit my house in Edinburgh, to enter some old house, where she was, as it were, transported to past times. In order to show that this is not an isolated fact, I shall here quote a very remarkable case, of a very similar kind, which occurred to Major Buckley, and which he has kindly allowed me to take from his notes.

Case 34.—What follows is transcribed from Major Buckley's note-book. B. denotes the patient, M. the mesmerist. Mr. B., the patient, was a young officer, whom Major Buckley mesmerised for his health, and who became lucid on the first occasion. He almost instantly acquired the power of visiting distant places, and of reading through opaque bodies. He used also to go into a deeper state, which he liked, probably because he had, in that state, very vivid and agreeable visions. On the 15th of Nov., 1845, Major Buckley, at his request, allowed him to go into that deeper state for about ten minutes, after which he awoke, so to speak, into his usual clairvoyant state, in which he could converse readily with his mesmerist. His first remark was: B. "I have had a strange dream about your ring (a medallion of Antony and Cleopatra). It is very valuable." M. "Yes; it is worth 60 guineas," B. "Oh! it is worth a great deal more." Placing the ring in his hand, Major B. said, M. "Can you tell me its history:" B. "Oh! now I see it all again. If what I say be true, it is very valuable. It has belonged to royalty." M. "In what country?" B. "I see, Mary Queen of Scots. It was given to her by a man, a foreigner, with other things, from Italy. It came from Naples. It is not the same gold (that is, the setting is not that which it once had). She wore it only once. The person who gave it to her was a musician." M. "Can you tell me his name?" B. "It begins with an R. Oh! I see his signature. After the R there is an I, then there is a letter which looks like Z, then another Z, then an I, then there is something which looks like an E, with a curious flourish over it. I can write it." (He went to a table and wrote the name, then added), "There is something more. All this is secret." He then wrote at long intervals, until the paper marked 1 was finished. Once when I looked over his shoulder, he said I had caused him to make a mistake. It was while he was writing to the left of the signature marked 2. B. "The writing (that which he saw, and was copying) is on vellum. Here (pointing to the middle) I see a diamond cross; the smallest diamond is larger than this (pointing to one of about four carats). It was worn, out of sight, by Mary. The vellum has been shown in the House of Lords (qu. ? of Scotland?—W. G.): not the cross. They were afterwards placed where I now see them, in the wall of a stone building, erected before the reign of Elizabeth. It is now in ruins, and used as a farmhouse." M. "Who are living in it?" B. "Only an old man. It is a place of concealment in the wall, opened by an iron spring. Oh! I see how to open it. You push in a small stone near it. There are many valuable things there. Nobody knows of them but myself. . . . The ring was taken off Mary's finger by a man." M. "Did he steal it?" B. "No; he took it off in anger and jealousy, and threw it into the water. When he

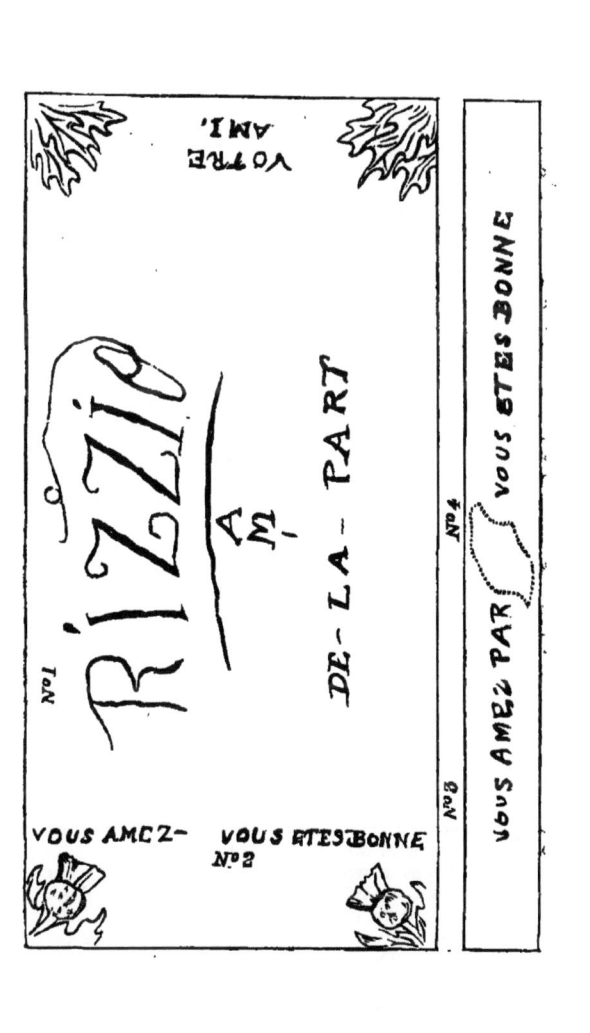

took it off, she was being carried in a kind of bed, with curtains (a litter). I now see the man who gave her the ring; he is in a room. I see many more men. There is a secret door. I see a man with a dagger." Here he shuddered very much; and added, " They have murdered him. There is a gash here (pointing to his throat). Oh, Mary is screaming dreadfully. That man (probably the one who took away the ring.—W.G.) has seized her by the hair." Here he was very much agitated. M. "Don't think any more about it." B. (after a pause) " I am looking back about 300 years." M. " Where are you ?" B. " In Scotland."

He was again mesmerised, three weeks afterwards. On placing the ring in his hand, he said : B. " You thought I would forget about the ring ?" M. " No; but I wish you to show me where you made a mistake in copying this" (producing the paper marked No. 1). B. ". It was here." He then re-wrote the words marked, separately, 3, adding the letters PAR after the word AMEZ. " Between PAR and VOUS some letters are covered with something green and wet." He dotted round the spot (of mould) marked 4. " I see some letters on the cross. There is an M, an S, then a small word, then a large R. The ornaments on the corners of the vellum are in gold." Major B. did not inquire what they represented. Those on the right of the signature resemble the leaves of the thistle, those on the left the flowers. Major Buckley appended to his letter a rough copy of the drawing or copy made by Mr. B. from the vellum he saw in his vision; to this sketch the numbers refer. It represents an oblong sheet, apparently of small size, in the copy sent to me about 5 inches by 2½.

1. The signature. 2. The words on the left, as first written. The second copy he made of these words, adding PAR, is given at the foot, No. 3. No. 4 is the spot of mould, concealing some letters. (The sentence most probably runs thus, "Vous amez (aimez) parceque vous êtes bonne." I do not know whether Mr. B. saw only a small bit of vellum, like that here sketched, or whether he saw only the end of a larger portion, to which the signature was attached. It would appear that Mr. B. saw the writing so distinctly as to be able to copy it, but the sketch here given only give a general idea of the style of it, as I have not seen the original drawing of Mr. B.—W. G.)

I regard this vision as a most remarkable one, because it was quite spontaneous, and nothing was known, even to Major Buckley, of the history of his ring, except that his father, in 1829, had had it for 60 years, having purchased it at the sale of the effects of a gentleman. The ideas in the vision, therefore, could not have been suggested by Major B., nor read in his thoughts. Then the very minute detail of the writing, and the intense agitation of the sleeper on seeing, acted before him, the murder of Rizzio, tend to show that the vision was at all events genuine. As to its accuracy, little can now be said; but it is conceivable that, if properly tried, the clairvoyant might be able to discover the place where the vellum lies. If Rizzio ever presented Mary with a ring, or cross (although so very valuable a present as the diamond cross, is more likely to have been from the Pope through Rizzio), accompanied by such a manuscript, it is probable enough, that it (the MS.) may have been produced against her by her enemies, in the House of Lords or Privy Council of Scotland, nay possibly even in England, and afterwards concealed. Unfortunately Major Buckley has not been able, on account of Mr. B.'s absence from England, to mesmerise him again. Other clairvoyants have been tried with the ring, and have, without the least knowledge of each other, or of what had been said, corroborated the main facts about the ring. This, as Major Buckley was the operator, might possibly depend on thought-reading, since he now had those ideas; but it would appear that this was not the case, as they have added new details of the history of the ring. Thus, one traced it from the time it was thrown into the water. It was fished up by a man in a boat, something like a wooden box (a punt); that he kept it for some years, and then gave it to his son, who lost it by shifting his stick from one hand to another as he walked. It remained lost for many years, till found by a man with two dogs, described by the clairvoyant; by him it was sold, and finally came into the posession of a gentleman who shot himself. The suicide was seen and described, both by this and another clairvoyant, who were much affected and agitated by it. (Indeed the sight of blood or death in their visions, almost invariably produces the most painful effects on clairvoyants, as was the case with Mr. B. on seeing Rizzio's murder.) They said it took place in his drawing-room, and that the ball passed through his body. One or both of these clairvoyants added that at the sale of the effects of the suicide, the ring had been bought by Major B.'s father, who had kept it for 60 years, and that Major B. had had it upwards of fifteen years. It must be admitted, then, that there is considerable probability that the vision of Mr. B. may have been a true vision of past events. Not only are the chief statements as to its early history confirmed by others, but the only part of its history that can at present be ascertained, proves to be as described by the clairvoyants. For, till they mentioned it, Major Buckley did not know that his father had obtained it at the sale of the effects of a gentleman who had committed suicide; but he has since ascertained that this was really the case. And although what Mr. B. said of the murder of Rizzio corresponds with the history of that tragedy, yet all the rest of the details are such as no history could suggest.

It appears to me highly probable, that if those who are fortunate enough to meet with clairvoyants of such a kind as Mr. B., and to be allowed to study them, were to try them with trinkets,

&c., which are known to have belonged to historical personages (as Mr. B. was tried with this ring by accident), many curious details might be discovered. Hair, handwriting and articles of dress worn by the persons, would probably answer equally well. In this way it is possible that missing historical documents might be recovered, where they still exist; and I know that the girl E. has been the means of recovering or finding three registers which had been sought for in vain.

Chapter XIV.

DIRECT CLAIRVOYANCE—CASES—TRAVELLING CLAIRVOYANCE—CASES—SINGULAR VISIONS OF MR. D.—LETTERS OF TWO CLERGYMEN, WITH CASES—CLAIRVOYANCE OF ALEXIS—OTHER CASES.

BEFORE passing on to another branch of the subject, I would point out, that those sensitives who exhibit clairvoyance in the forms hitherto described, are not always found accurate in their visions. I have already mentioned in part that there are many known and some unknown causes of confusion in the results, so that, where we cannot verify them, we must be cautious of attaching too much credit to these visions. One frequent source of error is, that the subject may see past events, and suppose that they are present. Sometimes, by urging the clairvoyant to attend very closely, we may discover that he is speaking, not of passing events, but of what has happened, a shorter or a longer time before. But notwithstanding this, and other sources of error, there are cases in which the clairvoyant sees so clearly, and sympathises so intensely with the person seen, as to be able to speak without hesitation. As a general rule, we ought to verify the vision, before admitting it as an instance of genuine clairvoyance.

E. DIRECT CLAIRVOYANCE IN THE MESMERIC SLEEP.

Under this head I include cases, in which the clairvoyant, without any means of communication whatever, such as contact with the person, or with his hair, or his handwriting, or any other object connected with him, exhibits the same powers as are mentioned under the preceding head, that of Sympathy. I include here also, those cases in which the sleeper is able to describe or read the contents of closed and sealed packets and envelopes, just as we have seen that some can do in the conscious state.

Case 35.—At the house of Dr. Schmitz, Rector of the High School here, I saw a little boy, of about nine years of age, put into the mesmeric sleep by a young man of seventeen. As the boy was said to be clairvoyant, I requested him, through his mesmerist, whom alone he heard, to visit, mentally, my house, which was nearly a mile off, and perfectly unknown to him. said he would, and soon, when asked, began to describe the back drawing-room in which he saw a sideboard with wine glasses, and on the sideboard a singular apparatus, which he described. In fact, this room, although I had not told him so, is used as a dining-room, and has a sideboard on which stood at that moment glasses, and an apparatus for preparing soda-water, which I had brought from Germany, and which was then quite new in Edinburgh. I then requested him, after he had mentioned some other details, to look at the front room, in which he described two small portraits, most of the furniture, mirrors, ornamental glasses, and the position of the pianoforte, which is very unusual. Being asked whom he saw in the room, he replied, only a lady, whose dress he described, and a boy. This I ascertained to be correct at that time. As it was just possible that this might have been done by thought-reading, although I could detect no trace of any sympathy with me, I then requested Dr. Schmitz to go into another room, and there to do whatever he pleased, while we should try whether the boy could see what he did. Dr. S. took with him his son, and when the sleeper was asked to look into other room, he began to laugh, and said that Theodore (Dr. S.'s son) was a funny boy, and was gesticulating in a particular way with his arms, while Dr. S. stood looking on. He then said that Theodore had left the room, and after a while that he had returned; then that Theodore was jumping about; and being asked about Dr. S., declined more than once to say, not liking to tell, as he said, but at last told us that he also was jumping about. Lastly, he said Dr. S. was beating his son, not with a stick, although he saw a stick in the room, but with a roll of paper. All this did not occupy more than seven or eight minutes, and when Dr. S. returned, I at once gave him the above account of his proceedings, which he, much astonished, declared to be correct in every particular. Here, thought-reading was absolutely impossible; for neither I, nor any one present, had the least idea of what Dr. S. was to do, nor indeed had Dr. S. himself, till I suggested it, known that such an experiment was to be tried. I am, therefore, perfectly satisfied that the boy actually saw what was done; for to suppose that he had guessed it, appears to me a great deal more wonderful; besides, his manner was entirely that of one describing what he saw. I regret much that I was unable to pursue further the investigation of this case, which would no doubt have presented many interesting phenomena. I have mentioned it as a recent one, and because Dr. Schmitz and others saw the facts, and can attest them.

Case 36.—After I had produced in Mr. D. (Case 1. and others) the deep mesmeric sleep, I found that he exhibited some forms of clairvoyance. Thus he often saw light flowing from my fingers, when my hand was held over his head, and his eyes were close shut. He also saw, in the same position, light from magnets, from a loadstone, and from many crystals. But the form in which clairvoyance was best developed in him, at least when I made these experiments, was that of visiting and describing distant places, both known and unknown to him. Having observed that he spontaneously described places which he said were quite unknown to him, I first tried him by asking him to look at my house in Prince's Street, he being then in the south side of the town; and although at first he saw it but dimly, owing to a thick mist of which he complained much, by degrees he came to see and describe it very plainly. He had several times been in the house, and might therefore be able to recall it in a general way; but I found that he could describe in detail any room, or part of a room, to which I directed his attention, and among these, some rooms that he had never seen. I next asked him to tell me whether he saw people in the rooms. Sometimes he did, at other times he said no one was there; and on some occasions, when he counted several persons, I found that, as near as the time could be ascertained, he was correct as to the number of visitors. I then desired him to look at a house, about two miles out of town, which he did not know, and which I did not farther describe. He found it, and said that it was of a peculiar form, describing especially the roof, which is unusual. He told me that he did so, because he saw it, as it were, from above. He said it stood in a garden, and had trees about it, but was not, at that time, able to see any persons about it, or to see the interior.—My next experiment was to ask him to visit Aix-la-Chapelle, which is quite unknown to him, but which I know well. My mind turned to the great place, in which is the Elisen-brunnen; but, to my surprise, he not only readily went to Aix, but began to describe what he saw so clearly, that I at once recognised the boulevard, or promenade outside the walls. This showed that he was not reading my present thoughts. When I asked, how he knew that the place he saw was that I named, he said, that an internal sense, like that which distinguishes right from wrong, told him that the first place he saw was the right one; but on subsequent occasions, he added, that sometimes another place would appear first, but that the same intuitive sense told him that it was not the place wanted. I now requested him to go to the great place, which, as I have stated, he accurately described, quite as accurately as I could have done. He also saw people moving in the place or streets, but every time he saw different people. He noticed soldiers in various uniforms, and said that some wore cloth caps, others helmets, such as he had never seen, but which he described correctly (the Prussian helmets or pickelhauben). He observed that many men wore beards and moustaches, which amused him much, and he described several fashions of beard that he saw. I asked him to look for several persons, known to me, naming them, and he found and described some of them, but not others. One gentleman, whom I thought of as sitting in his own room, he saw walking on the boulevard with another, an additional proof that he was not reading my thoughts. This gentleman, he said, wears neither beard nor moustaches, which was true. He saw, in an hotel, dinner going on at 2 P.M., and at our dinner-hour, without any questions being put in either case, he spoke of the saloon of the hotel as empty, and the tables uncovered. After trying him many times with Aix, I asked him to go to Cologne, and he soon told me that he saw it in a bird's-eye view, or as from a balloon, in which way I certainly never saw it, nor thought of it. He noticed the river, the bridge of boats, many spires, and one very large building, much higher than the rest. I begged him to go nearer to it, and he soon spoke of being in a street, where his attention was arrested by a fat, jolly-looking old boy, as he called him, standing in the doorway of his shop, without a hat, and with an apron on. At my request, he described the exterior of the large building, at one part, where he spoke of very tall windows, the shape of which he drew, and the buttresses and pinnacles between them. As he was much struck with the great size of the building, I conclude it was the Dom, and that he first saw the outside of the choir, and eastward part. He afterwards noticed a projecting part (transept), very high, with high windows, and going, at my request, to the west end, entered and saw inside many pillars and arches, and people kneeling on the floor; but, whether within or without, he could not see the roof, which a mist concealed from him. In the street he saw people, and remarked many soldiers. I next asked him to visit Bonn, when he found it, and gave me a most perfect description of it, as seen from the heights to the west, from a point on which he declared he was looking at the town. I need not go into detail, but his descriptions of the situation of the town, of the heights to the westward, of the course of the river, and of the heights on the east or right bank, were most graphic and accurate. In all these cases, he had the greatest pleasure in contemplating the new scenes, and particularly admired Bonn and the environs of Aix. But in every instance he called north south, and east west; telling me, for example, that the Rhine ran southwards, and that both Cologne and Bonn were on the east side, or what is actually the right bank of the river. I have no doubt he would have told me that the sun at noon was due north, had I tried him. He certainly persisted uniformly in the directions he gave for the position of places, and when I caused him to look, in his sleep, at Prince's Street, from a distance, declared that the street looked to the north, and that the Castle lay to the north of Prince's Street. Making the necessary corrections, his local descriptions, in these and in many other instances, which I cannot here detail, were not only correct, but strikingly graphic, and I could never discover the slightest reason to suppose that he was reading my thoughts; indeed, he very often spoke of one place or thing, when I was thinking of, and had perhaps asked him about another; and when he saw, as generally happened after a short interval, the particular house or street I asked for, he was sure to observe something not at all in my mind. Thus, in Cologne, he frequently returned to the street he had first seen, and very often saw the "jolly old boy," of whom I had no idea; but when I asked for him, it would frequently happen that he was not visible. Nor did that person always appear in the same way, for he sometimes stood in the doorway, at other times in the shop; sometimes with an apron on, at others without it. He declared he should know this man anywhere, and spontaneously added, that, in spite of his jolly appearance, he disliked him; why he could not tell.

I shall here mention some very singular facts that presented themselves in this case, in which Mr. D. saw and described, precisely as above, places quite unknown either to himself or to me. I had in my possession a so-called magic crystal, apparently of glass, certainly of some antiquity, but the history of which is unknown. I wished to try whether Mr. D. would feel any action from it while in the sleep, he being at that time singularly sensitive. I therefore placed it in his right hand, when he felt a strong current of cold up his arm; and, in the other hand, a strong current of heat. When I asked him if he saw it, his eyes being fast shut, he said it shone so brightly as to dazzle his eyes, and begged me to remove it. I did so, but found, in holding the glass over his head, that he saw it as well as before. Next day he was either in a different state, or at all events less sensitive, and saw it, his eyes being closed, whether in his hand, or held by me near his head, out of reach even of the open eyes. But this time it did not dazzle him, and he looked at it with extreme pleasure. He described it as traversed in every direction by broad bands of light, each of the bands exhibiting all the rainbow colours in great beauty. When held in his left hand, it caused a strong, but agreeable warm sensation, and appeared to produce a deeper sleep. All at once he said, spontaneously, "I see a man in a very strange dress." As it occurred to me that Mr. D. might be seeing some one in whose possession the crystal had been, or who was somehow connected with it, I encouraged Mr. D. to tell me all he saw; and as I found that on subsequent occasions he saw the same person, I very often got him to tell what the man was doing, and in fact for a time to trace his motions, day after day. I shall give the results as briefly as possible.

Mr. D. first found himself, after apparently traversing a large space, and seeing confusedly and dimly various objects (during the very short time that elapsed after the crystal was placed in his hand, before he began to speak), in a road, on one side of which ran a rapid and rough river, under high, perpendicular rocks, while on the other side of the road were also rocks not so high. On the road walked a man, above the middle height, between 40 and 50 years of age, very healthy and vigorous, with dark complexion, long face, prominent features, like a Spaniard, Italian, or Jew, black hair, long black beard, dressed in a black jacket, fur vest, and black knee breeches, striped stockings, and short boots folded over and furred at the top, below the calf of the leg. He had a belt or girdle round his middle, to which hung something, not a sword or dagger. Over all he wore a large cloak, which was open so as to allow his dress to be seen. His hat was a tall and conical cap of cloth, with fur round the lower part of it, and a broad brim, and there was a feather in it. In his hand he carried a long staff, taller than himself, with a crook at the upper end, and before him ran three or four sheep on the road. Mr. D. thought the shepherd's crook and the sheep so incongruous with the rest of the costume, which was handsome, and of fine materials, that he often expressed his surprise at these things. He followed the man along the road to an inn, into which he entered, and had refreshment in a public room, where some men, apparently peasants, were sitting. By them and by the landlord he was respectfully received. The inn did not look like one of this country, nor were the peasants English. Their dress was in form like that of the man, but of very coarse blue cloth, and they had cloaks of sheepskin. When Mr. D. next day saw the man, he was again walking on a road, but without the sheep or crook, and wrapped in his cloak. The road was now in a wild, bare country, in which, by and by, stunted trees appeared, along the course of the river. The valley gradually widened, the mountains receding on each side, and trees and uncultivated fields appeared. The road led to a town, lying near the foot of the hills on one side, and before entering the town, crosses the river by a bridge of many arches, elevated in the middle. The man first seen was apparently a mile or two from the town on the third or fourth occasion on which he was seen, and walking towards it. On the road were many peasants, some carrying baskets of eggs, &c., to the town, some with carts. The town, which appeared larger than Perth, lies on a slope rising up from the river to the hills behind; it is triangular in shape, the base of the triangle resting on the hills, the apex on the bridge. There are towers and spires in it, but Mr. D. could see no wall round it. The carts and waggons on the road are not like ours. As Mr. D. could not give a name to the man, and yet wished to have a name for him, and as he seemed to be a foreigner, and either a Jew, Italian, or Spaniard, in appearance, I suggested that he might be called Rafael. which name Mr. D. at once adopted. We shall call him R. Next day, Mr. D. found R. in a house in the main street of the town, which rises straight from the bridge, up a pretty steep slope. In passing the bridge, Mr. D. observed, first, that the water was very muddy; secondly, that towards the end of the bridge, next the town, where he conjectures there is a gate or archway, there appeared to him a dark space, in which he could see nothing, and he could only see clear again when he had reached to a point in the street, three or four houses from the bridge. This I cannot explain, but the same blank space has invariably presented itself. The house in which R. was now seen was a shop, his own apparently. A woman, much resembling him, was there with him; Mr. D. thought she was a sister. She seemed to listen attentively to what he told of his travels, as far as Mr. D. could judge. The shop appeared to be one in which are sold curiosities and second-hand jewellery; there were chains, crosses, &c., of gold and silver; but they did not look new. When describing these things for the first time, Mr. D. expressed much surprise, and some uneasiness, at his being, as he seemed to be, outside of the town, looking at the bridge, &c., and at the same time being able to see the inside of R.'s shop, without going there. Afterwards, he got accustomed to this, and made no observations about it. He saw R. and the woman at a meal in a small room behind the shop. The furniture was coarse and plain, the fare also plain. They did not say grace before meat. Their drink seemed to be wine. The woman in the shop took down from a shelf what seemed to be a small brass globe, to show to R. In the fire-place was a wood fire, burning on dogs, and some billets lay on the floor. Another day, he saw R. in his shop, and a ragged man offering something for sale in a small box, which R. would not buy. The main street, leading up hill from the bridge, has side pavements. It is crossed, half-way up, by another street. At one of the corners of this crossing, he saw a dragoon standing, with a green, long tailed coat, red facings, blue trousers with red stripe, cloth cap, with a shade in front, green, with a band of red, and a red tuft in the top; long sword, with long belt, boots and spurs. Saw many names on shops or sign-boards in the street; but could not read them, the letters being unknown to him, except on one large house, like an hotel, at the corner of the cross street already mentioned, on which he spelt out the name SCHULTZ, the only one that he could read of all that he saw. Another time, R. was not in the shop, but after seeking for him, he found him in a hut, in a mountainous district, along with a number of men. As some of these men were dressed as he had described the peasants whom he had seen, in coarse jackets and wide trousers, with sheepskin cloaks, while others wore a dress which seemed to him more Turkish than European, having cloths rolled round their heads, and very wide trousers, and wearing long beards, he spontaneously said that he believed the hut to be on the frontier between two countries.

I cannot help regarding these visions as very interesting. Supposing them to have been mere dreams, they are dreams of a very singular kind. Mr. D., from the time he first saw R., was requested to do so on not less than twenty different occasions.

I have only given a selection of a few of the more characteristic observations; had I given the whole, they would have filled many pages. Now, from the first observation to the last, he not only never saw the same vision twice, but as he was at that time mesmerised daily, and sometimes twice a day, there was an unmistakable connection between the separate visions. Thus, one day he saw R. in a wild country, travelling down the course of the river, by which the road ran. Next day, he was seen in a lower part of the valley, where the country became more cultivated; on the third day, he was observed approaching a considerable town, and on the fourth was found in his own home in that town. For about three weeks or a month, he was seen every day, often travelling on foot, at other times at home. The various localities were described with singular minuteness. Indeed, the town was so described, and that on many different occasions, that I am sure I should know it, were I ever to see it. It is obviously not in England, and from the costumes, and the often reiterated statement, that all the names on shops and signs, with one exception, a German name, were in strange letters, I conclude that it is not in Germany, but in the east of Europe, possibly near the frontiers of Russia and Turkey, or Transylvania and Turkey. It is possible that some reader may be able to specify the town. It is very difficult to see why Mr. D. should have seen these visions so clearly and so persistently; but I cannot help thinking that the crystal, which at first seemed to call them up, had really some connection with them. Mr. D. himself, in the sleep, thought that the crystal had at one time belonged to R., and we may suppose it to have acted on Mr. D. as R.'s handwriting might have done on some subjects. Supposing this to be the case, were the visions concerned with present or with past events? Mr. D. always said, that it appeared to him as if what he saw were then actually passing. But if the visions, as is possible, referred to past events, it is truly remarkable that the whole succession of these events should have been traced for about a month.

But this was not the only instance in which Mr. D. surprised me by very distinct visions of distant and unknown places. One day, while observing the town above mentioned, and describing it spontaneously, as I always encouraged him to do, he became suddenly silent, and after a short time told me, that he was travelling through air or space, to a great distance. I soon discovered that he had spontaneously passed into a higher stage; perhaps in consequence of the crystal, which he held in his hand, acting more powerfully than usual, he being then in a very susceptible state. As soon as he had come to the end of his journey, he began to describe a beautiful garden, with avenues of fine trees, of which he drew a plan. It was near a town, in which he could see no spires. At the end of one principal avenue was a round pond, or fountain, enclosed in stone and gravel, with two jets of water, and close to this fountain or pond stood an elderly man, in what, from the description, seemed to be the ancient Greek dress, the head bare, long beard, flowing white robes, and bare feet in sandals. He was surrounded by about a dozen younger men most of whom had black beards, and wore the same dress as their master. He seemed to be occupied in teaching them, and after a time, the lecture or conversation being finished, they all left the fountain, by twos and threes, and slowly walked along the avenues. Looking down these avenues, Mr. D. saw glimpses of the neighbouring hills, and of the town, which lay nearer to the garden than the hills, although still at some distance. This singular vision also recurred spontaneously two or three times; that is, Mr. D. saw the gardens and the localities, but not again the group at the fountain, although other persons were seen enjoying the walks, and on one occasion two ladies were noticed, whose dress seemed also to be ancient Greek. But what particularly struck me was, that this vision only occurred in a peculiar state, of which the consciousness was quite distinct, not only from his ordinary consciousness, but also from that in which he saw the former vision of the town, and of R. This peculiar, third consciousness was interpolated, and he always slept out his full time, as previously fixed, in the more common mesmeric state, while the time spent in this new state was added. On returning, which he always did of himself, to his first mesmeric state, he had not the slightest recollection of the new vision, nor did he ever remember it, except when he came into the new state. It certainly seems probable that, in that new state, he was transported to distant times and past events.

Another time he spontaneously passed into a similar state, but which I think had a fourth consciousness of its own, divided from all the others. He told me one day that he was travelling through the air or through space, as before, but all at once began to appear uneasy and alarmed, and told me he had fallen into the water, and would be drowned, if I did not help him. I commanded him to get out of the water, and after much actual exertion and alarm, he said he had got to the bank. He then said he had fallen into a river in Caffraria, at a place where a friend of his was born. But what was very remarkable was, that he spoke of the river, the fields, farmhouses, people, animals, and woods, as if perfectly familiar to him, and told me he had spent many years as a boy in that country, whereas he has never been out of Scotland. Moreover, he insisted he was not asleep, but wide awake, and although his eyes were closed, said they were open, and complained that I was making a fool of him, when I said he was asleep. He was somewhat puzzled to explain how I, whom he knew to be in Edinburgh, could be conversing with him in Caffraria, as he declared he was; and he was still more puzzled when I asked him, how he had gone to that country, for he admitted he had never been on board a ship. But still he maintained that he was in Caffraria, and had long lived there, and that he knew every man and every animal at a farm he described. It was evident that he had heard of Caffraria from his friend; but as he described all that he saw, precisely as a man would do who was looking at the place and the people, and as he maintained that all were familiar to him, I could hardly avoid supposing, that, his mind having been interested in what he had heard, he had, in some of his previous sleeps, visited Caffraria by clairvoyance, without telling me of it at the time; for it often happened, that he would sleep for an hour or half an hour without speaking; that when he had spontaneously passed into that state on this occasion, he not only saw, but recognised as well known, and as seen in previous portions of that peculiar consciousness, the localities, persons, &c., whom he described. Certainly his descriptions were such as to convey to me the impression that he actually saw these things as they exist. On two other occasions, he spontaneously got into the same state, and always then spoke as he had done the first time; but he retained not a trace of recollection of this South African vision in any other state but that one. Nay, when I asked him about Caffraria in his ordinary mesmeric sleep, he seemed not to understand me, and thought I was making fun of him when I asked whether he had ever been in Africa.

In these three distinct kinds of vision, that of R., that of the Greek garden and philosopher, and that of Caffraria, it is hardly possible to verify the visions; but when I reflect, that Mr. D. was able, in a certain state, to see and describe accurately towns, such as Aix and Cologne, countries, and 'persons, at a great distance, and quite unknown to him, I am disposed to think that in these visions also he saw the real places actually before him. It would have been most interesting to have studied more minutely the powers exhibited, or which might have been developed, in this very interesting case; but, as I have mentioned, Mr. D., whose extreme susceptibility at that time may have depended on the very unsatisfactory state of his health, was taken ill, and confined to bed with an affection of the chest, for five or six weeks; and when he had recovered, I found that his general health was far better than when he was first mesmerised, but his extreme susceptibility was gone. I can still mesmerise him, although with far more difficulty; and since his recovery, I have only once been able to get him to see the town formerly described and R. I intend, however, to pursue the investigation, and, perhaps, with patience, I may be able to bring him again into a state of decided and direct clairvoyance. Indeed, he several times told me, in the sleep, that he would acquire, after many operations, a high degree of lucidity, and as long as I was able to mesmerise him before his illness, he did continue to improve in lucidity.

Case 37.—Mr. T., formerly mentioned, occasionally showed,

quite spontaneously, some degree of clairvoyance. Thus, one day he told me that he saw my carriage arriving in the courtyard of the university, which I did not believe, as it had been ordered an hour later. But he said he not only saw it, but saw also the servant coming up stairs to announce it; and two minutes afterwards the servant appeared, the carriage having by mistake been brought an hour too soon. Another day, he spoke to me of seeing his uncle in Berwickshire, and said that his uncle was then sending off a letter to him. He had no reason to expect a letter, as he told me; on the contrary, he was at that time intending to write to his uncle. But with the first post from Berwickshire, the letter came. He several times saw and described what was going on in my house, from the college, and although I could not in all instances verify his statements, yet on several occasions I was able to do so, and found them correct as to the number, dress, &c., of the persons he saw in my drawing-room, at certain times, the distance being certainly upwards of a fourth of a mile in a straight line. One day he spontaneously visited Inchkeith and the lighthouse there, which he had never seen; but I found on inquiry, that his description, which was very minute, was accurate, and it was certainly given as by one seeing what he describes. He got at last alarmed, not seeing how he was to get home again, for he said he saw no boat that could bring him away. Another time, he, of his own accord, described very minutely a long avenue of fine trees, at the end of which was a large public building, and this appeared to be used as a barrack, for he saw lounging in front of it a number of cavalry officers in an undress uniform, which, from his very detailed account of it, was not an English one, but seemed to correspond with that of a regiment of Prussian hussars, which I have since heard of, but had never seen. Why he should have seen this vision, I cannot conjecture, for I do not know of any such barrack as he described, and he could not therefore, have read it in my thoughts. In this case, I could not always, when putting him to sleep, get him into the clairvoyant state. His clairvoyance was generally spontaneous; but sometimes I was able to get him to look where I wished, and he several times correctly told, at my request, what was doing in the room of a friend who lived at a distance of several hundred yards. He often told me that he saw places and persons very dimly, as if through a mist, like Mr. D., but I believe that, if I should be able to continue experiments with him for some time, he would gradually come to see more perfectly.

Case 38.—This case I have on the authority of a lady, who had it from the parties. Mr. B., of the E.I.C. civil service, being at Calcutta, and wishing to hear about Mrs. B., then on her voyage to England, applied to a clairvoyant at Calcutta, who being put into the mesmeric sleep, and asked where the ship *Queen* was at that moment, answered, that she was off the Western isles, and was then passing one of them, described as having a high peaked outline, resembling Madeira. The day, he said, was hazy and gloomy. Mr. B. asked him to enter Mrs. B.'s cabin, which he at first declined to do, for fear of intrusion, but finally agreed to do so, and said the cabin was in great confusion. Two ladies were sitting in it, conversing together, and his description of Mrs. B. was so exact and graphic that Mr. B. was much affected by it. Captain Macleod, of the *Queen*, afterwards compared the statement of the clairvoyant as to the position of the ship with his logbook, and found it perfectly correct. The *Queen* had left Calcutta on the 3rd of Feb., 1850, and having had an unusually long passage, did not reach England till the 6th of July. According to the average passage, she ought to have been in England when she was off the Western Isles. The clairvoyant had never seen Mrs. B., and it should be added, that a fellow passenger of that lady spoke from observation of the confused state in which her cabin usually was.

Case 39.—We have already seen that Major Buckley often produces waking or conscious clairvoyance, but he is also very successful in producing clairvoyance in the mesmeric sleep. The following case is extracted from a letter to me from Major B. A lady, who, after having been rendered clairvoyant in the sleep, could be rendered consciously clairvoyant, was found, in the sleep, to be able to read mottoes, &c., not in Major B.'s possession, and at a considerable distance. Another lady, having placed within a box, in her own drawing-room, a motto, requested Major B. to ask the clairvoyante to read it. This, while asleep, she did, she being in her own house, the motto in that of the lady, and the lady herself not being present. The motto was quite unknown to Major B. He then asked the clairvoyante to look into a shop of which he had heard, where mottoes were sold in nuts, but which neither he nor his subject had ever entered, and to tell him if she could perceive any new mottoes among those in the shop. She said she saw some new ones. "Many?" "No, only about three in an ounce of nuts." "Are you quite sure they are new?" "Quite. I see the one I have just read in the lady's house." "Were I," said Major B., "to buy an ounce there, should I have any new ones?" "Yes, the one just mentioned would be among them." "Will this happen if I purchase one ounce only?" "Yes. Mark them all before you bring them to me." Major B. left her asleep, went to the shop, purchased one ounce, eighteen nuts, marked them all with a file, and brought them to her. She instantly pointed to one, and directed him to open it. It contained the same words he had just before written down, and only two of the others contained new mottoes. Next day, Major B. called on that other lady, and saw the same motto taken from the box in which it had been put.

Case 40.—A lady who could read in boxes while awake, being

one day on her way with Major B., to leave a letter at the house of a professional singer, all at once exclaimed, "He has left his house, and the name plate has been taken off the door." On arriving at the house, this was found correct. As the people of the house could not give the new address, Major B. put the lady to sleep, when she said, truly, that the singer now lived out of town, and intended to come in to his duty by omnibus.

Case 41.—Miss G., a very intelligent young lady, was mesmerised by Mr. Lewis, and became clairvoyant at the second trial. In this state, she went to see some near relations in India, whom she found in a camp, and mentioned various details which cannot yet be verified. Mr. Lewis, while she was asleep, told her that he would mesmerise her next day from a distance, at 1 P.M., and that she must then go to sleep, and see him wherever he might be, so as to tell what he was doing. When awoke, she had not the slightest recollection of anything that had passed in her sleep, and she was not told of what Mr. Lewis had said. Next day, at 1 o'clock, while occupied in writing, she fell asleep, and after a time answered the questions of a gentleman who watched the experiment, and from whom I have the details. She soon saw Mr. Lewis, in a room, the furniture of which she described, and she also said he was writing at one time, but afterwards walked about the room gesticulating strangely, and making ludicrous grimaces. Everything that she said was correct, except that she took a travelling-desk for a large book, but could not say what book it was. Mr. Lewis was then in Dundee, the lady in Stirling. He was in the room she described, and had made gesticulations and grimaces after he thought she must be asleep, with the wish that she should repeat these gestures, which however she did not do. But at that distance she saw his gestures, and had been already put to sleep, whether by his direct influence at the time, or in consequence of the command given in his sleep the day before, of which, in her waking state, she knew nothing. And this was only the third time she had been mesmerised.

Case 42.—Mr. J. D., a plate-layer from Annan, was, I as have mentioned before, put to sleep by Mr. Lewis in my presence, and in that of several gentlemen. He exhibited beautifully the phenomena of suggestion in the sleep, as also detailed, Case 24 ; but after a time he spontaneously passed into the clairvoyant state, in which I examined him, he being transferred to me by Mr. Lewis, when he heard my voice, but not that of Mr. L., till I retransferred him. I have mentioned before that I asked him to find and describe my house, which he did most accurately, although he had only that day come to Edinburgh, and did not know me. In particular, after describing the street-door and steps, the lobby, the staircase, and the drawing-rooms, he said he saw a lady sitting in a particular chair, reading a new book. On returning home I found that Mrs. G. had at that time been sitting in the chair alluded to, which she hardly ever does, reading a new book, which had been sent to her just before, but of which I knew nothing. Besides, I found that J. D. did not, in describing the house, read my thoughts at all, but dwelt on many things, strange to him, which I never thought of, and omitted others which I did think of, and wished him to notice. I have now to add, that when he first said he would look for my house, and I did not even tell him the street in which it is, he very soon spoke of being in the Royal Infirmary. I found it impossible to divert him from this, till he had described what he saw. I cannot discover why his mind should have been led to the Infirmary, unless it be that it had been pointed out to him in the course of the day. But he had never entered it; yet he saw the interior, described two men putting a third into a bath on the ground floor, and afterwards, going upstairs, entered a ward, on the door of which he saw No. IV., counted the beds on one side of it, noticed the closets at the ends of the ward, and said that most of the patients were in bed, but that one man was smoking " up the lum." After leaving the Infirmary, he proceeded to look for my house, and very soon found it to be in Prince's-street.

On this occasion J. D. became only accidentally clairvoyant, and was not very highly lucid; but on other occasions Mr. Lewis found him to possess a rare degree of lucidity. At Mr. L.'s request, he once mentally visited St. John's, New Brunswick; told Mr. L. that his mother, of whom he, Mr. L., had not heard for years, was alive in that place. Also that on a certain day Mr. Lewis would receive a letter from that quarter, on business of importance, which was now on the way; that it was written by an agent or executor, who was then ill; that the mother of Mr. L. would also soon receive a letter which Mr. L. had written; and he added private information of much importance to Mr. Lewis. The whole proved quite correct. Mr. L. received the letter announced, from an agent, whose illness and death were mentioned in a later communication. Mr. L.'s mother proved to be living there, and she also received the last letter he had written, without a knowledge of her being alive, or of her address. I have since heard of various other instances of J. D.'s great lucidity.

Case 43.—L. W., a young woman, aged 25, of fair complexion and nervous temperament, servant in the family of Dr. M'Culloch, Dumfries, of excellent character in every respect, was mesmerised by Mr. Lewis, and became clairvoyant. On the 6th of October 1850, she was put to sleep in the evening, and asked to visit a school for young ladies at Boulogne, where Dr. M.'s daughter had been for several months, and then was. She said she saw her going to bed at a quarter to nine, and hanging up a brown dress which she had worn that day, because the day was wintry. It was supposed that this was a mistake, as that dress was not to be worn till winter. She said that the young lady was to return home in

June or July, and afterwards to go back to Boulogne. This also was supposed to be quite wrong. She described minutely the person of the English teacher, and said there were 25 scholars, a point not known to any one in Dumfries. She said they dined at half-past one, and drank no wine, but some light stuff out of a very large flask of a peculiar shape. She also described the bed and other furniture of the young ladies' bedroom, particularly a small carpet, of a stair-carpet pattern, a deep red colour, much faded. Miss M. heard from her mother last week, on the 4th, she thought, and intends to write home on Wednesday the 9th. She also described a lady, whom she took for the school-mistress, as a stout woman, dressed in black satin, wearing a cap, and with black hair. Miss M. generally sleeps alone, but sometimes one of the young ladies sleeps with her. One of the French teachers sometimes instructs in music. Miss M. was at the English Church in the forenoon, but not in the afternoon, because it was rather wet. The text of the sermon was from Luke xvii.

On inquiry, almost the whole of these statements were found correct. A few were wrong. Thus, Miss M. did go to church in the afternoon also, and the text in the forenoon was not from Luke. The lady L. W. took for the school-mistress was a friend on a visit to her. But in almost everything else L. W. was right. Thus, Miss M. had worn the brown dress that day, for the first time, sooner than she had intended, because the day was very cold. She had gone to bed that evening at a quarter to nine, and had hung the dress in her cupboard. Only the day before she and the other young ladies had been unexpectedly told that they would have to return home about the end of June or beginning of July, because the mistress was obliged to go at that time to Germany, which was entirely unknown to and unexpected by Miss M.'s family. The person of the English teacher was correctly described. The description of the dinner was generally correct, and the water flask was exactly as L. W. said. The furniture of the bedroom was also correctly described, as was the pattern and colour of the carpet before the bed. Miss M. did intend to write, and did write on the 9th. One of the young ladies sometimes slept with her; and one of the French teachers occasionally gave instructions in music. The number of scholars, which had varied much, was then 25.

Now it is impossible that these things could have been suggested to L. W., because no one in Dumfries knew them, and several of them, such as the unexpected return of Miss M., were quite opposed to what her parents understood, and had only then become known to herself. The girl, moreover, spoke and acted as if looking at what she described, and it cannot, I think, be doubted that, by some means, she did see it. The case is remarkable also because the girl mentioned correctly various points, which cannot here be given, as to what passed in the mind of Miss

M., and because there was some tendency to prevision of future events, as in regard to her writing on the 9th, and her return at a later period.

Case 44.—E., the clairvoyante of Dr. Haddock, formerly mentioned, is often clairvoyant without any means of communication, such as writing, &c., and sometimes spontaneously passes into a lucid state, without any artificial process. One day lately, Dr. H. received a letter, the writing of the address of which he did not recognise. E. requested him not to open it, till she had told him of a dream (that is, an act of spontaneous clairvoyance), she had had about it. She had seen, as it were, the form of its contents, but it had not appeared to her open. She said there were two sheets in it, one of which had a piece cut off, and a plain piece of paper, apparently that cut off, was also in the envelope. In connection with the letter she saw a man, and a funeral. On opening the letter, Dr. H. found everything just as E. had described, and the first sentence referred to the death of a gentleman.— This case I consider very interesting, as showing how true dreams, as this was, are, in all probability, very often acts of clairvoyance. It is remarkable also, because E., before the letter was opened, not only described its form, &c., but had ideas of death and a funeral connected with it, which ideas were in the mind of the writer when she began her letter. But the writer and E. were 200 miles apart. Whatever may have been the ultimate result, it appears that E. had found the trace of all the ideas concerning the letter in the writer's mind, at that great distance, and had then followed them up for herself.

It would be tedious here to go into detail; but I may briefly mention, that E., in the mesmeric sleep, as I saw more than once, could see perfectly what passed behind her, her eyes being closed; or anything placed in such a position, that had her eyes been open, she could not have seen it; she could also see very often all that passed outside of the door, and when I was there, told us how many of the servants of the hotel were listening at the door, in hopes of hearing wonders; she would also often tell what was doing in the room above or below her. In short she frequently exhibited direct clairvoyance in every form, not only in those just mentioned, but also in that of seeing prints or pictures shut up in boxes. Besides seeing various instances of direct clairvoyance, I was able to satisfy myself that Dr. Haddock's experiments were made with the greatest care and judgment; that he was particularly well acquainted with the various causes of error and confusion, very careful to avoid these, and that in short his accounts of such experiments as I had not seen were entirely trustworthy. I shall have occasion again to return to this case.

Case 45.—The next case is one which I regard as particularly valuable, because the observer, in whose words I shall give it, had never seen any mesmeric experiments whatever, and had only read a few works on the subject, when he resolved to try for himself, and succeeded, in the first case, in producing, not only the mesmeric sleep, but also clairvoyance. He is the Rev A. Gilmour, a highly respected clergyman, residing in Greenock, and well known to be a very able and highly accomplished gentleman. At my request, he wrote the following letter, which I give without abridgment, because it is an excellent example of what may be done by any of us, if we only take the trouble to experiment for ourselves. It will be seen also that this case illustrates many other phenomena, besides direct clairvoyance, such as sympathy and community of sensations. And I may here add, that if my space permitted, I could have published several similar accounts of the results obtained by various ladies and gentlemen of my acquaintance. As that cannot be done in this work, I must content myself with giving Mr. Gilmour's letter as a type of a numerous class of communications, and reminding you that many have been equally or even more successful, whose observations have never been published, and that the results of these numerous private experiments entirely confirm, as far as they go, the statements of professed mesmerists, and the multitudes of published cases to be found in the *Zoist*, and in numerous French and German periodicals devoted to mesmerism.

MY DEAR DR. GREGORY,

"I HAD read a good deal about mesmerism in the spring of 1843, but I had never seen any person put under its influence. The Rev. Mr. Townsend's works were my text-books upon the subject. The whole seemed to me to be a mystery, yet I felt that I was not warranted in rejecting the testimony of upright and honourable men, merely because I could not understand the subject in question.

"I resolved to make experiments for myself, following the directions of the Rev. Mr. Townsend. I asked one of my servants, V. R., May 27, 1843, if she was willing to be mesmerised; she consented. Her temperament is nervous, bilious, dark hair, and eyes, pulse 80 and small,* age 18, person thin and spare. I gazed steadily for about seven minutes upon the pupil of her right eye, directing her to look fixedly into mine. This I continued to do for about fourteen minutes, and was about to give it up, when she told me that she felt very strangely. I should have mentioned that she had never heard of mesmerism before this. On getting the hint that she felt very strangely, I persevered for ten minutes longer, when her eyes gently closed, and she was fast asleep.

"She appeared to be agitated; her hands and arms moved as if under the influence of irregular nervous twitches. Her head kept

---

* The pulse invariably lessened and softened under the mesmeric influence.

up a kind of rocking motion, and on being asked how she felt, answered 'Very funny.' I made a few reverse passes, when she said that she felt very happy. I kept her in this state for about 45 minutes. I tried to effect her phrenological developments, but could not. I tickled her nose and upper lip with a feather, but she was quite insensible to it. I also tried to render the arm cataleptic, but could not. I then demesmerised her, when she knew nothing of all that had taken place. I tried her with the feather, but she shrunk from the slightest touch. This was my first successful trial.

"After this I mesmerised her every night. She became more and more susceptible, and my power seemed to increase in proportion as it was exercised. At last I could throw her into the mesmeric sleep in 40 seconds. She is able to tell what I taste, such as soda, salt, sugar, milk, water, &c., though not in the same room with me. When my foot is pricked, or my hair pulled, or any part of my person pinched, she feels it, and describes it, unerringly.

"August 7th.—I found her in a state of clairvoyance. She went to my mother's, on being requested; described her cottage, her personal appearance, and her dress, with perfect accuracy.

"When in this state, I went into different rooms, leaving her in my study; and forming a strong wish that she should rise and come to me, she invariably did so. I also went into the garden, and on wishing her to come to me, she instantly did so, always proceeding in a direct line, slowly, but accurately. I observed that, as she came to me on such occasions, her two hands were slightly extended, and when they touched mine, it was with a sudden slightly jerking motion, the same as when a needle touches the magnet.

"Without giving you an historical detail of my proceedings, I may here mention that, on the 8th of March, 1844, one of our most intelligent physicians, his sister, two ladies, and one of our magistrates, dined with me, when we had a mesmeric *séance*. We requested her to visit the house of Mrs. P., one of the ladies present. This house was in Greenock, distant from my cottage about a mile and a quarter. She saw her servant in the kitchen, but said that another woman was with her. On being pressed to look earnestly at the woman, she said it was C——— M———. This, Mrs. P. declared to be true. We then asked her to see if any person was in Mrs. P.'s parlour, when she said that Miss Laing was there, a young lady from Edinburgh, who was boarding with Mrs. P. at the time; that she was sitting on the sofa; that she was crying, and that a letter was in her hand. On the party breaking up, I walked into Greenock with the ladies and gentlemen, in order to see if she was right about Miss L. It was true. Miss L. had received a letter by that evening's post from her father in Edinburgh, stating that her mother was not expected to live, and requesting her to come home by the first train in the morning.

"September 2nd.—I had made her follow the ship *Ellen* of Glasgow, Captain P., on a voyage from Glasgow to Ichaboe, which was towed down the Clyde with the ship *Chusan* on the 28th of August. She saw the *Chusan* lying becalmed the same evening, about seven miles down the river, but could nowhere see the *Ellen*. On being pressed to look out for her, she discovered her much farther down, a small boat at her stern, and the captain and a little man in the cabin taking their grog. A few days after, the pilot called, when she said (on being mesmerised) that was the man who was in the cabin with Captain P. The pilot stated that the steamboat threw off the *Chusan* opposite the Cloch Lighthouse; but that Captain P. had made them tow the *Ellen* eight or ten miles farther down the frith. This harmonises with what she saw. On the 2nd September she saw the *Ellen* in full sail, the sea a little rough, Captain P. in bed, and the mate on the quarter-deck. She is sure the ship is past Ireland, for there is no land to be seen ahead of her; but she cannot discover any more than one dog on board, there being two when the *Ellen* sailed. On Captain P.'s return from Ichaboe, I obtained the log of his voyage, which is still in my possession. The little dog had become so ill that it was thrown overboard in the channel, and at the above date the *Ellen* was by observation 53.25 north latitude, 17.41 west longitude, which you will see carries her far beyond Ireland.

"December 25th.—J. S., Esq., spending the evening with me, was anxious to test her clairvoyance accurately. She visited, at his request, his breakfast parlour at home, said that his father was reading *Blackwood's Magazine*, in his easy-chair by the fire; described the room with perfect accuracy, though, I need scarcely say, she had never been in it in her life; described the gasalier, and the number of burners lighted, and mentioned what Mrs. Scott was doing. Some of these statements, he felt perfectly sure, were incorrect; but, on going home, he found that she had been minutely accurate. On the same evening, he begged me, in writing, so that she might not hear the request, that I would send her along to our Provost's. On going into the room, she saw a great number of young ladies; but though she had seen some of them before, she could not name them. On entering another room, she saw a great many little misses. On being pressed to look earnestly at them, and see if she knew any of them, she discovered Mr. S.'s sisters, their governess, whom she named, and the Misses L. Mr. S. then told me that the Provost had a large party that evening, upwards of sixty young ladies; that his sisters, their governess, and the Misses L., were to his certain knowledge there. I may also mention, that while this was going on, I heard a knock at my door. On the person being admitted by the housemaid, I asked her to tell me if any one was there. She said, Yes, a lady had been taken into the parlour. On being pressed to look well at her, and tell me who she was, she named her. I went out and found that all this was true.

"I may also state, that during the summer, Dr. T———— of K————, Mrs. T————, and her two daughters, visited me. On the day that they left, I requested him to take notice of all that was doing in his house at 11 o'clock of that same night, and I would visit him, through my clairvoyant. I did so, and dispatched to the Doctor, by the next morning's post, my questions and her answers, stating that the Dr. and Mrs. T. were in a small parlour; that it was lighted by a gas jet from the mantelpiece; that Mrs. T. was sitting at the table with a book before her; that she had a turban on her head; that she had a dress of an uncommon kind, which she described; that the Doctor was standing in the room; she described his dress; that one little Miss was in a small bedroom off the parlour, and that another little Miss was in bed with the servant in a room at the head of the stair. I may state that she had never been in K. in her life. By return of post, the Doctor acknowledged the receipt of my letter; stated that Mrs. T. was dressed in the peculiar manner described, and that everything which I had stated was true; but he informed me that he was playing upon the flute, and expressed his unwillingness to believe in the possibility of any person telling what was doing at such a distance.

"These, my dear Dr. Gregory, are only a few of the many strange and startling statements which I could make upon this subject. I cannot comprehend the *modus operandi* of clairvoyance; but neither can I deny the evidence of my own senses; nor can I question the veracity of hundreds of upright and honourable men, who are far too clear-sighted to be imposed upon themselves, and much too honest to try to deceive others. Moreover, everything around me is a mystery, not opposed to reason, but far above all human comprehension. I cannot explain how I speak, or hear, or see; and yet I am compelled to admit the fact. Neither can I understand how a clairvoyant can tell me what is going on in any part of my house, or in any other house; and yet I know that this has been done hundreds of times, and with the most startling accuracy. I cannot tell how it is that the clairvoyant obeys all my volitions; but still I am compelled to admit that this is true. Many honest and upright men object to mesmerism, upon the principle that young women, when mesmerised, may be easily corrupted by unprincipled men. I am destitute of experience on this point, but express my firm conviction that it is an error. The young person that may be corrupted in this state, may be corrupted in any state; but the virtuous and the pure will neither indulge in an irregular thought, nor submit to an improper proposal, when mesmerised. Impure minds will indulge in impure actions in any state.—I am, my dear Dr. Gregory, with every sentiment of respect and affection, yours faithfully,

"AND. GILMOUR.

"MARTHABRAE COTTAGE,
"GREENOCK, 21*st, January*, 1851.
"PROFESSOR GREGORY, *Edinburgh*."

N.B.—In reference to this point, I would recall what I have said in one of the earlier letters, namely, that, as a general rule, the moral sense is exalted in the mesmeric state, and the so-called sleeper is fully awake to moral feeling. Mesmerism does not essentially change the character; but it does seem to give greater preponderance to certain feelings than is observed in the natural state of the subject. But so far as I have seen, when a change has been observed, the feelings exalted have been those of the love of truth and morality, and a very marked hatred of vice and falsehood. I have never observed any appearance of exaltation or excitement of the lower propensities, which, on the contrary, have seemed to be laid to sleep, while the higher sentiments were in full vigour. W. G.

Case 46.—The following letter, received when this sheet was in type, is particularly interesting as an example of the manner in which a man of sense and ability, like the reverend writer, although sceptical, investigates any case which may present itself. The case itself appears to be one of great interest, and well worthy of further investigation:—

"W——, 31st March, 1851.
"DEAR SIR,

"I VERY willingly furnish you with the information which you request from me. Till the month of August last, I regarded the whole subject of mesmerism with incredulity and dislike; and with respect to clairvoyance, in particular, I was a determined sceptic. I had not, indeed, at all investigated the matter; but I have now a somewhat uncomfortable consciousness that my scepticism was not a whit the less dogged and decided on *that* account.

"At the time referred to, I visited, in the company of another minister, some old friends in Shetland; and it was not long before we heard of certain wonderful performances in the way of mesmerism. We were told, that a man, whom I knew very well ten or twelve years ago, had been found to be an excellent mesmeric subject, and had enacted marvels as a clairvoyant. I listened to the narrative with cool and settled incredulity; not questioning, indeed, the veracity of the reporters, for that I could not question, but altogether disbelieving the correctness of the so-called clairvoyant's statements, and having no doubt that there must be rampant delusion or imposture somewhere. I made a pretty emphatic declaration to this effect, and expressed a wish to have the matter thoroughly tested, and to have the testing process put entirely into my own hands. This was at once agreed to ; and when I then intimated that I should not be satisfied with any mere *general* description of places or persons, nor any account whatever of Sir John Franklin, or such like, the accuracy of which I could not pronounce upon ; but that my tests should be such as would establish at the moment, and upon the spot, either the truth or the falsehood of the alleged revelations, I was told, in reply, to put what tests I pleased, as all would be left to myself. This was satisfactory ; and I had no doubt that I should speedily demolish the delusion. The individual was sent for, and mesmerised in my presence ; and when he was pronounced to be asleep, I furnished the mesmeriser with half-a-dozen questions to ask, not regarding Sir J. Franklin, but regarding *myself* ; and having requested that the answers to these queries should be accurately reported to me on my return, I left the room. The questions referred to my whereabouts after quitting the apartment ; my dress, in which I took care to make some rather material alterations ; my exact position and occupation, and suchlike particulars, which the clairvoyant could not *possibly* know by any ordinary means of intelligence, and which there were ten thousand to one against his hitting by so many consecutive guesses. On my return, I demanded a report ; and found, to my no small surprise, that all the questions had been correctly answered, and that the experiment by which I had intended to expose a hoax, was likely to lead to a very different result. In a word, I had taken such precautions, and had applied what seemed to me so fair a test, though a simple one, that my scepticism received a considerable shock. Resolved, however, to sift the matter still more thoroughly, we made a great number of very varied experiments ; and, as the result of the whole, I was compelled to admit the unquestionable reality of the facts, although utterly unable, as I still am, to account for them. The full detail would fill a volume, as we had the clairvoyant under our hands for a period of about two hours during each of three successive days. I may, however, state one or two facts which I witnessed, and for which I can vouch. Perhaps I should mention here, that the clairvoyant is a poor man—indeed *steeped* in poverty—not very robust in health ; but sufficiently active and intelligent, and, for his station in life, pretty well educated. He has never been out of Shetland, and has had little opportunity of acquainting himself in any way with the circumstances of 'the adjacent islands of Great Britain and Ireland.' One evening, after he had been thrown into a mesmeric sleep, my friend and fellow-traveller, whose name I shall send you, asked him to accompany him to a certain place which he was thinking of, but the name or locality of which he did not mention, nor in the least hint at. The clairvoyant described the house, first the outside, with 'big trees' round it, then several rooms in the interior ; and being directed to enter a particular apartment which was indicated to him by its position, he described the appearance and occupation of a gentleman and two ladies who were in it ; declared that he saw a picture over the mantelpiece ; and being further questioned, deposed that it was the picture of a man, and that there was a name below it ; and being urged to read the name, after experiencing some difficulty with the penmanship, he affirmed that the last word of the *name was ' Wood,' which he slowly but correctly spelt*. The house was near Edinburgh, and when we came to compare notes, on our return from Shetland, we found that the description of the individuals in the room at the time had been quite correct ; and we *saw* over the mantelpiece a print of the *Rev. J. J. Wood, of Dumfries*, with his name written below.

"I had the clairvoyant taken to other places, which were not named in his hearing, but which he described with great accuracy and minuteness ; and some of these experiments were, if possible, still more satisfactory to my own mind than that just mentioned, though I may not be able to present the evidence so palpably to others. For example, he accompanied me to my own house, without my naming it, though this indeed could have made no difference to him, and he gave a fuller and more detailed and accurate description of it than many who have spent hours in it could have done. He followed me, in thought, from place to place, and, with a momentary hesitation and confusion in one or two cases, he correctly described them all.

"He went in search of Sir John Franklin, and found the ships *Erebus* and *Terror*, spelling the name of each on the stern of the vessel. I am sorry now that I did not make such full and explicit inquiries upon this subject, as its importance and the interest attaching to it deserve, or as it would have been proper to institute, in order to compare the statements of this clairvoyant with those of others. During the time when I had him in hand, my experiments were almost entirely of a kind which were fitted to be conclusive upon the spot. However, I heard him declare that the *Erebus* was fast locked up ; that those on board were alive, but in low spirits, and that, in answer to his inquiries, they said they had little hope of making their escape. He affirmed that there was water for a certain distance round the *Terror*, but that she was not clear of the ice. Of course, I give no opinion as to the correctness of *these* revelations. The date when they were made was about the 22nd of August, 1850. When sent to these northern regions, and as long as he was kept there, he appeared to be shivering with cold, and declared the cold to be intense.

"I might mention other phenomena, which struck me as curious, but which are, I suppose, common enough in such cases, as, for instance, his insensibility to pinching and pricking, when applied to his own proper person, and then wincing and complaining when these applications were made to one of *my* hands, while I had hold of his with the other. And so, when a chain of five or six individuals was formed, and the pinching applied to the one farthest off, poor James got the full benefit of it. It was also amusing to notice his readiness in catching the air of a song, and chiming in with the singers—his aptitude in pronouncing Gaelic words, which it cost some of us not a little effort to master, and you know Shetland is too far north for Gaelic—his fluency in repeating German sentences, and in rolling forth after you Homer's hexameters in an *ore rotundo* style, which would have done credit to an advanced student at the university.

"I may state farther, that, when awakened, he had no recollection of what had passed while he was in the mesmerised state. On one of the three occasions on which I witnessed the proceedings, he was slightly sick when restored to consciousness, but immediately recovered on receiving a little cold water. Upon that occasion the room was warm, and the sun had been shining upon him while asleep. He declared that he had never experienced any bad effects, only his wife had once or twice said, that after he had been mesmerised, she thought him less talkative than usual.

"It is right to state, that I found him committing one or two mistakes; but it should also be stated, that they were of a kind which served rather to confirm than contradict the other evidence —exact instances, indeed, of the canon, 'Exceptio probat regulam' —as, for example, when he described a certain person, whom he named, as being in a certain place, and superintending labourers engaged in a particular employment, when it turned that another individual had been so engaged at the time, but who might easily be mistaken for the person in question, even by those who knew them both. It is right to mention another mistake, of which I have no explanation to offer. It seems that on the 21st. of Dec. last, James did me the honour of paying a visit, and inspecting my premises, and informed my friends in the North who were in my house, and how we were engaged. Now, it so happened that the individuals whom he named *were* here, and *were* engaged exactly as he described, but then this was certainly *not* on the day on which the clairvoyant professed to see us, in truth, not till nearly three weeks afterwards. How he should have been so correct with regard to persons and circumstances, and so much in error with respect to time, I do not pretend to explain.*

"I think I have not mentioned before that the name and address of this individual is James Smith, Whalsay, Shetland. He lives on the property of my friend, Mr. Bruce, of Symbister, in

---

* Here the clairvoyant would seem to have exerted prevision, as well as lucid vision. It was not retrovision, for the persons had never been there before. The fact illustrates what I have said as to sources of error.—W. G.

whose house the proceedings above related took place. If any one should wish to know more of the person who furnishes you with this statement than his initials will tell them, you may give them whatever amount of information they may desire.—I remain, dear sir, yours truly, "P. H."

Case 47.—Mr. C., with two daughters, travelling from Richmond, Yorkshire, to Cheltenham, stopped a night in York, where a son was to join them from Richmond. One of the young ladies, being put into the mesmeric sleep by a friend, was asked if she expected any one, when she said she saw her brother coming; but instead of describing the expected brother from the north, she spoke of another as coming from the south, whom she saw in a railway carriage, which she described, with a certain number of fellow-travellers, and reading a certain book. She also described his dress. This puzzled the party much, but their astonishment was much more increased when this brother, who lived in Manchester, came into the room, in the dress described, and confirmed every statement of the clairvoyante. His wife was ill, and having heard that his sisters were to be at York that evening, he had suddenly started, in order to see them, and take back one of them to nurse his wife. No communication had been possible, as he had only resolved to start an hour before doing so. There can be no doubt that, in this case, Miss C. saw her brother on his way to York. This interesting case was communicated to me by Alan Stevenson, Esq., C.E.

The same gentleman also communicated to me a brief notice of a remarkable instance of sympathetic warning, the details of which might be obtained from the parties if required. Two ladies, very intimate friends, had both been mesmerised, which proves that both were susceptible. On one occasion, one of them suddenly burst into tears, and declared that at that moment some great evil had befallen the other. This proved to be exact, although the friends were then one hundred miles asunder.

In the case of Miss C. above given, it would appear, that sympathy led her to find the trace of her brother, and that, having found it, she saw him and his companions in the railway carriage, by direct clairvoyance.

Case 48.—The following case is one in which the names of the parties are unknown to me, but I have no reason to doubt its accuracy, at all events in all the chief points, and, if necessary, I have no doubt it could be well substantiated. A young lady, on a visit away from her home, being mesmerised, was desired to visit her father's house. She did so, and said she saw the postman delivering a letter; that this letter was addressed to her; that it was from a brother or cousin, an officer, and dated from Cork, that her sister had opened it, &c. When she woke, she had no recollection of the vision. On being asked, incidentally, where the brother or cousin was whom she had spoken of as the writer of the letter, she said he had sailed a week or ten days before from Cork for the Colonies with his regiment. It was therefore supposed that she had been dreaming of some past letter. But on her return home, she found that at the time she saw it, a letter had arrived, addressed to her, the writer having been forced by stress of weather, to return to Cork, from which place he had written.

Such cases are very frequent, but a few are sufficient to illustrate the essential fact, that many persons, in the mesmeric sleep, possess some means, the nature of which we can only guess at, of perceiving absent persons and things, and correctly describing the occupation of the persons seen.

Case 49.—A young lady of 17, being mesmerised for the first time by Mr. Lewis, became clairvoyant, and when asked by him to do so, visited America. In New York, she accurately described the appearance and situation of the Astor House Hotel, then proceeded to Niagara, where she was at first much alarmed, but afterwards much delighted with the scenery, which she most correctly described, on both sides of the falls. She next took a view of the country, from the suspension bridge above the falls, describing the landscape on both sides. After this, she was taken to Buffalo, and immediately on entering a certain part of it, exclaimed at its extreme filthiness, which is true of that part. She was then made by silent volition to visit Louisville, and the slave market there, which horrified her much. She saw an open space, with something about slaves printed on a wall or building. She then visited the residence of a gentleman at Rochester, known to the family, and saw him seated in his parlour, looking over a newspaper, of which he is editor. This case might be regarded as one of thought-reading, although, on that view, it would not be the less interesting or difficult to explain. But I have chiefly noticed it, because it is an instance of clairvoyance, by thought-reading or otherwise, occurring in the very first trial, and because of the vividness of the images. Another young lady, mesmerised at the same time, was made to see the same things, and confirmed the statements of the first. This case I owe to the kindness of the father of the first-mentioned young lady.

Case 50.—Mr. Atkinson had mesmerised a young lady, the daughter of a medical man, who resided many miles from London, where the young lady was. She became clairvoyant, but her father, who came to see her, would not believe in her clairvoyance. Mr. A. then requested him, when he got home, to do anything he chose, not telling any one, at a certain hour and in a certain room. At the time appointed, Mr. A. mesmerised the young lady, and requested her to visit her father's dining-room. (It was at dinner-time.) She did so, and saw her father and the rest. But all at once she began laughing, and said, "What does my father mean? He has put a chair on the dinner-table, and the dog on the top of the chair!" Mr. A. sent by the first post an account of what his patient had seen, which was received next morning, and in answer he was informed that she had seen correctly, for that her father, to the amazement of his family, had put the chair on the table, and the dog on the chair, at the time agreed on.

Case 51.—Another lady, a patient also of Mr. Atkinson's, who, by a long and laborious mesmeric treatment, cured her of a most distressing complaint which had resisted all other treatment, became highly clairvoyant, and spontaneously, in the sleep, saw and described the house of a near relation, at a great distance in the country, with its inmates. She continued for a long time to visit his house, and one time saw that her relative was dangerously ill, and told how many persons were in the sick-room; saw the medical men, described the treatment, and pursued the case from day to day, involuntarily and even against her own wish, as it distressed her severely, till the distant patient died. After this, she was still involuntarily drawn to the scene of death, saw the corpse, described its appearance, and all the proceedings connected with the interment. Even after that, she felt compelled to visit the corpse in its grave, and described with horror the changes which took place in it. It was not for a long time that she was enabled to get rid of these painful visions. But everything that could be ascertained and verified was found exact.

The next case was communicated to me by Col. Gore Browne, 21st Fusiliers.

Case 52.—Two ladies, the Misses B., being in Paris, and about to have a *séance* with Alexis, wrote to their sister at Nottington, desiring her to send some question thence in writing. She, however, only wrote the following words, which were sealed up in an envelope, "Your letter came too late to answer in time for your *soirée*." This was given to Alexis, while in the sleep, by one of the Misses B., who knew neither the contents of the letter, nor who had written it. He took her hand and said, in French, "You do not know who wrote it. It is written very small, and is English." "The first word has four letters, it is 'your.' It is written so small that I can read no more. (It is very small.) It is a lady who sent it to me. She is not in her own house. She is living with another person. She is your sister. A man has had something to do with what is written. Your sister lives with a lady of about 50, in England, near a seaport, about two leagues from it. (Nottington is three miles from Weymouth.) The port is called Weymouth. I see your sister plainly. She has a high black dress, a cap, and her hair in bandeaux. She has a very slender figure; she is older than you; her shoulders are broader than yours. She is lively, and talks much, indeed continually. She has twice lately gone to visit this lady. She is very well, but she rose late this morning. At present she is about to read. She is reading. Besides her, I see two ladies and a gentleman. He has gone out. He is not in his own house; but he has a room to himself in the lady's house. He goes there often. The lady of the house has a high dress of green woollen stuff. (This was wrong.) She suffers in her limbs; and uses frictions. She is mesmerised; she does not sleep. Your sister does not believe in clairvoyance. She speaks very little French, but understands it. Present my compliments, and ask her to send me a line in French. Then I shall read it with much pleasure."

By mistake, a note addressed to a servant in Dorsetshire, had been put into the envelope shown to Alexis. He, being asked what it meant, said, "That is not your affair; it has been written by a lady in the house where your sister is." Observing the name of Mr. B. below the address on the letter, he exclaimed, "Ah! there is the name of the gentleman I saw in the house with your sister."

Col. Gore Browne informs me that members of his family were told by Alexis, whose powers of thought-reading are remarkable, of various private family matters, known only to themselves. And I have lately had a similar account, in the strongest language, from a very intelligent lady who lately saw him. I have heard of many instances of his power equally remarkable, but I shall only add the following, which are short and striking.

Case 53.—Mons. Sabine, Chief of the Station of the Havre Railroad, went a few days ago to consult Alexis, who, when in somnabulism, said, "You come about something lost in the service to which you belong?" "It is true," replied he. "You are employed on the Havre Railroad?" "It is likewise true. (Mons. Sabine not having previously stated his business to any one.). It is a basket that is missing, containing some little animals." "They are——they are——leeches. You sent to inquire about the basket at Rouen and at Havre, and you have received no news of it. This is what has taken place. A traveller, going to Havre by your carriages on the——the——the 11th November, was greatly annoyed, on arriving at his destination, to find only one basket instead of two, which he had on setting off." "This is wonderful!" said Mons. Sabine. "There were two baskets of leeches." "The train," continued Alexis, "on arriving at Rouen, left several travellers with their luggage, and one of the baskets was, by mistake, put on one of the omnibuses going into town, and the conductor was surprised to find no one claimed it. From fear of being scolded, he did not deposit it in the baggage warehouse, but hid it for some time in his stable; and while it was there you wrote to Rouen and Havre about it, the reply being that it could not be found. A few days ago the conductor put it in the goods depôt, near the entrance and beneath the first window on the right. You will find it if you set off to Rouen; only, on account of the length of time that has elapsed, you will find about 200 leeches dead." On the next day, Mons. Sabine returned from Rouen, having found the basket at the place indicated by Alexis, with 200 of the leeches dead. The directors of the railroad expressed themselves doubly obliged to the somnambulist and his mesmerism, inasmuch as the proprietor of the leeches, perceiving that they were not found after twenty-five days, had stated their value to be double what it actually was.

In the autumn of 1845, Alexis gave a series of mesmeric *séances* to the medical men of Havre, each of whom were permitted to bring one friend to witness the experiments. One of them took with him Mr. Featherstonhaugh, the Consul at Havre, who had come over the day before from California, and was a decided sceptic as to mesmerism. In order to test Alexis, Mr. Featherstonhaugh put in his pocket, enclosed in a box, a portion of a Japanese idol which he had picked up out of the wreck of a vessel from Japan, which had been lost on the coast of California during his stay there. On being asked by Mr. F., "What have I in my pocket?" Alexis answered, "It looks like a beetle; but it is not one, but part of a Japanese idol with an inscription on it: you picked it up during a walk on the sea-shore in California, and thought at first it was some curious stone, but you afterwards perceived it was an idol which had been washed up from the wreck of a Japanese vessel that was lost on the coast a few days before." The relater of this was Monsieur Paravet, of Havre, to whom it was told by one of the medical men present at the time.

———

At a *séance* which took place before the *élite* of the society at Versailles, Dr. Bataille, one of the principal physicians of this town, placed in the hands of Alexis a letter, and requested him to describe the residence of his son, who was living at Grandville. "Instead of giving you an account of the apartment of your son," said Alexis, "I am now occupied about his health, which is very bad." "How! bad?" replied his interrogator. "You have in your hand his last letter, dated six days ago, in which he states himself to be very well." "*To-morrow*," rejoined Alexis, "you will receive a letter from his wife, announcing to you that he is very ill. I recommend you on the receipt of this to set off *immediately*, for, knowing as you do the constitution of your son, there is only you who can save him. He is very ill." The next day the letter arrived, and Dr. Bataille immediately set off for Grandville, found his son very ill, and after a fortnight's sojourn, succeeded in restoring him to health. On his return to Versailles, this event produced a great sensation throughout the town.

I have given these few instances of the power of Alexis in thought-reading and clairvoyance, because from the reports of many friends who have seen him tried, on many different occasions, there can, I think, be no doubt that he does possess, at all events at some times, a very remarkable degree of power; while at the same time, I believe that he has often failed, and that his power varies very much at different times. But as I have explained before, failures can only prove the absence of the power when they occur, and have no weight in opposition to well-attested instances of success. In the case of Alexis, there is reason to think that failure has often been caused by over-exertion, and also by the influence exerted on him by the by-standers, to which he is peculiarly susceptible.

F. Introvision and Prevision in the Mesmeric Sleep.

Introvision, or the power of seeing and correctly describing the interior of the clairvoyant's own frame, is a tolerably frequent phenomenon, but it has not occurred in any of those persons whom I have myself mesmerised, none of whom, apparently, have yet reached a stage sufficiently advanced for that purpose. It is possessed, however, in a high degree, by the girl E., the clairvoyante of Dr. Haddock. She sees, in certain states, the whole of her frame bathed in light, transparent, and full of motion. At first, as often happens, she was much frightened and agitated at what she saw, but soon became reconciled to it, and described it in great detail. As I did not see her in this case, I shall merely say that Dr. Haddock's observations on this point in her case may be received with confidence, from the care and caution with which they are made. Many other cases of introvision are recorded, but space does not permit me to do more than thus briefly refer to these.

Case 59.—Allied to introvision is the power of seeing in the same way the interior of the frame of others *en rapport* with the clairvoyant, which I have already described. I may here adduce one instance of his power, as exercised by E. at a great distance. I have already stated (see Case 22) that E. in Bolton, described my son, then in Edinburgh, whom she has never yet seen. This was in October, and was done with the aid of his handwriting. In January he was attacked with the symptoms of inflammation of the membranes of the brain, such as usually precede hydrocephalus in children. While he was ill, I happened to mention, in a letter to Dr. Haddock, that he was suffering from illness, without giving any details. As E. had expressed a great liking for the boy when she saw him in her vision of October, Dr. H. asked her, when she was in the sleep, if she could see him and tell how he was. She had no writing to help her, but soon found him, and without having been told of his illness, at once said that he was very ill, and proceeded to described minutely the state of the cerebral membranes as they appeared to her. She gave a minute detail of the symptoms, which was as accurate as I could have given at the bedside. She also mentioned that he had studied too much, which had hurt him. In fact, I had carefully avoided giving him too much to do, because I had observed a tendency to the affection under which he then suffered; but even the very moderate amount of study allowed, had proved too much for him for some time before the inflammatory symptoms appeared. It was, in this case, quite evident to all who saw the case, and read E.'s account of it, that she had seen it as plainly as any of them had done, and her account of the state of the membranes was, although given in her own plain language, in all probability quite correct. At least such was the opinion of a medical friend who saw the boy, and who was very much struck with the accuracy of E.'s description.

I may here state, that a case is known to me at this time, to which I cannot more particularly refer, in which the sleeper, being mesmerised, has described the diseased part of his own frame, which is out of the reach of ordinary vision, in a manner not only very remarkable, but in all human probability correct.

With regard to prevision, I have not myself had opportunities of seeing much of it. It is comparatively rare, and I have mentioned various forms in which a certain degree of power of predicting the future occurs. The commonest is that of predicting the occurrence or recurrence of fits in the sleeper, often with the statement that the fit to occur at a specified time, will be the last, or the last but one, &c. Another form of prevision, which I have described, is that of predicting the duration of the sleep, which I have often seen; and that of predicting accurately the period at which the sleeper will become lucid, or will acquire certain powers In the case of Mr. D., he told me he should acquire a very high degree of lucidity after being mesmerised a number of times. I have already stated that he could not specify exactly the number of times, but that he went on improving in general lucidity, with occasional glimpses of a higher state, as long as I was able to mesmerise him. I had done so about 45 times, probably not much more than the half of the number required for full lucidity, and was in hopes of getting him at last to fix the precise number, and of ascertaining how far he would be correct, when his illness interfered, and on his recovery he had lost the extreme susceptibility he at first exhibited, so that I must once more begin from the beginning. But this I cannot now attempt for some time to come.

Some clairvoyants predict accidents to themselves, and one case of this kind is alluded to before, which I owe to Mr. Atkinson. In that case, the predicted accident was a fall on the steps on coming out of church, but it was only predicted generally, as to happen at a certain hour, and to produce certain effects. It is very common for such as go spontaneously into trance or extasis to predict these occurrences a long time before. I shall have to mention, briefly, a case of this kind under the head of extasis.

As to prevision of matters unconnected with the sleeper, and referring to other parties, this is much more rare, and I have not met with it in any of my own cases. But it has been frequently recorded, and I think must be admitted as a possible occurrence, beyond our power at present to explain. Major Buckley informs me, that it frequently occurs in his experience; and has communicated the following instances.

Case 55.—A young lady in London, being mesmerised, saw her family in the country, described their occupation, and added that her little brother had got the measles. Being asked, if her little sister had not also got the measles, she said, " No, but she will have them on Wednesday. Oh! my elder sister will have them too, but not until the Wednesday following." This came true.

Case 56.—A lady from Canada, who was present, asked the clairvoyante to go to Quebec. She declined then, but when next mesmerised, did so, and correctly described the house and its inmates that she was desired to see. She then said the lady would be able to read in nuts while awake, but not on that day; and that she herself would do so first. She was right in both predictions.

Case 57.—A clairvoyante told Major B., that if he would mesmerise a certain lady, who had never been tried, by making three passes round her head, the lady would be able to read three words, enclosed in boxes. The lady did sleep after the third pass, and read only three words, although there were four on the slip of paper enclosed in the boxes.

Major Buckley tells me that he finds similar predictions, as to the mode of mesmerising others, and the powers they will exhibit at certain periods, to be correct if made by good clairvoyantes.

Now it is obvious, that if prevision can exist to that extent, it may go still farther. It is quite as difficult to foresee when a person will take the measles, or when another will be able to read words enclosed in nuts, as to foresee any events whatever.

It appears to me also, as I have hinted before, that this fact furnishes the true key for the explanation of correct dreams of future events. That such dreams do occur, cannot, I think, be doubted; and I am disposed to regard them as instances of clairvoyant prevision occurring during sleep, or, as sometimes happens, in the waking hours, most probably in a state of reverie or abstraction. which resembles sleep in the circumstance that the mind is not dwelling on the impressions of the external senses. I shall have occasion to give a few instances, when I come to spontaneous prevision : here I have only to do with that which occurs in the mesmeric sleep.

In one or two cases I have known of predictions being made in the sleep, which may prove correct, but the accuracy of which remains to be ascertained, since the predicted events are still future.

Chapter XV.

TRANCE—EXTASIS—CASES—SPONTANEOUS MESMERIC PHENOMENA—APPARITIONS—PREDICTIONS.

Trance.

The state of Trance, as defined previously, has not yet occurred in my experience; but various cases are on record. I can only refer, here, to the celebrated Tinsbury case, recorded in the early transactions of the Royal Society, in which the trance lasted, with hardly an interruption, for seventeen weeks. In Mr. Braid's little work on Trance, recently published, the reader will find many interesting details of the very striking case of Col. Townsend, who could produce that state of apparent death at pleasure; and of various instances of the Faquirs in India, who are in the habit of doing so, and of allowing themselves to be buried in the trance, and awakened out of it days, weeks, or even months afterwards. A case, similar to the Tinsbury one, lately occurred in France.

Extasis.

This state is also one of which I have no experience. But it has been very often recorded, and I shall give a brief notice of its occurrence, on two distinct occasions, in the girl E. at Bolton. Both of these states were predicted accurately by E., and the second occurred only about a month ago. Dr. Haddock has been kind enough to let me see his notes, and from these I extract a brief notice of the phenomena, which will be fully described in Dr. H.'s new work.

Case 58.—In the summer of 1848, E. frequently went, spontaneously, and without any warning, into a state of extasis. This first happened on the 3rd of July, 1848. By degrees, she began, in the usual mesmeric sleep, to predict the occurrence of the extasis, and in one case did so two months before it occurred, which it did precisely at the time indicated. The same accuracy was observed in all her predictions of this kind, although she had no recollection, in her waking state, of having made them, and was never told that she had done so.

In the state of extasis she sometimes retained a recollection of the place she was in, and of the persons around her, but her mind was chiefly occupied with visions, apparently of another state of existence, and of what appeared to be spiritual beings. She always spoke of the state as of one to which she went away, or was taken away, and on returning to her usual mesmeric state, she would remember and describe what she had seen and felt. Her eyes were turned up, and she was entirely insensible to pain. At first, her limbs were flexible, but subsequently her whole frame was rigid. She could, when asked, perceive any concealed object by clairvoyance, but was usually too much engrossed with her spiritual perceptions to attend to such matters. On one occasion, when in her usual mesmeric state, she told Dr. H., that next night, a person long dead would come to her, and show her a book with some words in it, which she was to take to Dr. H. From her description of the book, Dr. H. conjectured that it was a small Bible, not then in the house; and he quietly procured it and placed it among other books. During the next night, she awoke in a kind of somnambulistic extasis, and in the dark went down two flights of stairs, selected the book, and brought it to Dr. H., opened at a certain page. In the darkness it fell, but she instantly found the passage by placing it on her forehead and turning over the pages. She said the passage had been shown to her in a similar but larger book by the person alluded to, and she added many circumstances connected with the history of the book, known to Dr. H. alone. She could not read, but explained, that, when looking for the passage by turning over the leaves, she found that, when she came to it, she could no longer turn them either way. This experiment was often repeated, in the dark, and for some months she could alway discover the passage, when mesmerised; but after a time she ceased to be able to do so. Here it is evident, that while her extatic vision somehow directed her to the book, she, who could not read, and was besides in the dark, had some means by which she saw and recognised the passage. When light was present, she never attempted to use her eyes, which were moreover turned up and closed, but always placed the book on her head, and then turned over the leaves. This vision was evidently connected with her states of spontaneous extasis, because the person seen was one who had always appeared to her in that state.

About the 11th of December, 1850, E., in the mesmeric state, predicted an extasis to occur on the 8th of January, 1851, and subsequently fixed the hour at 10 A.M. She had not gone into this state for some time, but as the day approached, she now and then went into a sort of partial extasis, and became almost insensible to what was passing around her, being much occupied with beings who came, as it appeared, to her. On the 8th, at 10 A.M., the predicted extasis occurred, and in it her visions were not only of another state and of spiritual beings, but obviously connected with all the former instances of extasis, of which the last well-marked one had occurred nearly two years before. I do not enter into the details of her visions, which will no doubt be given fully in Dr. Haddock's work, now in the press; but I may point out the remarkable clearness and consistency of these visions, which indicate a very peculiar and interesting mental state. In many points, her notions of the spiritual world, as derived from the visions, agree with those of the somnambulists or extatics of M. Cahagnet, alluded to previously; but it is remarkable, not only that these notions were not suggested to her, for Dr. H. most carefully avoided all suggestion, but were in many points directly opposed to the ideas that she had formed on such subjects from what she had been taught. It is singular that E. also, like the French extatics, spoke of Swedenborg as appearing to her, and as having possessed the power of seeing spirits. This is a subject on which it is impossible to form a decided opinion without far more extended investigation; but I may be allowed to observe, first, that whatever be the real nature of these visions or dreams, they appear to be genuine; and secondly, that their occurrence is always connected with a high degree of clairvoyant power, that is, with exalted perceptive faculties, acting through some unknown medium. Thus E., in gradually passing from her full state of extasis to her ordinary mesmeric state, was always found more clairvoyant than usual. It must therefore be admitted as possible, if we believe in the existence of a spiritual world at all, that in this state of exalted perception we may come into communication with it. Not having had any opportunity of examining a case of extasis, I can only judge from the reports of others; but it appears plainly, that when such cases occur, we ought to study them with care and attention. In this way alone, and not by rejecting the whole as imaginary, can we hope to ascertain the true nature of the phenomena. As far as I can judge at present, they do not appear to be suggested, at least directly, and there is a degree of harmony between the accounts of different observers, which is not easily reconcilable with the idea that they are altogether delusive. Certainly the clairvoyant visions of absent or even deceased persons (unknown to the clairvoyants, and often to any one present, but subsequently ascertained to be correct) are not delusive, although beyond our power to explain.

Spontaneous Mesmeric Phenomena.

We have already seen that many of the mesmeric phenomena occur spontaneously. Somnambulism is nothing else than the mesmeric sleep occurring during natural sleep. From the history of the numerous recorded cases, it appears that somnambulists can walk securely in the dark, or in dangerous situations, while the eyes are either closed or insensible to light. They are often deaf to the loudest sounds, and they can pursue their usual occupations, or read and write without waking. In short, they seem, like persons in the artificial mesmeric sleep, to possess a new sense or new senses. But I have had no opportunity of studying any case of sleep-waking.

Cataleptic rigidity is very frequent as a symptom of nervous disease, and is generally found associated with extreme susceptibility to the mesmeric influence.

Spontaneous Vision of Passing Events.

Sympathy with those who are absent is also very frequent as a spontaneous occurrence, and depends, in all probability, on the fact, that an impressible state is developed, either as a symptom of nervous disease, or in health, by deep concentration or abstraction. In this state the mind is affected by impressions which are usually overlooked. It is impossible to doubt, that in many instances, sympathetic warnings of the death or illness of absent friends have been experienced. Every one knows of some such instances, and it has been frequently stated, I believe correctly, that the persons who received the warnings, were those of whom the dying or sick person was earnestly thinking at the moment. I have already alluded to a case in which this sympathy was felt at the distance of 100 miles; and previously I have mentioned the case of a lady, related to me, who frequently had such warnings of the death of friends at still greater distances, which were always found true. The following cases are well attested. They are interesting, because, if we regard them as dreams, they are dreams occurring, at the same moment, to different persons, who all supposed themselves awake.

Case 59.—A gentleman of rank and property in Scotland served in his youth in the army of the Duke of York in Flanders. He occupied the same tent with two other officers, one of whom was sent on some service. One night, during his absence, this gentleman, while in bed, saw the figure of his absent friend sitting on the vacant bed. He called to his companion, who also saw the figure, which spoke to them, and said he had just been killed at a certain place, pointing to his wound. He then requested them, on returning to England, to call at a certain agent's house, in a certain street, and to procure from him a document of great importance for the family of the deceased. If the agent, as was probable, would deny the possession of it, it would be found in a certain drawer of a cabinet in his room. Next day, it appeared that the officer had been shot, as he had told them, in the manner, and at the time and place indicated. After the return of the troops to England, the two friends, walking together one day, found themselves in the street where the agent lived, and the request of their friend recurred to both, they having hitherto forgotten it. They called on the agent, who denied having the paper in question, when they compelled him, in their presence, to open the drawer of the cabinet, where it was found, and restored to the widow. Such is, briefly, the story, which illustrates the effect of sympathy in producing the vision of the absent man at the moment of his death. It may be ridiculed as a ghost story; but as I first heard it from a relative of my own, who was a neighbour in the country to the gentleman to whom it happened, and had often heard it from himself; as I subsequently heard it confirmed by the gentleman himself; and as I know that many others had it from him, I am satisfied that the facts are true. Even if we suppose the persons who saw the vision to have heard the details given by the figure from him when alive, and to have forgotten them, which is not likely, how are we to account for both seeing the same vision at the same time? I may add, that a lady of rank, who had often heard the story, told me, that when she or some one else once spoke of it in a company, a gentleman present said, "Perhaps you are not aware that the officer whose figure was seen was my father, and that I hold certain property by virtue of the very document which was so recovered."

Case 60.—An officer occupied the same room with another officer in the West Indies. One night he awoke his companion, and asked him if he saw anything in the room, when the latter answered that he saw an old man in the corner, whom he did not know. That, said the other, is my father, and I am sure he is dead. In due time, news arrived of his death in England at that very time. Long afterwards, the officer took his friend who had seen the vision to visit the widow, when, on entering the room, he started, and said, That is the portrait of the old man I saw. It was, in fact, the portrait of the father, whom the friend had never seen, except in the vision. This story I have also on the best authority; and every one knows that such stories are not uncommon. It is very easy, but not satisfactory, to laugh at them as incredible ghost stories; but there is a natural truth in them, whatever that may be.

Case 61.—The next case is one widely known, but interesting from the fact that the vision was seen by many persons. At a mess-table in America, the whole of the officers present saw the door open, and the figure pass through the room to an inner room. It was that of an absent comrade. As the figure did not reappear, and there was no other issue from the inner room, the company, surprised, looked into the inner room, and found it empty. It appeared that the person seen, died or was killed at the same time. Here it is very remarkable, that the sympathy to which we must, in all probability, ascribe the vision, affected so many persons. But the most striking fact was, that one officer, who had never seen the absent man, saw the figure. Some years afterwards, this officer (a very distinguished one), being in the streets of London, along with another who had also been present when the figure was seen, exclaimed, on seeing a gentleman, "There is the man whose figure I saw!" "No," replied his friend; "it is not he, but his twin brother." So that the officer who had never seen the dead man, except in the vision, recognised his brother by his strong likeness to the figure he had seen.

Case 62.—A lady saw, in a spontaneous vision or dream, a hand taking a brooch from her desk, where she had shortly placed it before. She saw the hand so plainly, that she could have pointed it out among a hundred, and recognised it as the hand of one of her servants. When morning came, the desk was examined, and the brooch was gone. As it was not ascertained, in this case, by whom it had been taken, we have not the full proof of the accuracy of the vision; but it is nevertheless interesting. The same lady exhibited many remarkable phenomena; but was not found to be clairvoyant in the mesmeric sleep. On one occasion, she fell into a state of trance, resembling death, and was supposed to be dead; but, as happens in many similar cases, she was conscious, and quite aware of what was passing, without the power of making the slightest movement, or of uttering a sound. For this case I am indebted to Mr. Atkinson.

SPONTANEOUS RETROVISION.

In the preceding cases, there has always occurred the vision of a dying person, then at a distance, and it is this which I am disposed to ascribe to what we call sympathy between certain persons in regard to events then passing. But it would appear that a similar sympathy may exist in reference to past events, just as we have seen retrovision occuring in clairvoyance. It is well known, that in all ages, and in every country, the belief has prevailed that certain places are visited by visions of the former inhabitants, and, according to popular belief, this occurs chiefly where crimes have been committed, It is easy to see how many such stories have originated in the minds of the timid, the ignorant, and the superstitious, perhaps from dreams, or from waking visions; and how they may have come to be much exaggerated and distorted. But there appear to be cases in which such an explanation is hardly admissible, and in which the same vision has presented itself, at long intervals, to many different persons, many of whom had never heard of them. Every one must have heard of many such cases, apparently well attested; and I confess that I find it difficult to convince myself, that this universal belief is altogether destitute of a true natural foundation. On the contrary, just as Reichenbach has shown, that one kind of apparitions is caused by luminous odylic emanations from dead bodies, or from other objects, I am inclined to think that a careful study of mesmerism would clear up other kinds of visions or spectres. And since persons in the mesmeric state may have distinct visions of past events, so we may imagine persons, not actually in that state, but in one approaching to it, in which they are intensely susceptible to mesmeric influences, to possess, occasionally, the power of seeing past events imaged before them. The chief difficulty is to understand how any vision comes to be attached to any special locality. But since highly lucid and sensitive subjects often perceive the traces of persons in places where they no longer are, we may suppose such traces to adhere to places, and to act on or be perceived by the sensitive, at a great distance of time. The following case will illustrate the phenomena alluded to. I have it on very good authority, and have every reason to believe that the facts occurred as here stated.

Case 63.—In an old house near Edinburgh, long since pulled down, resided a family, in which was a negro servant, who slept in a wing of the house. The first time he slept there, he saw a vision, which nearly terrified him out of his senses. (It is to be remembered that negroes are very sensitive to the mesmeric influence.) In the middle of the night, he saw the figure of a lady, richly dressed, but without her head, and carrying a child, pass through his room. When he spoke of it, the family, who had only recently rented the house, ridiculed him, and said he must have been drunk or mad. But he declared he had been quite sober. He was, however, compelled to sleep there again, and again saw the vision, and this went on till it was considered cruel to insist on his sleeping there, from the distress and terror it occasioned. When the vision was mentioned to persons belonging to the family who were the proprietors of the house, it was instantly recognised as one which had often appeared to those who slept in that room. Many years afterwards, the house was pulled down, and in a hollow in the wall of that room was found a box, shorter than a coffin, in which were the remains, the bones, of a female and of a child, the head of the female being at the side, and not in its natural place. A lady who heard of this, and had also known of the negro's vision, found him, now an old man, living in a remote part of the country, and without telling him what had been found, got him to tell the story of his vision, which he did, word for word, as above stated, and, even at that period, with every sign of terror. The only idea I can form of the cause of the vision, which was known in the family to have appeared to various persons, is this, that some influence, of a mesmeric or odylic nature, proceeding from the remains, acted on sensitive persons within its reach, so as to cause them to have a clairvoyant vision of the past.

The following case, which was mentioned to me by Sir David Brewster as well attested, is of a similar nature. I do not remember the names of the parties, but I believe the case has been published.

Case 64.—A nobleman one day went to hear a very distinguished Professor of Philosophy lecture in Berlin. During the lecture, which turned on apparitions, he observed the Professor to be much agitated; and after it was over, he mentioned this to the Professor, and begged for an explanation. This he gave, observing that the subject was one on which he could not dwell without deep emotion. It appeared that he had once been appointed to a living in East Prussia, where his predecessor was a priest generally respected. The first time he slept there, he saw, as he awoke in the morning, the figure of a priest cross the room, leading two children by the hand, and disappear behind the stove. If I remember rightly, he recognised the priest he saw to be the late incumbent, from a portrait in the room. Having discovered it was a vision he had seen, he made some inquiries, and was informed that two children, supposed to be natural children of the late incumbent, and who lived with him, had disappeared. For some time nothing more was discovered; but when it became necessary to light a fire in the stove, behind which the figures had vanished, a most offensive smell was observed to proceed from the fire, which would not burn properly, and, on examination, and taking down the wall of the stove, the remains of two children were found concealed in it.

I shall here mention another case, kindly communicated to me by Sir David Brewster. It is that of the "Tower Ghost." Unlike the two preceding cases, the origin of it is quite obscure.

Case 65.—"At the trial of Queen Caroline, in 1821, the guards at the Tower were doubled, and Col. S., the keeper of the Regalia, was quartered there with his family. Towards twilight, one evening, and before dark, he, his wife, son and daughter, were sitting, listening to the sentinels, who were singing and answering one another, on the beats above and below them. The evening was sultry, and the door stood ajar, when something suddenly rolled in through the open space. Col. S. at first thought it was a cloud of smoke, but it assumed the shape of a pyramid of dark thick grey, with something working towards its centre. Mrs. S. saw a form. Miss S. felt an indescribable sensation of chill and horror. The son sat at the window, staring at the terrified and agitated party, but saw nothing. Mrs. S. threw her head down upon her arms on the table, and screamed out, 'Oh! Christ! it has seized me!' The Colonel took a chair, and hurled it at the phantom, through which it passed. The cloud seemed to him to revolve round the room, and then disappear, as it came, through the door. He had scarcely risen from his chair to follow, when he heard a loud shriek and heavy fall at the bottom of the stair. He stopped to listen, and in a few minutes the guard came up and challenged the poor sentry, who had been so lately singing, but who now lay at the entrance in a swoon. The sergeant shook him rudely, declared he was asleep on his post, and put him under arrest. Next day, the soldier was brought to a court-martial, when Col. S. appeared on his behalf, to testify that he could not have been asleep, for that he had been singing, and the Colonel's family had been listening, ten minutes before. The man declared that, while walking towards the stair entrance, a dreadful figure had issued from the door-way, which he took at first for an escaped bear, on his hind legs. It passed him, and scowled upon him with a human face, and the expression of a demon, disappearing over the Barbican. He was so frightened, that he became giddy, and knew no more. His story, of course, was not believed by his judges, but he was believed to have had an attack of vertigo, and was acquitted and released on Col. S.'s evidence. That evening, Col. S. went to congratulate the man; but he was so changed that he did not know him. From a glow of rude health in his handsome face, he had become of the colour of bad paste. Col. S. said to him, 'Why do you look so dejected, my lad? I think I have done you a great favour in getting you off; and I would advise you in future to continue your habit of singing.' 'Colonel,' he replied, 'you have saved my character, and I thank you; but as for anything else, it little signifies. From the moment I saw that demon, I felt I was a dead man.' He never recovered his spirits, and died next day, forty-eight hours after he had seen the spectre. Col. S. had conversed with the serjeant about it, who quietly remarked, 'It was a bad job, but he was only a recruit, and must get used to it, like the rest.' 'What!' said Col. S., 'have you heard of others seeing the same?' 'Oh, yes,' answered the serjeant; 'there are many queer, unaccountable things seen here, I assure you, and many of our recruits faint a time or two, but they get used to it, and it don't hurt them.' Mrs. S. never got used to it. She remained in a state of dejection for six weeks, and then died. Col. S. was long in recovering from the impression, and was reluctant to speak of it; but he said he would never deny the thing he had seen."

It is evident that, in this case, the fatal results were chiefly caused by terror; but the essential fact is, that something was seen, and that too, by different persons, and that similar things had been seen before. It is worthy of notice, that what to some appeared as a horrible human form, appeared to others as a mass of grey smoke, and was not visible to others. Now, this is one of the characters of odylic light. Some cannot see it; others see only a faintly luminous grey smoke; others a more luminous and less dense vapour; and others a bright light; the differences being caused by various degrees of sensitiveness, and by the different distances at which the light is seen, as well as its various degrees of intensity. If it be asked, Could a mere shapeless luminous cloud or emanation assume a human form merely by the force of imagination? we must admit this to be possible. But we do not know that all odylic emanations are shapeless, when viewed at the proper distance; nor do we know that they may not, in some cases, possess the form of that from which they are derived.

Spontaneous Prevision.

We now come to spontaneous prevision. This has been recorded, as occurring in all ages, sometimes in the form of dreams, at other times in that of waking visions or second sight. By far the most remarkable, because the best attested, instance in modern times, is the celebrated prediction of M. Cazotte, concerning the events of the Reign of Terror. I shall give it entire, and I shall only premise, that it was well known, in all its details, both in Paris and London, at a time when every one thought it a mere dream. I have seen persons who heard of it very soon after it was delivered, and who remembered hearing it ridiculed in society as absurd. It is particularly worthy of notice, that Cazotte, who was a man of a very peculiar turn of mind, and much addicted to the study of occult science, was also subject to fits of abstraction, reverie, or dreaming, in which he seems to have been clairvoyant, and that this was far from being the only occasion in which he uttered predictions which were verified. He is to be considered as a man subject to fits of spontaneous lucidity, which, in his case, often took the form of prevision. The following account is extracted from the posthumous memoirs of La Harpe.

Case 66.—"It appears but as yesterday; yet, nevertheless, it was at the beginning of the year 1788. We were dining with one of our brethren at the academy—a man of considerable wealth and genius. The company was numerous and diversified—courtiers, lawyers, academicians, &c.; and, according to custom, there had been a magnificent dinner. At dessert, the wines of Malvoisin and Constantia added to the gaiety of the guests that sort of license which is sometimes forgetful of *bon ton;*—we had arrived in the world, just at that time when anything was permitted that would raise a laugh. Chamfort had read to us some of his impious and libertine tales, and even the great ladies had listened without having recourse to their fans. From this arose a deluge of jests against religion. One quoted a tirade from the *Pucelle;* another recalled the philosophic lines of Diderot,—

'Et des boyaux du dernier prêtre,
Serrer le cou du dernier roi,'

for the sake of applauding them. A third rose, and holding his glass in his hand, exclaimed, '*Yes, gentlemen, I am as sure that there is no God, as I am sure that Homer was a fool;*' and, in truth, he was as sure of the one as of the other. The conversation became more serious; much admiration was expressed on the revolution which Voltaire had effected, and it was agreed that it was his first claim to the reputation he enjoyed:—he had given the prevailing tone to his age, and had been read in the antechamber, as well as in the drawing-room. One of the guests told us, while bursting with laughter, that his hairdresser, while powdering his hair, had said to him—'*Do you observe, sir, that although I am but a poor miserable barber, I have no more religion than any other.*' We concluded that the revolution must soon be consummated,—that it was indispensable that superstition and fanaticism should give place to philosophy, and we began to calculate the probability of the period when this should be, and which of the present company should live to see the *reign of reason*. The oldest complained that they could scarcely flatter themselves with the hope; the young rejoiced that they might entertain this very probable expectation;—and they congratulated the academy especially for having prepared the *great work*, and for having been the great rallying point, the centre, and the prime mover of the liberty of thought.

"One only of the guests had not taken part in all the joyousness of this conversation, and had even gently and cheerfully checked our splendid enthusiasm. This was Cazotte, an amiable and original man, but unhappily infatuated with the reveries of the illuminati. He spoke, and with the most serious tone. 'Gentlemen,' said he, 'be satisfied; you will all see this great and sublime revolution, which you so much admire. You know that I am a little inclined to prophecy: I repeat, you will see it.' He was answered by the common rejoinder, '*One need not be a conjuror to see that.*' 'Be it so; but perhaps one must be a little more than conjuror, for what remains for me to tell you. Do you know what will be the consequence of this revolution,—what will be the consequence to all of you, and what will be the immediate result,—the well-established effect,—the thoroughly recognised consequence to all of you who are here present?' 'Ah!' said Condorcet, with his insolent and half-suppressed smile, 'let us hear,—a philosopher is not afraid to encounter a prophet.' 'You, Monsieur de Condorcet, you will yield up your last breath on the floor of a dungeon; you will die from poison, which you will have taken, in order to escape from execution,—from poison, which *the happiness* of that time will oblige you to carry about your person.'

"At first astonishment was most marked, but it was soon recollected that the good Cazotte is liable to dreaming, though apparently wide awake, and a hearty laugh is the consequence. 'Monsieur Cazotte, the relation you give is not so agreeable as your *Diable Amoureux*'—(a novel of Cazotte's).

"'But what diable has put into your head this prison and this poison, and these executioners? What can all these have in common with philosophy and the reign of reason?' 'This is exactly what I say to you; it is in the name of philosophy,—of humanity,—of liberty;—it is under the reign of reason, that it will happen to you thus to end your career;—and it will indeed be *the reign of reason*; for then she will have her temples, and indeed, at that time, there will be no other temples in France than the temples of reason.' 'By my truth,' said Chamfort, with a

sarcastic smile, '*you* will not be one of the priests of those temples.'
'I do not hope it; but you, Monsieur de Chamfort, you will be one, and most worthy to be so; you will open your veins with twenty-two cuts of a razor, and yet you will not die till some months afterwards.' They looked at each other, and laughed again. 'You, Monsieur Vicq d'Azir, you will not open your own veins, but you will cause yourself to be bled, six times in one day, during a paroxysm of the gout, in order to make more sure of your end, and you will die in the night. You, Monsieur de Nicolai, you will die upon the scaffold;—you, M. Bailly, on the scaffold;—you, Monsieur de Malesherbes, on the scaffold.' 'Ah! God be thanked,' exclaimed Roucher, 'it seems that Monsieur has no eye but for the academy—of it he has just made a terrible execution, and I, thank Heaven. . .' 'You! you also will die upon the scaffold.' 'Oh, what an admirable guesser,' was uttered on all sides; 'he has sworn to exterminate us all.' 'No, it is not I who have sworn it.' 'But shall we then be conquered by the Turks or the Tartars? Yet again. . . .' 'Not at all; I have already told you, you will then be governed only by philosophy, —only by reason. They who will thus treat you, will be all philosophers,—will always have upon their lips the selfsame phrases which you have been putting forth for the last hour,—will repeat all your maxims,—and will quote, as you have done, the verses of Diderot, and from *La Pucelle*.' They then whispered among themselves,—' You see that he is gone mad,'—for he preserved all this time the most serious and solemn manner. 'Do you not see that he is joking? and you know that, in the character of his jokes, there is always much of the marvellous.' 'Yes,' replied Chamfort, 'but his marvellousness is not cheerful,— it savours too much of the gibbet;—and when will all this happen?' 'Six years will not have passed over before all that I have said to you shall be accomplished.'

"'Here are some astonishing miracles' (and this time it was myself who spoke), 'but you have not included me in your list.' 'But you will be there, as an equally extraordinary miracle; you will then be a Christian.'

"Vehement exclamations on all sides. 'Ah,' replied Chamfort, 'I am comforted; if *we* shall perish only when La Harpe shall be a Christian, we are immortal.'

"'As for that,' then observed Madame la Duchesse de Grammont," 'we women, we are happy to be counted for nothing in these revolutions: when I say for nothing, it is not that we do not always mix ourselves up with them a little, but it is a received maxim, that they take no notice of us, and of our sex.' 'Your sex, ladies, will not protect you this time; and you had far better meddle with nothing, for you will be treated entirely as men, without any difference whatever.' 'But what, then, are you really telling us of, Monsieur Cazotte? You are preaching to us the end of the world.' 'I know nothing on that subject: but what I do know is, that you, Madame le Duchesse, will be conducted to the scaffold, you and many other ladies with you, in the cart of the executioner, and with your hands tied behind your backs.' 'Ah! I hope that, in that case, I shall have a carriage hung in black.' 'No, madame; higher ladies than yourself will go like you in the common car, with their hands tied behind them.' 'Higher ladies! what, the princesses of the blood?' 'Still more exalted personages.' Here a sensible emotion pervaded the whole company, and the countenance of the host was dark and lowering: they began to feel that the joke was become too serious. Madame de Grammont, in order to dissipate the cloud, took no notice of the reply, and contented herself with saying, in a careless tone, '*You see that he will not leave me even a confessor.*' 'No, madame, you will not have one, neither you, nor any one besides. The last victim to whom this favour will be afforded, will be . . . .' He stopped for a moment. 'Well! who then will be the happy mortal, to whom this prerogative will be given?' ''Tis the only one which he will have then retained—and that will be the king of France.'

"The master of the house rose hastily, and every one with him. He walked up to M Cazotte, and addressed him with a tone of deep emotion: 'My dear Monsieur Cazotte, this mournful joke has lasted long enough. You carry it too far, even so far as to derogate from the society in which you are, and from your own character.' Cazotte answered not a word, and was preparing to leave, when Madame de Grammont, who always sought to dissipate serious thought, and to restore the lost gaiety of the party, approached him, saying, 'Monsieur the prophet, who has foretold us of our good fortune, you have told us nothing of your own.' He remained silent for some time, with downcast eyes. 'Madame, have you ever read the seige of Jerusalem, in Josephus?' 'Yes! who has not read that? But answer as if I had never read it.' 'Well, then, madame, during the siege, a man, for seven days in succession, went round the ramparts of the city, in sight of the besiegers and besieged, crying unceasingly, with an ominous and thundering voice, *Woe to Jerusalem;* and the seventh time he cried, *Woe to Jerusalem, woe to myself*—and at that moment an enormous stone, projected from one of the machines of the besieging army, struck him, and destroyed him.'

"And, after this reply, M. Cazotte made his bow and retired.

"When, for the first time, I read this astonishing prediction, I thought that it was only a fiction of La Harpe's and that that celebrated critic wished to depict the astonishment which would have seized persons distinguished for their rank, their talents, and their fortune, if, several years before the revolution, one could have brought before them the causes which were preparing, and the frightful consequences which would follow. The inquiries which I have since made, and the information I have gained, have induced me to change my opinion. M. le Comte A. de Montesquieu, having assured me that Madame de Genlis had repeatedly told him that she had often heard this prediction related by M. de la Harpe, I begged of him to have the goodness to solicit from that lady more ample details. This is her reply :—

"'November, 1825.

"'I think I have somewhere placed, among my *souvenirs*, the anecdote of M. Cazotte, but I am not sure. I have heard it related a hundred times by M. de La Harpe, before the revolution, and always in the same form as I have met with it in print, and as he himself has caused it to be printed. This is all that I can say, or certify, or authenticate by my signature.—COMTESSE DE GENLIS.'

"I have also seen the son of M. Cazotte, who assured me that his father was gifted, in a most remarkable manner, with a faculty of prevision, of which he had numberless proofs; one of the most remarkable of which was, that on returning home on the day on which his daughter had succeeded in delivering him from the hands of the wretches who were conducting him to the scaffold, instead of partaking the joy of his surrounding family, he declared that in three days he should be again arrested, and that he should then undergo his fate; and in truth he perished on the 25th of Sept., 1792, at the age of 72.'

"In reference to the above narrative, M. Cazotte, jun., would not undertake to affirm that the relation of La Harpe was exact in all its *expressions* but had not the smallest doubts as to the reality of the *facts*.

"I ought to add, that a friend of Vicq d'Azir, an inhabitant of Rennes, told me, that that celebrated physician, having travelled into Brittany some years before the revolution, had related to him, before his family, the prophecy of Cazotte. It seemed that, notwithstanding his scepticism, Vicq d'Azir was uneasy about this prediction.

"Letter on this subject addressed to M. Mialle by M. le Baron Delamothe Langon :—

"'You inquire of me, my dear friend, what I know concerning the famous prediction of Cazotte mentioned by La Harpe. I have only on this subject to assure you *upon* my honour, that I have heard Madame la Comtesse de Beauharnais many times assert that she was present at this very singular historical fact. She related it always in the same way, and with the accent of truth;—her evidence is fully corroborated by that of La Harpe. She spoke thus, before all the persons of the society in which she moved, many of whom still live, and could equally attest this assertion.

"'You may make what use you please of this communication.

"'Adieu, my good old friend. I remain with inviolable attachment, yours,
"'BARON DELAMOTHE LANGON.
"'PARIS, *Dec.* 18*th*, 1833.'"
—*La Harpe: Posthumous Memoirs*, Paris, 1806, Vol. I. p. 62.

I have mentioned the case of a lady who had fits of spontaneous lucidity. Mr. Atkinson, to whom I am indebted for the instances of this power in that lady, already given, has kindly furnished me with some further details of her case, from which it appears that it also presented some phenomena of prevision.

Case 67.—The lady in question, is one possessed of the highest qualities both of mind and person, and has enjoyed the esteem and respect of many distinguished men. She has always had the power of clairvoyance, both as to present events, and sometimes as to future events, occurring spontaneously, and generally when she has been sitting alone and quiet in the evening (that is in circumstances favourable to abstraction or concentration of thought). Her visions do not always relate to important events, but frequently refer to some trifling occurrence in the neighbouring street. At other times she will see clearly all the circumstances connected with the death-bed of a friend, the persons present, with other details, the whole facts perhaps not to be realised for some years, and then occurring as foreseen. Sometimes she sees what appear to be mere optical delusions, as, for example, an empty arm-chair where no chair exists. But it is possible that even these visions, if understood, or properly interpreted, might be found to have a meaning. The case previously narrated is an excellent example of her power of seeing present or passing events, and although I am not permitted to give in detail any instances of her prevision, I have no doubt that she has repeatedly possessed that power. It is indeed impossible to form a satisfactory theory to explain this, but neither can we explain the power of seeing passing events. One point of this case is very interesting, namely, that the lady, besides being subject to fits of spontaneous clairvoyance, is also, as might be expected, of an exceedingly sensitive and impressible nature. Thus, on one occasion, when a gentleman visited her house, she experienced a very uncomfortable sensation so long as he was present, and observed a spot or sore on his cheek. Two days after, a similar spot or sore appeared on her own cheek, in precisely the same situation, and with the same characters. It is evidently in such idiosyncrasies that spontaneous clairvoyance is mostly likely to appear.

To Mr. Atkinson, who has profoundly and acutely studied the whole of this subject, I am also indebted for the following instance of spontaneous prevision, in the shape of a dream, in the words of the gentleman who had the dream.

Case 68.—"My brother, who was an officer in the Royal Engineers, and to whom I was tenderly attached, died in the West Indies in the autumn of 1826. As well as I can recollect, about a month before the news of his death arrived, I had the following dream concerning him. I was then pursuing my studies in the University of Dublin, and used generally to spend my evenings at a friend's house. I dreamed that, on returning to my lodgings one night, I received a message from my uncle, who resided in Dublin, to come to him directly; that I accordingly went, and was ushered into his private room; that he was seated at his desk in a particular corner, and asked me to take a chair at the fire. He then told me, he was sorry to say that he had bad news to communicate to me respecting my brother, and that, in fact, he was no more. I thought that I then immediately replied, Is there any evidence to show in what state of mind he died? to which my uncle replied there was, and then handed to me the letters which he had received, upon which I took my departure. Such was the dream, and it made so strong an impression on my mind that I was greatly distressed, and could not, as I had always hitherto done, make mention of him in my prayers. I related the dream, at the time, to the lady to whom I was married, and she has a perfect recollection of all the circumstances. After a little time, the impression wore off, and I had nearly forgotten it, when returning to my lodgings one evening, I was informed that my cousin had called, and had left an urgent request for me to proceed to his father's house as soon as I came home. I accordingly went, and was shown into his room; he was seated in the same spot in which I had seen him in my dream; the desk, papers, and even candles, were in exactly the same position. He invited me to a chair at the fire, and the same conversation took place, *verbatim*, as in my dream. He made the communication to me precisely in the same words, and I made exactly the same reply, as related above. He then handed me the letters, and I took my leave, being too much agitated and shocked to continue the conversation. But strange to say, I did not recollect the dream till the interview was over, when it suddenly recurred to me, with very startling effect." Mr. Atkinson adds: "The subject of this dream is a clear and sober-minded clergyman, greatly respected by all who know him, and on the accuracy of whose statements you may place the fullest reliance."—It is, I think, evident that mere coincidence is not sufficient to account for the accuracy of this prevision, even were it a solitary case. And if the facts be admitted, they are quite as marvellous and inexplicable as any recorded prediction whatever.

Case 69.—A lady, who had left her only child in Edinburgh, and was then in Germany, told me at the time that she had seen a vision or dream of her son seriously ill in bed, and of his nurse standing in a particular spot, where he could not see her, in great distress, watching the sick child. On returning home, she pointed out the spot where she had seen the nurse, who had stood for a long time there, watching her patient. She was then informed that he had been seriously ill, which had not been mentioned before, as he had soon recovered. But while abroad, she had often told me that, from what she saw, she felt sure he had been very ill, although her letters had only alluded to a very slight indisposition. I cannot now ascertain whether this vision occurred precisely at, or before or after the time at which the child was so ill. It was certainly very near to the time.

Case 70.—Major Buckley, twenty-three years ago, before he had heard of mesmerism, was on the voyage between England and India, when one day a lady remarked, that they had not seen a sail for many days. He replied, that they would see one next day at noon, on the starboard bow. Being asked by the officers in the ship how he knew, he could only say that he saw it, and that it would happen. When the time came, the captain jested him on his prediction, but at that moment a man who had been sent aloft half-an-hour before, in consequence of the prophecy, sung out, "A sail!" "Where?"— "On the starboard bow."—I consider this case interesting, because it tends to show a relation between mesmeric power, which Major Buckley possesses in an eminent degree, and susceptibility to the mesmeric or other influences concerned. The same combination is found in Mr. Lewis.

Case 71.—A soldier in a Highland regiment, than in America, named Evan Campbell, was summoned before his officer for having spread among the men a prediction that a certain officer would be killed next day. He could only explain that he had seen a vision of it, and that he saw the officer killed, in the first onset, by a ball in the forehead. He was reprimanded, and desired to say no more about it. Next day, an engagement took place, and in the first attack the officer was killed by a ball in the forehead. I am told that this instance of second sight may be entirely depended on.

The above cases are only a few out of many that might be adduced, and tend to show that, by some obscure means, certain persons, in a peculiar state, may have visions of events yet future. And indeed it is only by admitting some such influence, that we can at all account for the fulfilment of prophetic dreams, which, it cannot be doubted, has frequently taken place. Coincidence, as I have before remarked, is insufficient to explain even one case, so enormously great are the chances against it; but when several cases occur, it is absolutely out of the question to explain them by coincidence.

Chapter XVI.

CURATIVE AGENCY OF MESMERISM—CONCLUDING REMARKS, AND SUMMARY.

Therapeutic Use of Mesmerism.

This part of the subject may be considered under several heads. First, the use of mesmerism in relieving pain and curing disease; secondly, its use in preventing pain in surgical operation; thirdly, the use of magnets, crystals, and other inanimate objects, as well as of mesmerised water, or mesmerised objects of any kind; and, lastly, the use of clairvoyance in diagnosis. My own experience in these matters is very limited indeed, so that I cannot enter into minute details; but I have seen enough to convince me, that mesmerism is a most powerful and valuable ally to the physician. I have already mentioned that the persons mesmerised by me have generally been healthy; but I have also described one case, that of the blind Mr. H. W., in which, without intention on my part, I cured him of an obstinate and annoying discharge of the nose, by a few operations. A remarkable improvement also took place in his general health, insomuch that, after fifteen or twenty operations, his whole aspect was changed for the better, and his strength and spirits very much improved. At the same time, a slow but steady improvement in his eyesight has continued to appear, he having now, in about three months, been operated on nearly fifty times. It is, of course, impossible to say whether his sight will ever be restored; but the results, up to the present time, are such as to yield some hope, and the state of the eye, externally, has undergone a most marked change, so that it is now natural in colour and moisture, which was not the case when I began.

Every one who tries it will find that, by means of mesmerism, various pains, especially neuralgic and rheumatic pains, may be very often and easily relieved. Cases of this kind occur every day. You will find, in the *Zoist*, and in the foreign journals, innumerable cases recorded, most of them by medical men, of cures effected by mesmerism, and very often in cases which had baffled all other treatment. It appears that some individuals possess much greater curative virtue than others, although all healthy persons probably possess the power of curing in this way in a greater or less degree. The celebrated Valentine Greatrakes obviously possessed an unusual share of mesmeric power; but, at the present day, we have a remarkable instance of it in Mr. Capern, a gentleman of Devonshire, who has just published a small work, giving an account of his mesmeric practice. This work, which is a simple, unaffected statement of fact, is well worthy of persual, and I cannot do better than refer you to it. I might quote also many other instances of mesmeric cures, performed by persons known to me, and among them a large number by Mr. Lewis, whose power is very great; but my space forbids me to do this, and it would lead to a tedious repetition of details. In connection with the use of mesmerism as a curative agent, I prefer, therefore, to give some extracts from a communication kindly made to me by Mr. H. G. Atkinson, who not only has great mesmeric power, but very great experience, and has studied the whole subject profoundly. It will be seen that his observations illustrate some very important points in mesmeric treatment, and more particularly the very curious question of the transference of pain or disease. Mr. A. describes his first experiment on the conveyance of mesmeric influence to a distance, by means of inanimate objects, as follows:—

Case 72.—"I was requested by a physician to try the effect of mesmerism on a lady who was suffering fearfully from tic, a complaint to which she had been subject for many years. The trial was most successful, but before a cure could be effected, the lady was obliged to go to Paris with her husband. Now, as my peculiar influence had so good an effect upon her, it seemed most desirable to continue the process, if possible; and as I had already tested the fact that mesmeric power could be conveyed by water, cotton, leather, and other substances, I suggested the plan of sending her mesmerised gloves by post to Paris. The experiment succeeded perfectly; the glove put on her hand always sent her into mesmeric sleep, and relieved her intense suffering, which all other means had wholly failed to do. The mesmerised glove by use gradually lost its property, and then failed to cause sleep, after a third time; so that I had to send newly mesmerised gloves every week, and the old ones were from time to time returned, to be charged afresh. This led to the observation of a very striking fact. I found that, before I could renew the healthy power, I had to remove the unhealthy influence or contagion, which the glove had absorbed from the patient. I felt in my hand, on approaching the old gloves, the same unpleasant sensations as I have from touching a diseased individual, besides absolute pain from the tic. The sensations were as clear and unmistakable as those of heat from a flame, or of the roughness or smoothness of objects. The pain was the same in character as that of the patient. After I had mesmerised the glove for two or three minutes, the sensation ceased, and the glove was now cleansed from the influence it had absorbed and brought with it. The sensation now was the same as I have when I relieve pain by mesmerising; when I can tell at once that the pain is relieved. I was in fact a complete mesmerometer, and had within myself the most convincing proof of the mesmeric fact, and of its relation to ordinary contagion. It might be supposed that the influence of the gloves on the patient was due only to the imagination; but 1 tested this, by sending sometimes unmesmerised gloves, and at other times such as had been used by the patient, without doing anything to them; and always found that the unmesmerised gloves had no effect, and the used gloves a most disagreeable one. I have made the experiment in a great many cases, and with the same results. The perception I had of pain and other states appeared at first very strange to me, as if it were a new sense; but I soon became familiar with it, so that it ceased to attract notice. My patients used to try to deceive me as to their pains, but could never succeed, and they used to remark that I knew their sensations better than they did. When mesmerising nervous patients, I have felt a prickling sensation in my hand, but as soon as the sleep came on, I felt a slight shock, as it were; all disagreeable feelings ceased, and I experienced an agreeable influence in their place. On the occurrence of any decided change in the patients, as from trance to somnambulism, I felt the same slight shock. I have found that one's own peculiar mesmeric power may be in a measure conveyed to another, and also that the peculiar mesmeric state or sleep may be conveyed from one patient to another. This accounts for the occasional contagion of fits, and for such phenomena as the contagious preaching mania in Sweden. I have experienced the same sympathetic influence as in the case of gloves, from letters, especially if the paper were glazed; and I could thus tell the state of the patient before reading the letter. Sometimes the heat and prickling have been so strong, that I have laid the letter on the table, to read without touching it. The influence from a feverish state would cause my hand to feel hot and feverish, even to others, the whole day. On one occasion, on reading a letter from a distance, I had the sensation of tears. It was so strong, that I felt sure the writer had been in tears while writing it, although nothing in the case, or in the letter, led to this conclusion. It proved, however, on enquiry, that the writer had been in tears, and that the tears had fallen on the paper. In one case, in which the patient, a lady, was too sensitive to be treated in the usual way, I gave her mesmerised water, which immediately caused her to sleep, and she was thus cured of sleeplessness. Once, when she sent for the water, I could not venture to mesmerise it, as I had just been mesmerising a diseased object, so I sent ordinary water, without any remark. In a few days, I received a note to say, that the water had lost its power, and no longer caused sleep. On one occasion I breathed a dream into a glove, which I sent to a lady; the dream occurred. One of the ladies above mentioned, with her whole family, are sensitive to the approach of iron, which recalls the faculty possessed by some, of discovering veins of ore, or springs of water under ground. I found that if, when engaged in mesmerising for pain, my mind was bent on what I was doing, I received little or no influence. I could at will either impress the patients, or absorb their condition. In either case, if the will was active, the influence never affected me beyond the hand used; but if I were thinking of other things, I experienced the whole symptoms of the patients, so far as pain was concerned; and I thus approached to the state of the somnambulist who detects the diseases of others.—The following is a remarkable instance of sympathy. I had mesmerised a young lady who was living with my brother and sister, twenty miles from town. She proved an excellent clairvoyante. One Sunday I was walking with a lady, after church, in her garden, in St. John's Wood, when I found the dead body of a new-born infant, wrapped up in a clean cloth; it must have been thrown over the wall. Next morning I received a note from my sister, telling me that my patient had on the Sunday, after church, insisted on taking her all about the garden to look for a baby, because she was sure she should find one. Once when I had been mesmerising the same young lady before a company, and was making a cross pass to wake her, a lady standing close by received the influence from my hand, and ran away screaming like one possessed. From that time, whenever I mesmerised my patient, which was at very uncertain times, when I could get away from London, this lady, living four miles off, fell into mesmeric sleep at the same time, and her case exhibited the same phenomena of clairvoyance. Two of my patients fell into a dead sleep or trance, so deep that I believe you might have cut them to pieces, and they would have felt nothing. No ordinary means could arouse them; yet if a drop of water fell, even on their dress, it immediately set them a-trembling; the touch of a piece of silver convulsed them with laughter, which the touch of another metal instantly stopped. The trembling and laughing seemed wholly without consciousness, a kind of life in death, or merely spasmodic. I have seen and heard some patients writhe and groan and scream as if in agony, and yet on awaking declare they had had delightful dreams. Here a conscious and an unconscious state seemed to exist together so that the persons may be truly said to be *beside themselves.* One lady whom I know, suffers acutely from the contact of metals. She feels a pricking sensation, with general discomfort, and at times even sickness. A brass thimble caused her finger to swell, and she was forced to leave it off. She is compelled to eat with wooden spoons, and to use her handkerchief to open the door when the door-handle is of metal. This sensitiveness is not uncommon, and should be carefully attended to as it may be the unsuspected cause of much suffering. In general, medical men pay no attention to such things, and call them nervous or hysterical, and thus save the trouble of investigation. Sir C. Bell said that the eternal answer of the indolent is, ' It is hysterical.' But supposing them hysterical, what then ?   They are not the less real nor the less interesting."

The above extracts contain a number of facts well worthy of consideration, in reference to the curative agency of mesmerism; and may serve to show how much there is for us to learn, and how much any of us might contribute to the advancement of knowledge in these obscure subjects, if we availed ourselves of our opportunities with the same zeal and the same patience as Mr. Atkinson. I ought to mention that Mr. A. is not a medical man, but has often generously devoted his time for weeks and months to the relief of suffering; and it is in this benevolent occupation that he had met with so many interesting facts, of which a few are briefly sketched above.   It will be observed, that several of the facts above related tend to throw light on the propagation of disease by contagion, and if duly investigated, may lead to results of great importance. We may also see, in the very powerful effects produced by inanimate bodies, such as metals, a glimpse of the principle on which the effects of minute doses of medicines are produced in certain constitutions. It cannot be doubted, that in some cases the infinitesimal doses used in homœopathy do produce strong and marked effects, which cannot be explained on the ordinary principles of medicine; but we can see how, in susceptible subjects, not only a minute dose, but the mere approach of certain substances, may have a decided action. Many persons in the mesmeric sleep, or in a certain stage of it, exhibit a singular degree of sensitiveness to the influence of different substances; and Reichenbach has proved that the same sensitiveness, in various degrees, is found in many persons in the ordinary conscious state, so that, by feeling the substances, or even the bottles which contain them, these persons can infallibly distinguish one from the other.

It is quite unnecessary to give details of cases concerning the possibility of mesmerising water, and other substances, so that they shall act strongly on the susceptible. Nothing, as Reichenbach justly remarks, can more satisfactorily prove the existence of an influence, capable of being transferred from the hand of the operator to the patient, or to any substance, than this simple, fundamental experiment. Any one may easily satisfy himself of the fact, as soon as he finds a case in which he can produce the mesmeric sleep.

As to the action of magnets, crystals, and metals, numerous cases occcur, and are daily to be met with in which pain is relieved by the contract or approach of these bodies. I know of one lady, subject to severe nervous headaches, who is relieved at once by holding in her hand a large crystal of fluor-spar, which generally thnows her into mesmeric sleep. The effect is so well marked, that when she suffers, her children always beg her to use the crystal. But, in exact correspondence with that Reichenbach has observed, the position of the poles of the crystals must be reversed if it be shifted from one hand to the other. The action of magnets, and even of galvanic rings, in relieving rheumatic pains, is very far from being imaginary in many cases. It has been generally rejected by medical men, because they could not explain it ; and it has been said, that since the galvanic rings could not cause a current of galvanic electricity, they could have no effect. But this is a *non sequitur*. Not only rings of two metals, but rings or other masses of one metal, often produce strong effects, relieve pain, and cause sleep, as do magnets also ; and on the principles developed by Reichenbach, they act, not by electricity, nor by ferro-magnetism, but by their odylic force. Instead of rejecting the facts, therefore, on theoretical grounds, or because we cannot explain them, we ought rather to multiply our observations, and from them, in process of time, deduce our theory or explanation.

With regard to the use of mesmerism or clairvoyance in diagnosis, I have already given several instances of it, and from what I have seen, I am satisfied that, with a good subject, much may be done in this way. Did my space permit, I could give very many instances, in the practice of others, in which the clairvoyant, either by contact with the patient, or with the aid of his hair or handwriting at a distance, has most accurately described the whole symptoms, and often has detected the true, though unsuspected cause of these. As my space, however, is exhausted, I

For some remarks on the use of mesmerism in Insanity, I am compelled, too, by want of space, to refer to what has been said before.

I had collected a number of cases to illustrate the therapeutic action of mesmerism, but I find I must omit them. This I do with the less regret, that this branch of the subject is sure to be worked out by many persons, while the works of Dr. Esdaile and Mr. Capern, besides others, and the long lists of cases published in the *Zoist* by Dr. Elliotson, and many other medical men, as well as in the foreign Journals, has sufficiently established the great value and importance of mesmeric treatment. Moreover, the cases which I had collected, being from the practice of others, and very much of the same character as the published cases, would merely have added to the size of the work, without the advantage of anything new. No one who has attended to the subject can doubt the value of mesmerism as a therapeutic agent, and for the present we must admit this application of mesmerism to be the most important, and to be worthy of the attentive study of medical men.

Dr. Esdaile's admirable work on Mesmerism in India, has firmly established the value of mesmerism as a means of preventing pain during surgical operations; it has also been used extensively for that purpose, both in this country and in France. The fact, recently demonstrated here, to the satisfaction of many medical and scientific men, by Mr. Lewis and by Dr. Darling, that complete insensibility to pain, either general or local, may be produced in many persons in the conscious state, will tend to dissipate the incredulity which still lingers in some quarters as to the possibility of rendering patients insensible to pain by mesmerism, even although some should still persist in ascribing the result to the imagination. In the conscious state, it is certainly done by suggestion, but it is also, without any suggestion whatever, a frequent feature of the mesmeric sleep. The essential point is, that such a state can be produced in various ways, and this can no longer be doubted by any one who has examined the subject. I have myself often seen it produced by others, and also produced it myself, both in the conscious state and in the mesmeric sleep. In fact, this was done, and in many instances, long before the powers of ether and chloroform were discovered; and although at that time many at once rejected the alleged fact, without inquiry, who afterwards at once admitted the corresponding fact in regard to ether, the evidence for both is, and always was, the same in kind. The only difference is, that in the case of the ether, the existence of a drug, a tangible agent, is supposed by many to enable us to understand or explain the result, while, because in the mesmeric power the agent is not tangible, they reject the fact. Nothing can be more illogical; for in the case of ether or chloroform, we only know *that they act*, but not *how they act*, which no one can explain; and, in the case of mesmerism, whether we can explain it or not, the fact is equally certain. Besides, the mesmeric sleep and insensibility to pain may be produced by means as tangible as ether, namely, by magnets, crystals, and other bodies, such as mesmerised water. The following case will show, in addition to those already given, that in susceptible persons, ether or chloroform may be dispensed with, and the full effects produced by suggestion. The experiment was performed in my house, in the presence of Sir D. Brewster and Sir W. C. Trevelyan, besides others.

Case 74.—Mr. B., a student of medicine, was found to be easily thrown either into the mesmeric sleep, or in the conscious impressible state, by Mr. Lewis. Mr. L., when the patient was quite conscious, gave him a handkerchief moistened with ordinary water, and told him that it was chloroform, and that he was to breathe it, and to become unconscious and insensible to pain. Although Mr. B. knew it to be water, he could not resist the suggested impression; he breathed exactly as one does with chloroform, and in about a minute became unconscious, when he was found utterly insensible to pain. On waking, he had no recollection of what had past. Some time afterwards, he put the wet handkerchief in his pocket, and as long as it remained there, he fell asleep every few minutes, till Mr. Lewis removed it, when that effect ceased. Nothing can show more strongly the power of suggestion in certain states.

On the same evening I had occasion to observe a fact which is of some importance, as showing that causes, hitherto unsuspected, may interfere with our experiments.

Case 75.—There were present on that evening, in all, ten persons, including Sir D. Brewster, Sir Walter C. Trevelyan, Mr. B., Mr. Lewis, Mr. W., the blind German formerly mentioned, myself, besides four ladies. Mr. H. W., whom I was then in the habit of mesmerising daily in the forenoon, told me, a few minutes after he entered the room, that he felt the influence of Mr. Lewis very strongly, and soon afterwards, while Mr. L. was trying to put Mr. B. to sleep in the front room, Mr. W., who was engaged in conversation in the back room, fell asleep. At the same time, Mr. B., who had gone into the sleep, suddenly woke. Mr. L. then showed his power of controlling the muscular motions of Mr. B., which he did for a time very completely, but all at once Mr. B' said, " I feel that you have no longer any power over me ;" and at the same instant Mr. W.; who was now nearer than before, fell into a deep mesmeric sleep. Mr. L. was not aware of his having previously fallen asleep, and had not thought of affecting him; but it appeared as if he, being more susceptible, had twice carried off the influence from Mr. B., without knowing it. But this was not all; for it soon appeared, that all the four ladies were more or less affected, although Mr. L. had never tried to affect them. It was proposed that he should put one of the ladies fully to sleep, with the view of then awaking her, and thus removing her unpleasant sensations. This she declined; but Mr. L., at my request, tried to put her to sleep by silent volition. This had just begun to act, when Mr. B. came, not knowing what was going on, and sat down, nearer to Mr. L. than the lady; and almost instantly he fell asleep. The lady then allowed Mr. L. to put her to sleep, which he did, and then awoke her, now free from all unpleasant sensations; but as she awoke, Mr. B. again fell asleep. Another lady was then put to sleep and was soon in a very deep sleep; but when I, observing Mr. B. asleep, spoke to him, and, finding that he would not answer me, took hold of his hand, he suddenly awoke, and at the same instant the sleeping lady, about 14 feet off, suddenly awoke also. All this was before the handkerchief was tried; and it plainly showed, that where a powerful mesmeriser is in the same room with several susceptible persons, the results are very apt to be very much confused, and experiments may fail which would succeed perfectly if only one susceptible subject were present.

I must now conclude, and I would do so by once more pointing out, that my object has not been to explain the facts I have described, but rather to show that a large number of facts exist, which require explanation, but which can never be explained unless we study them. I am quite content that any theoretical suggestions I have made should be thrown aside as quite unimportant, provided only the facts be attended to; because I consider it too early for a comprehensive theory, and because I believe that the facts are as yet but very partially known.

But I think we may regard it as established: first that one individual may exercise a certain influence on another, even at a distance; secondly, that one individual may acquire a control over the motion, sensations, memory, emotions, and volition of another, both by suggestion, in the conscious, impressible state, and in the mesmeric sleep, with or without suggestion; thirdly, that the mesmeric sleep is a very peculiar state, with a distinct and separate consciousness; fourthly, that in this state, the subject often possesses a new power of perception, the nature of which is unknown, but by means of which he can see objects or persons, near or distant, without the use of the external organs of vision; fifthly, that he very often possesses a very high degree of sympathy with others, so as to be able to read their thoughts; sixthly, that by these powers of clairvoyance and sympathy, he can sometimes perceive and describe, not only present, but past, and even future events; seventhly, that he can often perceive and describe the bodily state of himself or others; eighthly, that he may fall into trance and extasis, the period of which he often predicts accurately; ninthly, that every one of these phenomena has occurred, and frequently occurs, spontaneously, which I hold to be the fundamental fact of the whole inquiry; Somnambulism, Clairvoyance, Sympathy, Trance, Extasis, Insensibility to pain, and Prevision, having often been recorded as natural occurrences. Tenthly, that not only the human body, but inanimate objects, such as magnets, crystals, metals, &c., &c., exert on sensitive persons an influence, identical, so far as it is known, with that which produces mesmerism; that such an influence really exists, because it may act without a shadow of suggestion, and may be transferred to water and other bodies; and lastly, that it is only by studying the characters of this influence, as we should those of any other, such as Electricity or Light, that we can hope to throw light on these obscure subjects. Let us in the meantime observe and accumulate facts; and whether we succeed or not in tracing these to their true causes, the facts, if well observed and faithfully recorded, will remain, and, in a more advanced state of science, will lead to a true and comprehensive theory.

Address. — Manager of The
PSYCHOLOGICAL PRESS
4 Ave Maria Lane E.C.

☞ *This List has been considerably enlarged and revised.*

A NEW

LIST OF STANDARD BOOKS

ON

𝔓𝔰𝔶𝔠𝔥𝔬𝔩𝔬𝔤𝔶, 𝔐𝔢𝔰𝔪𝔢𝔯𝔦𝔰𝔪, 𝔐𝔦𝔫𝔡-ℜ𝔢𝔞𝔡𝔦𝔫𝔤, ℭ𝔩𝔞𝔦𝔯𝔳𝔬𝔶𝔞𝔫𝔠𝔢,
𝔖𝔭𝔦𝔯𝔦𝔱𝔲𝔞𝔩𝔦𝔰𝔪, 𝔗𝔥𝔢𝔬𝔰𝔬𝔭𝔥𝔶; 𝔱𝔥𝔢 𝔒𝔠𝔠𝔲𝔩𝔱 𝔖𝔠𝔦𝔢𝔫𝔠𝔢𝔰,
𝔓𝔥𝔶𝔰𝔦𝔬𝔩𝔬𝔤𝔶, 𝔏𝔦𝔟𝔢𝔯𝔞𝔩 𝔞𝔫𝔡 ℜ𝔢𝔣𝔬𝔯𝔪 𝔖𝔲𝔟𝔧𝔢𝔠𝔱𝔰.

PUBLISHED AND SOLD BY

THE PSYCHOLOGICAL PRESS ASSOCIATION,

LONDON.

☞ *All Communications to be addressed to the Psychological Press Association. All books sent post free on receipt of published price except where specially indicated.*

REMITTANCES may be made either in Penny or Halfpenny Stamps, Postal Orders, Post Office Orders, or Cheques. If made in the two last forms, they should be made payable to "The Psychological Press Association," and crossed "Birkbeck Bank."

THE PSYCHOLOGICAL PRESS ASSOCIATION supplies all the latest publications upon Spiritual Science and Psychic Research, etc., issued at home or abroad. Customers can be supplied, in most cases, through the post without extra charge. Subscriptions received for the various Spiritual periodicals.

THE PSYCHOLOGICAL PRESS ASSOCIATION undertakes the PRINTING and PUBLISHING of books, tracts, pamphlets, etc., upon advantageous terms to authors and others; also general printing for Psychological and other Societies at very reasonable rates. For specimens of work see "Ghostly Visitors," "Psychography," "More Forget-me-Nots," and all Mr. J. S. Farmer's books. Apply to Manager.

WORKS BY JOHN S. FARMER.

A New Basis of Belief in Immortality. This book was specially mentioned by Canon B. Wilberforce at the Church Congress. He said:—"The exact position claimed at this moment by the warmest advocates of Spiritualism is set forth ably and eloquently in a work by Mr. J. S. Farmer, published by E. W. Allen, and called 'A New Basis of Belief,' which, without necessarily endorsing, I commend to the perusal of my brethren." Mr. S. C. Hall, F.S.A., and Editor of the *Art Journal*, says:—"Your book is both useful and interesting; a very serviceable addition to the literature of Spiritualism." "One of the calmest and weightiest arguments, from the Spiritualist's side, ever issued. . . . Those desirous of knowing what can be said on this present-day question, by one of its ablest advocates, cannot do better than procure this volume."—*Christian World*. "This is an exceedingly thoughtful book; temperate, earnest, and bright with vivid and intelligent love of truth. Mr. Farmer is no fanatic, if we may judge of him by his book, but a brave seeker after the truth. . . . We commend his book to the attention of all who are prepared to give serious attention to a very serious subject."—*Truthseeker*. "Mr. Farmer writes clearly and forcibly."—*Literary World*. Printed on Superior Paper, Cloth, Bevelled Edges, with Portrait of Author. Price Three Shillings. Paper, One Shilling.

How to Investigate Spiritualism. A Collection of Evidence showing the possibility of Communion between the Living and the so-called Dead, with Hints and Suggestions to Inquirers, and other Useful Information. INTRODUCTION — What is Spiritualism? — The Rise of Modern Spiritualism—Its Progress—Theories—The Argument for the Spiritual Hypothesis—The Two Classes of Phenomena. Physical: Spirit Raps—Altering the Weight of Bodies—Moving Inanimate Objects without Human Agency—Raising Bodies into the Air—Conveying Objects to a distance out of and into Closed Rooms—Releasing Mediums from Bonds—Preserving from the Effects of Fire—Producing Writing or Drawing on Marked Papers placed in such Positions that no Human Hand can Touch them—Musical Instruments of Various Kinds Played without Human Agency—The Materialisation of Luminous Appearances, Hands, Faces, or Entire Human Forms—Spirit Photographs, etc., etc. Mental: Automatic Writing—Clairvoyance—Clairaudience—Trance Speaking—Impersonation—Healing—Concessions to Sceptics—Postulata—The Weight and Value of the Testimony—List of Names—Testimonies of Professors Challis, De Morgan, Wagner—Zöllner, Butlerof, Fechner, Scheibner, Weber, Hare, Crookes, Mapes, Gregory, Barrett—Testimonies of Sergeant Cox; Alfred Russel Wallace, Dr. Chambers, Dr. Robertson, Dr. Elliotson, Camille Flammarion, Leon Favre, Cromwell F. Varley, Lord Brougham, Nassau Senior, The Dialectical Committee—Thackeray, Archbishop Whately, and many others—Conclusions—The Literature of Spiritualism—Spiritualism not Conjuring—Advice to Inquirers. Price Sixpence, post free.

WORKS BY "M.A. (OXON.)"

Psychography. Second edition, with a new introductory chapter and other additional matter. Revised and brought down to date. Illustrated with diagrams. A collection of evidence of the reality of the phenomenon of writing without human agency, in a closed slate or other space, access to which by ordinary means is precluded. Cloth, Demy 8vo., Price 3s., and Paper, 1s.

Works Published by the Psychological Press Association. 3

Higher Aspects of Spiritualism: A Statement of the Moral and Religious Teachings of Spiritualism; and a Comparison of the present Epoch with its Spiritual interventions with the Age immediately preceding the Birth of Christ. "The work bears throughout those indications of careful investigation, a cordial admission of newly discovered truths, and an appeal to the loftiest sentiments of humanity, that have characterised the previous writings of its distinguished author."—*Banner of Light*, Boston, United States of America. The *Chicago Times* of June 19, 1880, which highly commends the tone and style of the book in a long review of nearly two closely-columns, says:—"The author does not weary the reader with spiritual communications conveyed to himself alone; he writes with exceptional clearness, candour, and cogency; he is a master of strong and graphic English; his logic is unassailable, and his spirit extremely suave, manly, and straightforward. He is a high authority among Spiritualists." Price 2s. 6d.

Spirit Teachings. First series, revised and corrected, forming a volume of some 300 pages, uniform with the second edition of "Psychography," Price 10s. 6d.

Spirit Identity. An argument for the reality of the return of departed human spirits, illustrated by many narratives from personal experience; together with a discussion of some difficulties that beset the inquirer. 5s.

The Day-Dawn of the Past: being a Series of Six Lectures on Science and Revelation. By "An Old Etonian." Contents.—1. Divine Truths and Human Instruments—2. In the Beginning: God—3. From Chaos to Cosmos—4. Foot-prints of the Creator—5. Man or Monkey?—6. The Cradle and the Grave. 200 pp., Crown 8vo., Profusely Illustrated, printed on good paper, and handsomely bound in thick bevelled cloth boards. Price 3s. 6d.; post free, 4s. "This excellent little volume."—*Public Opinion*.

Ghostly Visitors: a Series of Authentic Narratives. By "SPECTRE-STRICKEN." With an Introduction by "M.A. (Oxon.)." Printed on Superior Paper, Cloth Boards, Demy 8vo. Price 3s.; post free, 3s. 4d. Contents.—A Mother's Warning—A Mysterious Visitor—The Spectral Candle—The Spectral Carriage—Nugent's Story—Spalding's Dog—Gascoigne's Story—Anne Boleyn's Ghost at the Tower—A Prophetic Dream—The Spectre of Huddleston—Gordon's Story—The Fifeshire Story—The Wrecked Major—A Story of Second Sight—The Phantom Butler—The Haunted Convent—The Ghost of the Carmelite Friar—Footsteps on the Stairs—The Walled-up Door—The Butler's Ghost—The Mission Laundry—The Brown Lady of R.—The Mystery of Castle Caledonia—The Ghost Dressed in Blue—The Ayah's Ghost—The Supposed Burglar—A Considerate Ghost—Billy, the Ostler—The Old Eight day Clock—The Hidden Skeleton—The Headless Sentry—The Spectral Cavalcade—The Haunted Glen—Another Ghost who Nursed a Baby—The Old Clergyman's Ghost—The Haunted Rectories—The Haunted Chest—The Ghosts of Dutton Hall—The Death Secret—The Death Summons: A Remarkable Incident—A Haunted Billiard Room—"The Old Oak Chest"—Stories of Second Sight in the Island of Skye: Mrs. M——'s Story; Mr. M'K——'s Story; Mr. N——'s Story; Mrs. M'D——'s Story; Major C——n's Story; Miss M'A——r's Story—The Spectre Maiden—A Weird Story.

Immortality: Its People, Punishments, and Pursuits, with Five other Trance Addresses, delivered through the Mediumship of J. J. MORSE, at Goswell Hall, London. Among the subjects dealt with are:—Spiritualism: its Consolations—Deeds *versus* Dogmas—Concerning Angels—A Coming Creed—The Day of Judgment. It is a work that can be advantageously distributed among inquirers, and is well suited to Spiritualists, being the utterances of a Spirit upon the topics discussed. Cloth, 1s. 6d. Paper, 1s.

4   *Works Published by the Psychological Press Association.*

The Psychological Review. Bound Volumes can be had as follows:—
Vol. I. Old Series, 4 Parts, price 2s. 6d. each; or 10s. the set, unbound.
Vol. II. Only 4 complete sets of this vol. remain. Price, per set, unbound, £2. New Series. Vols. I. and II. Cloth Boards, 4s. 6d. each; Morocco, 10s. Vol. III., Cloth Boards, 7s. 6d.; Morocco, 12s.

"SIMPLE—EFFECTIVE."

The Psychograph (Registered). Supersedes the old-fashioned Planchette. Moves easily. Writes rapidly. Is better suited to its work than the more expensive instrument. Invaluable for Writing Mediums. Price 2s. 6d. Plain; or 3s. Polished. Post free, 3d. extra.

Recently Published. Price 12s. 6d.

The Perfect Way; or, The Finding of Christ. A New Gospel of Interpretation, solving the great problems of existence, and meeting the need of the age by reconstructing religion on a scientific, and science on a religious basis. "A grand book, by noble-minded writers, keen of insight, and eloquent in exposition; an upheaval of true spirituality. Were every man in London, above a certain level of culture, to read it attentively, a theological revolution would be accomplished."—*The* (Bombay) *Theosophist* (Buddhist Organ).

The Dervishes; or, Oriental Spiritualism. By JOHN P. BROWN, Secretary of the Legation of the United States of America at Constantinople. 12s.

Son, Remember; an Essay on the Discipline of the Soul beyond the Grave. By the Rev. JOHN PAUL, B.A. 3s. 6d.

Third Edition, 592 pp., Price 6s., by post, 6s. 6d.

Hafed Prince of Persia: His Experiences in Earth-Life and Spirit-Life. being Spirit Communications received through Mr. DAVID DUGUID, the Glasgow Trance-Painting Medium. With an Appendix containing Communications from the Spirit Artists, Ruisdal and Steen. Illustrated by many fac-similes of Drawings and Writings, the Direct Work of the Spirits. "'Hafed' is a book that will excite severe criticism and receive great praise. It furnishes intensely interesting reading, and at the same time requires patient and cultured study for its complete understanding. The sceptic will find it a weird and strange story; the Spiritualist will be charmed with its facts and philosophy."—HUDSON TUTTLE in *Religio-Philosophical Journal*. "We know something of the subjects treated of in this volume, and we know personally David Duguid; and of this fact we are certain, that, unaided by any power outside his own mind, he could no more have answered the questions put to him in the form in which they are answered in this volume than he could have written Bacon's *Organon*, Newton's *Principia*, or Shakespeare's Plays. . . . Every Spiritualist who can afford it ought certainly to purchase a copy." —*Spiritual Magazine.* The evening oration was entitled "What is Death?" given by Mrs. Nosworthy in that literary, poetical, and dramatic style which has long made her one of the most famous elocutionists in the kingdom; and the various quotations from "Hafed" on that important subject, proved a source of intense delight to the audience, as well as demonstrating that beautiful and charming work to be one of the greatest treasures yet vouchsafed to humanity. The subject-matter of Mrs. Nosworthy's lectures on Spiritualism is both original and select. Especially attractive are the quotations from "Hafed." In fact she has done much to make this truly charming book a universal favourite in Liverpool, Birkenhead, Rochdale, Southport, and other towns, from whence visitors have come to hear its beautiful teaching, elegant composition, and spiritual dignity publicly set forth with historical grace and dramatic splendour.—*Medium.*

Works Published by the Psychological Press Association.

The Koran, commonly called the Alcoran of Mahommed. Translated into English immediately from the original Arabic by GEORGE SALE. 10s. 6d.

The Spirit's Book. Containing the Principles of Spiritist Doctrine on the Immortality of the Soul, etc., etc., according to the Teachings of Spirits of High Degree, transmitted through various Mediums, collected and set in order by ALLEN KARDEC. Translated from the 120th thousand by ANNA BLACKWELL. Crown 8vo, pp. 512, cloth. 7s. 6d.

The Medium's Book; or, Guide for Mediums and for Evocations. Containing the Theoretic Teachings of Spirits concerning all kinds of Manifestations, the Means of Communication with the Invisible World, the Development of Medianimity, etc., etc. By ALLEN KARDEC. Translated by ANNA BLACKWELL. Crown 8vo., pp. 456, cloth. 7s. 6d.

Mythology and Popular Traditions of Scandinavia, North Germany, and the Netherlands. By BENJAMIN THORPE. In three vols. 18s.

Heaven and Hell; or, The Divine Justice Vindicated in the Plurality of Existences. By ALLEN KARDEC. Translated by ANNA BLACKWELL. Crown 8vo, pp. viii. and 448, cloth. 7s. 6d.

Childhood of the World. By EDWARD CLODD, F.R.A.S. Special Edition. 1s.

Footfalls on the Boundary of Another World. With Narrative Illustrations. By ROBERT DALE OWEN. An enlarged English Copyright Edition. Post 8vo, pp. xx. and 392, cloth. 7s. 6d.

The Debatable Land between this World and the Next. With Illustrative Narrations. By ROBERT DALE OWEN. Second Edition. Crown 8vo, pp. 456, cloth. 7s. 6d.

Apparitions; An Essay Explanatory of Old Facts and a New Theory. To which are added Sketches and Adventures. By NEWTON CROSLAND. Crown 8vo, pp. viii. and 166, cloth. 2s. 6d.

Pith; Essays and Sketches, Grave and Gay; with some Verses and Illustrations. By NEWTON CROSLAND. Crown 8vo, pp. 310, cloth. 5s.

Evidences of Spiritualism, Hints for the. Second Edition. Crown 8vo, pp. viii. and 120, cloth. 2s. 6d,

An Essay on Spiritual Evolution, considered in its bearings upon Modern Spiritualism, Science, and Religion. By J. P. B. Crown 8vo, pp. 156, cloth. 3s.

Evenings at Home in Spiritual Seance. Prefaced and welded together by a species of Autobiography. By Miss HOUGHTON. First and Second Series. Crown 8vo, cloth. 10s. 6d.

Chronicles of Spirit Photography. By Miss HOUGHTON. Crown 8vo, cloth. 10s. 6d.

Life after Death, from the German of GUSTAV THEODORE FECHNER. By HUGO WERNEKKE. 2s. 6d.

The Childhood of Religions, including a Simple Account of the Birth and Growth of Myths and Legends. By EDWARD CLODD, F.R.A.S. Crown 8vo. 5s.

Religion and Science, The Reconciliation of. Being Essays on Immortality, Inspiration, Miracles, etc. By T. W. FOWLE, M.A. Demy 8vo. 10s. 6d.

The Devil: His Origin, Greatness, and Decadence. By A. ROVILLE. Second Edition. 12mo, cloth. 2s.

Mind; A Quarterly Review of Psychology and Philosophy, contributions by Mr. Herbert Spencer, Professor Bain, Mr. Henry Sidgwick, and others. Vols. I. to VI.—1876 to 1882. Each 12s.; cloth, 13s. 6d. 12s. per annum, post free.

Principles of Psychology. Fourth Thousand. 2 vols. 36s.

New Life of Jesus for the People. The Authorised English Version. 2 vols., 8vo, cloth. 24s.

The Psychonomy of the Hand; or, the Hand an Index of Mental Development. With thirty-one Tracings from Living and other Hands. By R. BEAMEST, F.R.S. Demy 4to, cloth. 7s. 6d.

Our Eternal Homes. Contents: What is Heaven?—Guardian Angels—Heavenly Scenery—Death the Gate of Life—Do the Departed Forget us?—Man's Book of Life—Infants in Heaven. Crown 8vo. 2s. 6d.

Astrology, Introduction to Selby's. A new and improved edition by ZADKIEL, with his Grammar of Astrology and Tables of Nativities. 5s.

Religious System of the Amazulu. By the Rev. Canon CALLAWAY, M.D. 8vo. Part I. Unkulunkulu; or, the Traditions of Creation as existing among the Amazulu and other tribes of South Africa, in their own words, with a translation into English, and Notes. Pp. 128, sewed. 1868. 4s.—Part II. Amatongo; or, Ancestor Worship, as existing among the Amazulu, in their own words, with a translation into English, and Notes. Pp. 197, sewed. 1869. 4s.—Part III. Izinyanga Zokubula; or, Divination, as existing among the Amazulu, in their own words, with a translation into English, and Notes. Pp. 150, sewed, 1870. 4s.—Part IV. Abatakati, or Medical Magic and Witchcraft. Pp. 40, sewed. 1s. 6d.

Esoteric Buddhism. By A. P. SINNETT. Cloth, second edition, 7s. 6d. *Contents:* Esoteric Teachers—The Constitution of Man—The Planetary Chain—The World Periods—Devachan—Kama Loca—The Human Tide-Wave—The Progress of Humanity—Buddha—Nirvana—The Universe—The Doctrine Reviewed.

Reflections on the Character and Spread of Spiritualism. By BENJAMIN WILLS NEWTON. Second edition. 1s. 6d.

Trance and Muscle-Reading. By C. M. BEARD, M.D. 2s.

Modern Christianity a Civilised Heathenism. 2s.

Natural Law in the Spiritual World. By HENRY DRUMMOND, F.R.S.E., F.G.S. 7s. 6d.

Lancashire Folk-Lore: The Superstitious Beliefs and Practices, Local Customs and Usages. By JOHN HARLAND and T. T. WILKINSON. 5s.

The Legends and Myths of Guiana. 5s.

The Probable Effects of Spiritualism upon the Social, Moral, and Religious Condition of Society. Two prize essays by Anna Blackwell and Mr. G. F. Green. These essays won the first and second gold medals of the British National Association of Spiritualists. Price 6d.; post free, 7½d.; cloth.

Mesmerism, with Hints for Beginners. By Captain JOHN JAMES. Price 2s. 6d. Crown 8vo. Cloth. Red edges.

The Science of the Stars. By ALFRED J. PEARCE, author of "The Text Book of Astrology," etc. A popular epitome of Astrology in all its branches. Containing Tables of Houses for London, Edinburgh, Calcutta, and New York. Also horoscopes of eminent personages. Price 5s. Bound.

Works Published by the Psychological Press Association. 7

Heaven Opened, a Series of Communications from Children in the Spirit-World to their Brothers and Sisters on Earth. By F. J. T., with Observations by Mrs. DE MORGAN. Paper 4d., cloth, 1s.

The Nature of Spirit and of Man as a Spiritual Being. Lectures by the Rev. CHAUNCEY GILES. Lecture I.—"The Nature of Spirit and of the Spiritual World." Lecture II.—"Man essentially a Spiritual Being." Lecture III.—"The Death of Man." Lecture IV.—"The Resurrection of Man." Lecture V.—"Man in the World of Spirits." Lecture VI.—"The Judgment of Man." Lecture VII.—"Man's Preparation for his Final Home." Lecture VIII.—"The State of man in Hell." Lecture IX.—"Man in Heaven." Appendix—"The Lord's Resurrection not Material." 186 pp., boards, 1s.; cloth 1s. 6d.

The Report of the London Dialectical Society's Committee on Spiritualism. Together with a full Account of the Proceedings of Committee, the Reports and Minutes of the Experimental Sub-committees, and the Evidence, *pro and contra*, of the following eminent Persons:—"Lord Lytton, Lord Lindsay, Lord Borthwick; the Countess de Pomar; Professor Huxley, Professor Tyndall; Drs. W. B. Carpenter, Chambers, Davey, Dixon, Edmunds, Kidd, Robertson, Garth Wilkinson; Mr. Serjeant Cox; Messrs. Edwin Arnold, Henry G. Atkinson, Laman Blanchard, Chevalier Damiani, Leon Favre, Camille Flammarion, Hain Friswell, D. D. Home, William Howitt, H. D. Jencken, George Henry Lewes, Hawkins Simpson, J. Murray Spear, T. Adolphus Trollope, Cromwell Varley, A. R. Wallace, W. M. Wilkinson; Mesdames Anna Blackwell, Hardinge, Houghton, etc., etc., etc. 5s.

Now in Press.

19th Century Miracles, or, Spirits and their Work in every Country of the Earth. The plan of the work includes:—Introduction, Spiritualism in Germany, France, Great Britain, Australia, New Zealand, Polynesia and West Indian Islands, Cape Town, South America, Mexico, China, Japan, Thibet, India, Holland, Java, Dutch Colonies, Russia, Sweden, Switzerland, Scandinavia, Spain, Italy, Austria, Belgium, Turkey, &c., &c., and America. 15s.

Rivers of Life; or, Sources and Streams of the Faiths of Man in all Lands, showing the Evolution of Religious Thought from the Rudest Symbolisms to the Latest Spiritual Developments. By MAJOR GENERAL J. G. R. FORLONG, F.R.G.S., F.R.S.E., M.A.I., A.I.C.E., F.R.H.S., F.R.A. SOCY., etc., etc. In two volumes, demy 4to, embracing 1278 pages, with maps, plates, and numerous illustrations, cloth; and large separate Chart in cloth case on roller, price £6 6s.; separate Charts in case or on roller, £2 each.

Good Health and how to Secure it. By R. B. D. WELLS. Bound in cloth with "Food," by the same author. Price 3s. 6d.

Swedenborg's Heaven and Hell, being a Relation of Things Heard and Seen, 8vo, whole bound morocco extra, full gilt. Price 10s. 6d.

Spirits before our Eyes. By WILLIAM H. HARRISON. This book shows that one section at least of the phenomena of Spiritualism is produced by the spirits of departed human beings, who have passed over the river of Death. It contains a great number of well attested facts, proving that the said spirits are the persons they say they are. The work, from beginning to end, is full of evidence of Spirit Identity. The Author attempts to prove the Immortality of Man by strictly scientific methods, giving well-proved facts first, and conclusions which naturally flow from them afterwards. Price 5s. 6d., post free.

The Autobiography of Satan. By Rev. Dr. JOHN R. BEARD. 1872. Crown 8vo, cloth, 7s. 6d.

The Pioneers of the Spiritual Reformation. Life and Works of Dr. Justinus Kerner (adapted from the German). William Howitt and his Work for Spiritualism. Biographical Sketches. By ANNA MARY HOWITT WATTS. *Contents*—LIFE OF DR. JUSTINUS KERNER : Birth and Parentage—At College—In the Black Forest and in the Welzheim Forest—At Weinsberg—The Last Days of Justinus Kerner—Kerner's Home—Specimens from the Works on Psychology of Dr. Kerner—Periodicals Edited by Dr. Kerner, with other later Works on Psychology—Some Researches after Memorials of Mesmer in the Place of his Birth—Mesmer's first practical career as a Physician, together with strange Experiences in Hungary—Statement made by the Father of. the Blind Girl Fraulein Paradis, regarding whom Mesmer suffered Persecution—Mesmer's Twenty-seven Aphorisms, containing in brief the Substance of his Discoveries—Mesmer's Departure from Vienna : Journey to Munich, and sojourn in Paris—Regarding Mesmer's Followers and Opponents in Germany ; also Regarding the Gradual Development of Animal Magnetism, and the Publication of Mesmer's Collected Writings—Mesmer's Last Years.——WILLIAM HOWITT AND HIS WORK FOR SPIRITUALISM : His Childhood—Early Manhood—Noon-Day of Life—Pioneer in the Gold-Fields of the Visible and Invisible Worlds—His Psychological Experiences—His Work for Spiritualism—Bright Sunset of Life. Price 10s.

On Exalted States of the Nervous System: an (alleged) Explanation of the Mysteries of Modern Spiritualism, Dreams, Trance, Somnambulism, Vital Photography, Faith, Will, Origin of Life, Anæsthesia, and Nervous Congestion. By ROBERT H. COLLYER, M.D. 2s.

Lectures on the Origin and Growth of Religion, as illustrated by the Religion of Ancient Egypt. By P. LE PAGE RENOUF. (Hibbert Lectures, 1879.) 8vo, cloth, 10s. 6d.

The Devil: his Origin, Greatness, and Decadence. By Rev. Dr. A. REVILLE. Translated from the French. 2nd Edition. 1877. 12mo, cloth, 2s.

WORKS BY MISS F. POWER COBBE.

The Peak of Darien, and other inquiries touching concerns of the soul and the body. 1882. Crown 8vo, cloth, 7s. 6d.

The Duties of Women. A course of lectures delivered in London and Clifton. 2nd Edition. 193 pp. 1882. Crown 8vo, cloth, 5s.

Alone to the Alone. Prayers for Theists, by several Contributors. 3rd Edition. 1882. Crown 8vo, cloth, gilt edges, 5s.

Broken Lights. An inquiry into the present condition and future prospects of Religious Faith. 3rd Edition. 1878. Crown 8vo, cloth, 5s.

Dawning Lights. An inquiry concerning the Secular Results of the New Reformation. 1868. 8vo, cloth, 5s.

Re-echoes. 1876. Crown 8vo, cloth, 7s. 6d.

Darwinism in Morals, and (13) other Essays (Religion in Childhood, Unconscious Cerebration, Dreams, the Devil, Auricular Confession, etc., etc.) 400 pp. 1872. 8vo, cloth, 5s.

Religious Duty. 1864. 8vo, cloth, 5s.

Studies, New and Old, of Ethical and Social Subjects. 1865. 8vo, cloth, 5s.

Italics. Brief notes on Politics, People, and Places in Italy in 1864. 8vo, cloth, 5s.

Hours of Work and Play. 1867. 8vo, cloth, 2s. 6d.

Thanksgiving. A chapter of Religious Duty. 1863. 12mo, cloth, 1s.

The Other World: or Glimpses of the Supernatural, being Facts, Records, and Traditions relating to Dreams, Omens, Miraculous Occurrences, Apparitions, Wraiths, Warnings, Second Sight, Necromancy, Witchcraft, &c. 2 vols. A new edition, by F. G. LEE, D.C.L. 15s.

Ancient Pagan and Modern Christian Symbolism Exposed and Explained. By THOMAS INMAN, M.D. Second Edition. With Illustrations. Demy 8vo, cloth. Pp. xl. and 148. 1874. 7s. 6d.

Oriental Religions, and their Relation to Universal Religion. By SAMUEL JOHNSTON. First Section—India. In 2 vols., post 8vo, cloth. Pp. 408 and 402. 21s.

Demonology and Devil Lore. By MONCURE D. CONWAY, M.A. Two vols. royal 8vo, with 65 Illustrations. 28s.

The Rosicrucians: Their Rites and Mysteries. With Chapters on the Ancient Fire and Serpent Worshippers. By HARGRAVE JENNINGS. Crown 8vo, cloth extra, with five full page plates, and upwards of 300 Illustrations. 7s. 6d.

Hopes of the Human Race, Hereafter and Here. Essays on the Life after Death. Second edition, crown 8vo, cloth. 5s.

Origin and Growth of Religion, Lectures on the. 8vo, cloth. Price 10s. 6d.

Lost and Hostile Gospels. An Account of the Toledoth Jesher, two Hebrew Gospels circulating in the Middle Ages, and extant fragments of the Gospels of the first three centuries, of Petrine and Pauline origin. By S. BARING GOULD, M.A. Crown 8vo, cloth. 7s. 6d.

Spirit People. A scientifically accurate description of Manifestations produced by Spirits, and simultaneously witnessed by the Author and other observers in London. By WILLIAM H. HARRISON. Limp cloth, red edges. Price 1s.; post free, 1s. 1d.

Rifts in the Veil. A collection of choice poems and prose essays given through mediumship, also of articles and poems written by Spiritualists. A useful book to place in public libraries, and to present or lend to those who are unacquainted with Spiritualism, and is one of the most refined and elegant works ever published in connection with the subject. *Rifts in the Veil* contains articles or poems by Florence Marryat, Prince Emille of Sayn-Wittgenstein, Baroness Adelma Von Vay, Mr. William White, Mr. Gerald Massey, Mrs. Makdougall Gregory, Mrs. Eric Baker, Mr. Alexander Calder, President of the British National Association of Spiritualists, Mr. Alfred Russel Wallace, F.R.G.S., and other able writers. Price 5s., richly gilt, post free.

Researches into the Phenomena of Spiritualism. By WILLIAM CROOKES, F.R.S., &c. I. "Spiritualism viewed by the Light of Modern Science," and "Experimental Investigations in Psychic Force."—II. "Psychic Force and Modern Spiritualism:" a reply to the *Quarterly Review* and other critics.—III. "Notes on an inquiry into the phenomena called Spiritual, during the years 1870-1873." 16 Illustrations. Price 5s.

Lectures. By Col. ROBERT INGERSOLL, of America. Ghosts 4d. Mistakes of Moses, 3d. What must I do to be saved? 3d. The Christian Religion, 3d. The Religion of the Future, 2d. Farm Life, 1d. Any of the above post free 1 halfpenny extra. All the above post free for 16 penny stamps.

Clairvoyance, Hygienic and Medical. By Dr. DIXON. 1s. "The sight being closed to the External, the Soul perceives truly the affections of the body."—HIPPOCRATES.

Clairvoyance. By ADOLPHE DIDIER. Remarkable facts from thirty-five years' personal exercise of the Clairvoyant faculty.

Rationale of Spiritualism. By F. F. COOK (of Chicago). Second edition, 24 pages, price 2d., by post, 2½d. "'Spiritualism,' as Mr. F. F. Cook points out in his very able paper, 'is Revolution, not simply Reform.' There is very little conservative about it; little that is orderly, any more than there was in the Great Revolution that left us Christianity. It is an upheaval, and it is attended with all the apparent disorder and chaotic confusion of an earthquake."—"M.A. (Oxon.)" in "Higher Aspects of Spiritualism."

Life and Works of Mencius. Translated into English from the Chinese Classics by JAMES LEGGE, D.D., LL.D. 12s.

A Forecast of the Religion of the Future. By W. W. CLARK. New edition, price 3s. 6d. *Contents:* The Philosophy of Evil and Suffering—Conscience: its Place and Function—Religion and Dogma—Psychism and Spiritualism—The Philosophy of Inspiration and Revelation—Christianity: its Divine and Human Elements. "The series forms a chain of condensed reasoning and argument—so far as we know, unique of the kind."—*Light.*

Awas-I-Hind; or, a Voice from the Ganges. By an Indian Officer. 5s.

Philosophy of Mesmerism and Electrical Psychology. Eighteen Lectures by Dr. JOHN BOVEE DODS, including the Lecture on "The Secret Revealed, so that all may know how to Experiment without an Instructor." Price 3s. 6d. "Mesmerism is a stepping-stone to the study of Spiritualism."—GEORGE WYLD, M.D., in his Evidence in the Case of Dr. Slade at Bow Street.

Myths and Myth-makers. Old Tales and Superstitions interpreted by Comparative Mythology by JOHN FISKE, M.A. 10s. 6d.

Anacalypsis: an Attempt to draw aside the Veil of the Saitic Isis; or, An Inquiry into the Origin of Languages, Nations, and Religions. By GODFREY HIGGINS, Esq., F.S.A., F.R. Asiat. Soc. (late of Skellow Grange, near Doncaster). This magnificent work has always been scarce, but is now out of print. To be completed in 16 Parts. Parts I. to V. now ready, 2s. 6d. each. Vol. I. 12s. 6d.

Life and Teachings of Confucius. Translated into English with Preliminary Essays and Explanatory Notes by JAMES LEGGE, D.D. 10s.

Miracles and Modern Spiritualism. By ALFRED R. WALLACE, F.R.G.S., F.Z.S., Author of "Travels on the Amazon and Rio Negro," "Palm Trees of the Amazon," "The Malay Archipelago," etc., etc. Embracing: I.—"An Answer to the Arguments of Hume, Lecky, and others against Miracles." II.—"The Scientific Aspects of the Supernatural," much enlarged, and with an Appendix of Personal Evidence. III.—"A Defence of Modern Spiritualism," reprinted from the *Fortnightly Review*. Price 5s.

Theosophy and the Higher Life; or, Spiritual Dynamics, and the Divine and Miraculous Man. By G. WYLD, M.D., Edin., President of the British Theosophical Society. Contents: I. The Synopsis. II. The Key to Theosophy. III. Spiritual Dynamics. IV. Man as a Spirit. V. The Divine and Miraculous Man. VI. How best to become a Theosophist. VII. Can Anæsthetics Demonstrate the Existence of the Soul? VIII. The British Theosophical Society. 138 pp. cloth, 3s.

Nursery Tales, Traditions, and Histories of the Zulus. By the Rev. HENRY CALLAWAY, M.D. In six parts. 16s.

Works Published by the Psychological Press Association. 11

The Occult World. By A. P. SINNETT. Contents: Introduction. Occultism and its Adepts. The Theosophical Society. Recent Occult Phenomena. Teachings of Occult Philosophy. 172 pp., cloth, price 3s. 6d.

The Alpha; or, The First Principles of the Human Mind. A Revelation but no Mystery. By EDWARD N. DENNYS. With Spiritual Advent given through the Mediumship of Mr. J. J. Morse, and steel engraving of Author. Fourth edition, cloth, 360 pp. 8vo, 3s. 6d. "We can call to mind few books fit to be its fellow."—*Athenæum.* "It contains more truth, poetry, philosophy, and logic, than any work we have ever read; it is a new revelation, and one of the most remarkable productions ever given to the world."—*Cosmopolitan.*

Life Lectures. By EDWARD N. DENNYS. With steel engraving of Author. Cloth, 460 pp. 8vo, price 3s. 6d.

The Religion of Jesus compared with the Christianity of To-Day. By FREDERICK A. BINNEY. Reduced to 6s:, post free; published at 7s. 6d. "Well worthy of the attentive consideration of the clergy of all denominations, as showing in what direction a strong current of opinion is unmistakeably setting among a large class of earnest and thoughtful men. The author must be credited with a more than average share of candour, reasonableness, and love of truth."—*The Scotsman*, May 26th, 1877. "He thanks Mr. Greg for a large portion of his iconoclasm, but when that pervervid opponent of orthodoxy seeks to shatter the long-cherished hopes of immortality, Mr. Binney gives him a powerful thrashing."—*Newcastle Daily Chronicle.*

Back to the Father's House. A Parabolic Inspiration. Parts 1 to 6, 1s. each. Parts 7 to 12, 1s. 6d. each.

Christ and Buddha Contrasted. The new Leek Bijou reprint (Buddhistic Spiritualism). By an Oriental who visited Europe. 150 pages; price 3d., post free, 4d. "It will do an immense deal of good. I am no Buddhist, but some of their philosophy is very fine. Any way, it is of great service to have Buddhism and Christianity compared in this popular and intelligible way. Some of the Author's pithy definitions of the Christian belief are unsurpassed and terribly true."—*A Spiritualist.*

History of the Council of Nice, A.D. 325. By DEAN DUDLEY. 4s. 6d.

Private Practical Instructions in the Science and Art of Organic Magnetism (Third Edition, just published). By MISS CHANDOS LEIGH HUNT. Being her original *Three* Guinea *private* Manuscript Instructions, printed, revised, and greatly enlarged. Valuable and practical translations, and the concentrated essence of all previous practical works. Numerous illustrations of passes, signs, etc.—*Contents:* Organic Magnetism, its Nature. Development of the Magnetic Power. Processes of Magnetising, Mesmerising, Electro-Biologising, Psychologising, Hypnotising, Statuvolising, Comatising, Fascinating, Entrancing, etc., etc. How to become a Professional Public and Private Demonstrator of Magnetic Somnambulism. Thought-Reading, Clairvoyance, and Phreno-Magnetism. How to become a Professional Healer. How to conduct an Institution for Performing Surgical, Dental, and Midwifery cases Painlessly. How to Magnetise Animals and Plants. Instructions to Sensitives. How to Induce and Develop their Powers. The Dangers of Magnetism. Thirty-three Miscellaneous Fragments of Recapitulatory Cautions, Curiosities in Magnetism, etc. List of over one hundred English Works upon Organic Magnetism, and where to obtain them.—Price One Guinea, paper. French Morocco, with double lock and key, 5s. extra; best Morocco, ditto, 7s. extra.

Oriental Religions. JOHNSON. 24s.

Mechanism of Man: Life, Mind, Soul. A popular introduction to Mental Physiology and Psychology. By the late Serjeant Cox, S.L., Pres. Psychological Society of Great Britain. Vol. I. 10s. 6d; Vol. II. 12s. 6d.

Practical Instructions in Animal Magnetism. By J. P. F. DELEUZE; translated by THOMAS HARTSHORN. Cloth, 12mo., 8s. 6d.

Bob and I; or, Forget-me-Nots from God's Garden. By F. J. THEOBALD. Price 1s. 6d. "One of the most beautiful little stories I ever read."—Dr. MAURICE DAVIES in *Kensington News*.——"Conventional Christianity will be alternately shocked and bewildered by the broad and mystical things in this singular but fascinating little book."—*Hastings News*.

More "Forget-me-Nots from God's Garden" (Sequel to "Bob and I"). By F. J. THEOBALD. Author of "Heaven Opened." Price 2s., post free, 2s. 3d.

Keys of the Creeds. 5s.

The Wheel of the Law. ALABASTER. 14s.

Between the Lights. By LISETTE EARLE. Price 3s. 6d. This is a charming book which every Spiritualist should read. We have a strong suspicion that it is not altogether the *conscious* work of the lady whose name it bears, but that it has been given "inspirationally" by a Spirit of high and noble character. The stories have nothing in common with the ordinary novels of the day, but are full of sweet and elevating thoughts, at the same time that they abound in incidents which sustain the reader's interest to the end.

Three Lectures on Buddhism. EITEL. 5s.

Esoteric Theosophy. The entire subject, *pro* and *con*, argued out in a masterly style by an eminent writer, in a handsomely printed pamphlet of 108 pages, which is published for the benefit of the Theosophical Society. This work is even more remarkable and interesting than Mr. Sinnett's "Occult World." 3s.

History of American Socialisms. NOYES. 18s.

The Light of Asia, or the Great Renunciation. Being the Life and Teaching of Gautama, Prince of India and Founder of Buddhism (as told in verse by an Indian Buddhist). By EDWIN ARNOLD, C.S.I., Officer of the Order of the White Elephant of Siam, etc., etc. Sixth Edition. 2s. 6d.

Transcendental Physics. By Professor ZÖLLNER; translated by C. C. MASSEY. Price 3s. 6d., or post free, 4s. Containing all the original illustrations, and perhaps the most valuable book at the price ever issued in connection with Spiritualism.

A Philosophy of Immortality. By the Hon. RODEN NOEL, author of "A Little Child's Monument," etc. 7s. 6d.

Spirits before our Eyes, Vol. I. A book on Spontaneous Apparitions in private families. By W. H. HARRISON. 5s. 6d.

A Glance at the Passion Play. By Captain R. F. BURTON. With a frontispiece. 5s. 6d.

A Clergyman on Spiritualism, with a Preface by LISETTE MARK-DOUGALL GREGORY. 1s.

Psychic Facts; A Collection of Authoritative Evidence demonstrating Psychical Phenomena. 5s.

Epes Sargent, Portrait of. *The Psychological Review* Series, Carte-de-Visite, or unmounted 1s. Postage and packing in thick boards, 2d.

Religions of the World. LEIGH. A well-written little book. 2s. 6d.

Threading my Way: an Autobiography. By ROBERT DALE OWEN. 7s. 6d.

Travels of Fah-Hian, and Sun-Yun, Buddhist Pilgrims, from China to India (400 A.D. and 518 A.D.) Translated from the Chinese by SAMUEL BEAL, B.A., Trin. Coll. Cam. 10s. 6d.

Romantic History of Buddha. BEAL. 12s. 6d.

Attempts at Truth. By St. GEORGE STOCK. *Contents*: The Two Schools of Thought—Why must I do what is Right?—What is Right?—Hume on Miracles—The Mediumship of the Emperor Vespasian—Positive View of Spiritualism and the Philosophy of Force—Value of à priori Reasoning in Theology—Theism—A New Religion—The Bearings of Spiritualism—A "Test" from the Delphic Oracle—Materialism and Modern Spiritualism—What is Reality? Berkeley and Positivism—Illusion and Delusion—Where is Heaven? Price 5s.

Catena of Buddhist Scriptures. BEAL. 15s.

The Man of the Future. By A. CALDER. *Contents:* Preface—The Goal of Life—Blind Guides—Stumbling Blocks in the Way—Health of Body—Health of Soul—The Morality of Individuals—The Morality of Societies—Prospect and Retrospect. Price, cloth boards, 10s. 6d.

The Coming Era. By A. CALDER. 10s. 6d.

History of Magic (ENNEMOSER'S). Translated by WILLIAM HOWITT. To which is appended some most remarkable and well-authenticated stories of Apparitions, Dreams, Second Sight, Somnambulism, Predictions, Divinations, Witchcraft, Vampires, Fairies, Table Turning, &c. 2 vols., 10s.

Fireside Stories, Told by my Grandmother. 1. The Stroke of Fortune—2. A Life's Experience—3. Dora; or, the Evil Eye—4. Truth *versus* Grundy—5. Eva's Inheritance. By EDITH SAVILLE. Price 3s. 6d.

The New Earth: a Spiritual Essay. All who are interested in tracing the fulfilment of prophecy should read the little book. Price 1s.

Physiology for the Young; or the House of Life: Human Physiology with its application to the Preservation of Health. With numerous illustrations, by Mrs. FENWICK MILLER. 2s. 6d.

A Woman's Work in Water Cure and Sanitary Education. By Mrs. MARY S. G. NICHOLS. Foolscap 8vo., 158 pages. Treating of: My Education and Mission—Health and Hydropathy—The Processes of Water Cure—Water Cure in Acute Diseases—In Chronic Diseases—The Disease of Infancy—Female Diseases—Gestation and Parturition—Consumption—Cholera—Miscellaneous and Malvern Cases—Health Mission to American Convents—Sanitary Education. Cloth, 1s. 6d.; paper, 1s.

Autobiography of Satan. By JOHN R. BEARD. Price 7s. 6d.

The Unseen Universe; or, Physical Speculations on a Future State. By B. STEWART and P. G. TAIT. Eighth Edition. Price 6s.

Esoteric Anthropology (The Mysteries of Man). A comprehensive and confidential treatise on the Structure, Functions, Passional Attractions and Perversions, True and False Physical and Social Conditions, and the most Intimate Relations of Men and Women. Anatomical, Physiological, Pathological, Therapeutical, and Obstetrical; Hygienic and Hydropathic. 5s.

Human Physiology the Basis of Sanitary and Social Science. Crown 8vo., 496 pages, 70 engravings. In six parts—The Actual Condition of Humanity; Matter, Force, and Life; The Human Body; The Laws of Generation; Health, Disease, and Cure; Morals of Society. 6s.

(Just added. Never before translated into English.)

Genesis: The Miracles and Predictions according to Spiritism. By ALLAN KARDEC, author of "The Spirits' Book," "Book on Mediums," and "Heaven and Hell." The object of this book is the study of three subjects—Genesis, Miracles, and Prophecies—and the work presents the highest teachings thereon received during a period of several years by its eminent author through the mediumship of a large number of the very best French and other mediums. The books of ALLAN KARDEC upon Spiritualism attained an immense circulation throughout France, and were received with great favour by all classes. In this work, here for the first time presented in English, it is conceded by everyone he has far surpassed all his previous efforts, and effectually cleared up the mystery which has long enshrouded the history of the progress of the human spirit. The ground taken throughout is consistent, logical, and sublime; the ideas of Deity, human free agency, instinct, spirit-communion, and many other equally profound and perplexing subjects incomparably grand. The iconoclasm of KARDEC is reverential; his radicalism constructive, and his idea of the divine plan of nature a perfect reconciliation of scientific with religious truth; while his explanation of miracles and prophecy in harmony with the immutable laws of nature carries with it the unmistakable impress of an unusually exalted inspiration. Whatever view may be taken of the author's conclusions, no one can deny the force of his arguments, or fail to admire the sublimity of a mind devoting itself through the best years of an earthly existence to intercourse with the denizens of the spirit-world, and to the presentation of the teachings thus received to the comprehensions of all classes of readers. The book will be hailed by all Spiritualists, and by those as well who, having no belief in Spiritualism, are willing to consider its claims and to read what may be said in support of their truth, as a valuable addition to a literature that embraces the philosophies of two worlds, and recognises the continuity of this life in another and higher form of existence. Price 6s. 6d.

Lives of the Necromancers. By WILLIAM GODWIN. Post 8vo., cloth limp, 2s.

Elizabethan Demonology: An Essay in Illustration of the Belief in the Existence of Devils, and the Powers possessed by Them. By T. ALFRED SPALDING, LL.B. Crown 8vo., cloth extra, 5s.

A Little Pilgrim: In the Unseen. Eighth Thousand. Crown 8vo. 2s. 6d.

Religion and Science, History of the Conflict between. Fourth Edition. By Professor DRAPER. 5s.

Phrenological Pamphlets. By L. N. FOWLER. All 1d. each.

Formation of Character.	Utility of Phrenology.
Tact and Talent.	Temperaments.
Self Knowledge.	How to Train a Child.
Language of the Faculties.	Temperance in a Nutshell.
Love, Courtship, and Marriage.	Self-Made Men.
Education.	John Bull and Brother Jonathan.
Proofs of Phrenology.	Memory.
Objections to Phrenology.	How to Succeed in the World.
Thinkers, Authors, and Speakers.	Perfection of Character.
Health, Wealth, and Happiness.	How to Read Character.
The Moral Laws and Duties.	

Library of Mesmerism and Psychology: Mesmerism, Clairvoyance, Electrical Psychology, Fascination, Science of the Soul, etc., etc. 15s.

The Philosophy of Mesmerism and Electrical Psychology. By Dr. J. B. Dods. A New Edition, cloth, 3s. 6d.

Clairvoyance, Hygienic and Medical. By Dr. Dixon. 1s.

Clairvoyant Travels in Hades: or the Phantom Ships. By A. Gardner. 3d.

Mesmerism in Connection with Popular Superstition. By J. W. Jackson. Stiff paper, 1s.

Mental Cure: illustrating the Influence of Mind on the Body in Health and Disease, and the Psychological Mode of Treatment. By R. F. Evans. Cloth, 3s.

Statuvolism: or, Artificial Somnambulism, hitherto called Mesmerism or Animal Magnetism. Containing a brief historical survey of Mesmer's operations, and the examination of the same by French Commissioners. Phreno-Somnambulism; or, the Exposition of Phreno-Magnetism and Neurology—a new division of the Phrenological Organs, etc., etc. By Wm. Baker Fahnestock, M.D. 6s.

Psychopathy, or The True Healing Art. By Joseph Ashman. With Photograph of the Author, by Hudson, showing a halo of healing aura over his hands. Second Edition, cloth, 2s. 6d.

Vital Magnetic Cure: being an Exposition of Vital Magnetism, and its Application to the Treatment of Mental and Physical Disease. By a Magnetic Physician. Cloth, 7s. 6d.

The Electric Physician: or, Self-Cure through Electricity. A Plain Guide to the use of Electricity, with accurate directions for the Treatment and Cure of various Diseases, chronic and acute. By Emma Hardinge-Britten, Electric Physician. 2s. 6d.

Health Hints; showing how to acquire and retain bodily symmetry, health, vigour, and beauty. A New Book for Everybody. Now ready, in neat cloth, eighty pages, price 1s.—Table of Contents: Chap. I., Laws of Beauty. Chap. II., Hereditary Transmission. Chap. III., Air, Sunshine, Water, and Food. Chap. IV., Work and Rest. Chap. V., Dress and Ornament. Chap. VI., The Hair and its Management. Chapter VII., The Skin and Complexion. Chap. VIII., The Mouth. Chap. IX., The Eyes, Ears, and Nose. Chap. X., The Neck, Hands, and Feet. Chap. XI., Growth, Marks, etc., that are enemies of Beauty. Chap. XII., Cosmetics and Perfumery.

Healing by Laying on of Hands. An Exposition of the Art of Healing by Manipulation. By Dr. J. Mack. Cloth, 6s. 6d.

Night-Side of Nature; or, Ghosts and Ghost-Seers. By Catherine Crowe. Cloth, 6s. 6d.

Old Truths in a New Light; or, An Earnest Endeavour to Reconcile Material Science with Spiritual Science and with Scripture. By the Countess of Caithness. 18s.

The Four Gospels explained by their writers. Edited by J. B. Roustaing. 3 vols. Price 15s. This work is a further development of the religious philosophy of which the first principles are laid down in the writings of Allan Kardec.

The Spirit Disembodied. "When we die we do not fall asleep: we only change our place." By Herbert Broughton. Price 5s.

Mystical Philosophy and Spirit Manifestations. St. Martin. Price 7s. 6d.

Francis of Assisi. By Mrs. Oliphant. 10s. 6d.

Hamartia: An Inquiry into the Nature and Origin of Evil. Price 1s. In this treatise the relation of the lower animate creation to man, and the final perfection and immortality of each individual life in him, are deduced from the teachings of Scripture, and confirmed by Reason.—"A very thoughtful pamphlet, well deserving the attention of all students of the Bible, as well as of all students of their own hearts. Whether the author of this pamphlet hit the mark or not, he will certainly aid others in their attempts to understand why they so often miss it" (*Spectator*).—"We have read this thoughtful and fine-toned essay with much interest and sympathy, and can heartily commend it to all students whose eschatological views have not yet run and hardened into the 'orthodox' form" (*The Expositor*).

Home Cure and Eradication of Disease. Tastefully bound in cloth, price 3s. 6d., by post, 3s. 8d.

Visibility Invisible, Invisibility Visible. A New Year's Story, founded on fact. 78 pages. Paper cover, price 6d.

The Christian Saints: their Method and their Power. By GEORGE WYLD, M.D. Price 6d.

Emanuel Swedenborg: His Life and Writings. By WILLIAM WHITE. Two vols., demy 8vo, of upwards of 1200 pages, with four Steel Engravings, price 24s. This work was some years in preparation, and forms a sort of Swedenborg Cyclopædia.

Songs of the Spirit. By H. H. Price 3s. 6d.

After Death; or, Disembodied Man. The Location, Topography, and Scenery of the Supernal Universe: Its Inhabitants; their Customs, Habits, Modes of Existence; Sex after Death; Marriage in the World of Souls; the Sin against the Holy Ghost; its Fearful Penalties, etc. Being the Sequel to *Dealings with the Dead.* 6s.

Glimpses of a Brighter Land. A Series of Communications through a Writing Medium. 2s. 6d.

A Twenty Years' Record of Modern American Spiritualism, in 49 chapters, 1 vol., large octavo, of 600 pages, cloth, bevelled edges, superbly and profusely illustrated, with fine portraits, etc. on steel, wood in tint, lithography, etc. 15s.

Strange Visitors: A Series of Original Papers, embracing Philosophy, Science, Government, Religion, Poetry, Art, Fiction, Satire, Humour, Narrative, and Prophecy, by the Spirits of Irving, Willis, Thackeray, Byron, Bronté, Richter, Humboldt, Hawthorne, Wesley, Browning, and others. 6s. 6d.

The Theory of Germs. Showing the Origin, Career, and Destination of all Men, Spirits and Angels. By J. EMMETT, B.A., Author of the *Solar Paradises of God*, etc. Price 1s.

Letters and Tracts on Spiritualism. By JUDGE EDMONDS. Containing 338 pages, in wrapper. Price 3s. 6d. This is one of the cheapest, most useful, comprehensive and instructive works in the whole literature of Spiritualism. It elucidates some of the most difficult problems in Spirit Communion.

The Gates Ajar; or, a Glimpse into Heaven. By ELIZABETH STUART PHELPS. The only complete English Edition. Price 6d., cloth, 1s. This is a remarkable book, and has created more sensation than any work issued outside the ranks of Spiritualism, on account of its beautiful delineation of the Spiritual Philosophy. It cheers and blesses all who read it, and has thrown rays of light into many a darkened soul. Everyone should own this little gem.—"In that little work entitled *Gates Ajar*, more is being done for the race than was accomplished by the Colleges of Andover for two hundred years."—*Rev. Mr. Spalding*, Salem, Mass.

AMERICAN DEPARTMENT.

COMPLETE WORKS OF A. J. DAVIS.
COMPRISING THIRTY VOLUMES, ALL NEATLY BOUND IN CLOTH.

Principles of Nature: Her Divine Revelations, and a Voice to Mankind. In Three Parts. Thirty-fourth edition, with a likeness of the author. 15s.

Great Harmonia: Being a Philosophical Revelation of the Natural, Spiritual, and Celestial Universe. In five volumes, in which the principles of the Harmonial Philosophy are more fully elaborated and illustrated.

Vol. I. THE PHYSICIAN. In this volume is considered the Origin and Nature of Man; the Philosophy of Health, of Disease, of Sleep, of Death, of Psychology, and of Healing. 6s. 6d.

Vol. II. THE TEACHER. In this volume is presented "Spirit and its Culture;" the "Existence of God;" My Early Experience; My Preacher and his Church; the True Reformer; Philosophy of Charity; Individual and Social Culture; the Mission of Woman; the True Marriage; Moral Freedom; Philosophy of Immortality; the Spirit's Destiny; Concerning the Deity. 6s. 6d.

Vol. III. THE SEER. This volume is composed of twenty-seven Lectures on Magnetism and Clairvoyance in the past and present; Psychology, Clairvoyance, and Inspiration are examined in detail, and the conclusions obtained are believed to be consistent with the principles of Nature, and are the author's personal experience. 6s. 6d.

Vol. IV. THE REFORMER. This volume treats on "Physiological Vices and Virtues, and the Seven Phases of Marriage," the uses of the conjugal principle, which tend directly either to demolish or upbuild man's moral and physical nature; views of marriage and parentage; woman's rights and wrongs; laws of attraction and marriage; transient and permanent marriage; temperaments; the rights and wrongs of divorce, etc. 6s. 6d.

Vol. V. THE THINKER. Part First is a description of the Truthful Thinker, and an analysis of the natures and powers of mind. Part Second—the Pantheon of Progress, comprising psychometrical delineations of Egyptian, Chaldean, Persian, Greek, Pagan, Jew, Christian, Roman, and Protestant characters, illustrating the philosophy of universal progress. Part Third—the Origin of Life and the Law of Immortality. 6s. 6d.

A Stellar Key to the Summer-Land. Illustrated with diagrams and engravings. Revised edition, cloth binding, 3s. 6d.

Magic Staff. An Autobiography of Andrew Jackson Davis. A history of the domestic, social, physical, and literary career of the author, with his experience as a Clairvoyant and Seer, including the autobiographical parts of "Arabula" and "Memoranda," which enter largely into the author's personal experiences. 8s.

Views of our Heavenly Home. A sequel to "A Stellar Key." Illustrated. Contents; Statements in regard to Individual Occupation—Progress after Death—Eating and Breathing in the Spirit-Life—Disappearance of the Bodily Organs at Death—Domestic Enjoyments and True Conjugal Unions—Origin of the Doctrine of the Devil, etc. Cloth, 3s. 6d.

Arabula; or, The Divine Guest. Pre-eminently a religious and spiritual volume. To some extent a continuation of the author's autobiography, but chiefly a record of deeply interesting experiences, involving alternations of faith and scepticism, lights and shades, heaven and hades, joys and sorrows. 6s. 6d.

Harmonial Man; or, Thoughts for the Age. 3s. 6d.

18 Works Published by the Psychological Press Association.

Children's Progressive Lyceum. A manual, with directions for the organisation and management of Sunday Schools, adapted to the bodies and minds of the young, and containing rules, methods, exercises, marches, lessons, questions and answers, invocations, silver-chain recitations, hymns, and songs. 3s.

Approaching Crisis; or, Truth v. Theology. This is a close and searching criticism of Dr. Bushnell's Sermons on the Bible, Nature, Religion, Scepticism, and the Supernatural. 4s. 6d.

Answers to Ever-Recurring Questions from the People. A sequel to "Penetralia." 6s. 6d.

History and Philosophy of Evil. With suggestions for more ennobling institutions, and philosophical systems of education. The question of Evil—individual, social, national and general—is analysed and answered. 3s. 6d.

Death and the After-Life. The "Stellar Key" is the philosophical introduction to the revelations contained in this book. Some idea of this little volume may be gained from the following table of contents: 1—Death and the After-Life; 2—Scenes in the Summer-Land; 3—Society in the Summer-Land; 4—Social Centres in the Summer-Land; 5—Winter-Land and Summer-Land; 6—Language and Life in Summer-Land; 7—Material Work for Spiritual Workers; 8—Ultimates in the Summer-Land; 9—Voice from James Victor Wilson. 3s. 6d.

Morning Lectures. A series of twenty popular discourses. The subjects are as follows: Defeats and Victories; the World's True Redeemer; the End of the World; the New Birth; the Shortest Road to the Kingdom of Heaven; the Reign of Anti-Christ; the Spirit and its Circumstances; Eternal Value of Pure Purposes; Wars of the Blood, Brain, and Spirit; Truths, Male and Female; False and True Education; the Equalities and Inequalities of Human Nature; Social Centres in the Summer-Land; Poverty and Riches; the Object of Life; Expensiveness of Error in Religion; Winter-Land and Summer-Land; Language and Life in the Summer-Land; Material Work for Spiritual Workers. 6s. 6d.

Harbinger of Health. Containing medical prescriptions for the human body and mind. 6s. 6d.

Memoranda of Persons, Places and Events. Embracing authentic facts, visions, impressions, discoveries in Magnetism, Clairvoyance, and Spiritualism. Also, quotations from the opposition. With an appendix, containing Zschokke's great story, "Hortensia," vividly portraying the difference between the ordinary state and that of Clairvoyance. 6s. 6d.

The Diakka, and their Earthly Victims. Being an explanation of much that is false and repulsive in Spiritualism. 2s. 6d.

Special Providences, Philosophy of. 2s. 6d.

Free Thoughts Concerning Religion. 3s. 6d.

Penetralia, Containing Harmonial Answers. The topics treated in this work are mainly theological and spiritual. 8s.

Spiritual Intercourse, Philosophy of. Contents: Guardianship of Spirits; Discernment of Spirits; Stratford Mysteries; Doctrine of Evil Spirits; Origin of Spirit Sounds; Concerning Sympathetic Spirits; Formation of Circles; Resurrection of the Dead; A Voice from the Spirit-Land; True Religion. This work has been translated into the French and German. It contains an account of the very wonderful spiritual developments at the house of Rev. Dr. Phelps, Stratford, Conn., and similar cases in all parts of the country. 6s.

The Inner-Life; or, Spirit Mysteries Explained. This is a sequel to "Philosophy of Spiritual Intercourse," revised and enlarged. It presents a compend of the Harmonial Philosophy of "Spiritualism," with illustrative facts of spiritual intercourse, both ancient and modern, and an original treatise upon the laws and conditions of mediumship. 6s. 6d.

The Temple: On Diseases of the Brain and Nerves. 6s. 6d.

Tale of a Physician; or, The Seeds and Fruits of Crime. In Three Parts—complete in one volume. Part 1—Planting the Seeds of Crime; Part 2—Trees of Crime in Full Bloom; Part 3—Reaping the Fruits of Crime. 4s. 6d.

The Fountain: With Jets of New Meanings. Illustrated with 142 Engravings. 4s. 6d.

Conjugal Love, Genesis and Ethics of. This book is of peculiar interest to all men and women. It treats of the delicate and important questions involved in Conjugal Love; is straightforward, unmistakably emphatic, and perfectly explicit and plain in every particular. 3s. 6d.

Sacred Gospels of Arabula. 4s. 6d.

WORKS OF HUDSON TUTTLE.

Arcana of Nature: or, the History and Laws of Creation. 2 Vols. First Vol. A philosophical work, aiming to show How the Universe was Evolved from Chaos by Laws Inherent in the Constitution of Matter; the Origin of Life on the Globe; the Various Divisions of the Living World; Origin of Man from the Animal World, and History of His Primitive State; How Mind Originated and is Governed by Fixed Laws; Man proved Immortal, and Controlled by as Immutable Laws as His Physical State. Second Vol. Intensely interesting, offering Evidences of Man's Immortality drawn from Ancient History and from Modern Spiritualism; showing the objects of Modern Spiritualism; the Distinction of Spiritual Phenomena from those not Spiritual; Philosophy of Imponderable Agents, and of Spiritual Elements; Origin, Faculties, and Power of Spirit; a Clairvoyant's View of the Spirit Sphere; Philosophy of Spirit-World and Spirit-Life. Cloth, 12s., complete.

The Arcana of Spiritualism. A Manual of Spiritual Science and Philosophy. 450 pp., handsome cloth. 5s. "In taking leave of the author, I desire to express my conviction that his book is one which all Spiritualists may read with advantage, and from which even the most advanced may learn much. It would be an excellent text-book for societies to read at meetings gathered for mutual instruction. I have always regretted that such meetings are not more widely held, that there is not an attempt to study the philosophy of the subject, more mutual counsel and interchange of thought among us. A suggestive work of this kind read aloud, and criticised by those who are capable of so doing, or commented on by those who can confirm and elucidate its statements from personal experience, would be extremely useful."—*Opinion of* "M. A. (Oxon.)" *in* "*Human Nature*" *for April, 1877.*

Ethics of Spiritualism. A System of Moral Philosophy, founded on Evolution and Continuity of Man's Existence beyond the Grave. Cloth, 2s. 6d.

Origin and Antiquity of Physical Man, Scientifically Considered. This is an original and startling book, proving man to have been contemporary with the mastodon, detailing the history of his development from the domain of the brute, and dispersion by great waves of emigration from Central Asia. Cloth, 6s. 6d.

Revivals; Their Cause and Cure. The demand for this able article has induced the publishers to print it in tract form of eight pages. 6d.

WORKS OF ALLEN PUTNAM.

Witchcraft of New England Explained by Modern Spiritualism. 6s. 6d.

Bible Marvel-Workers, and the Power which helped or made them perform Mighty Works, and utter Inspired Words; together with some Personal Traits and Characteristics of Prophets, Apostles, and Jesus, or New Readings of "The Miracles." 6s.

Agassiz and Spiritualism: Involving the Investigation of Harvard College Professors in 1857. 1s. 6d.

Natty, a Spirit: His Portrait and his Life. 3s. 6d.

Mesmerism, Spiritualism, Witchcraft, and Miracle. A Treatise, showing that Mesmerism is a Key which will Unlock many Chambers of Mystery. 1s. 6d.

Spirit Works; Real but not Miraculous. A Lecture read at the City Hall in Roxbury, Mass. Paper, 1s. 6d.

WORKS OF EPES SARGENT.

Scientific Basis of Spiritualism. The author takes the ground that since natural science is concerned with a knowledge of real phenomena, appealing to our sense-perceptions, and which are not only historically imparted, but are directly presented in the irresistible form of daily demonstration to any faithful investigator, therefore Spiritualism is a natural science, and all opposition to it, under the ignorant pretence that it is outside of nature, is unscientific and unphilosophical. All this is clearly shown; and the objections from "scientific," clerical, and literary denouncers of Spiritualism, ever since 1847, are answered with that penetrating force which only arguments, winged with incisive facts, can impart. Third edition. Post free, 6s. 6d.

Planchette; or, The Despair of Science. Being a full account of Modern Spiritualism, its phenomena and the various theories regarding it. With a survey of French Spiritism. The work contains chapters on the following subjects:—What Science says of it—The Phenomena of 1847—Manifestations through Miss Fox—Manifestations through Mr. Home—The Salem Phenomena, etc.—Various Mediums and Manifestations—The Seeress of Prevorst—Kerner—Stilling—Somnambulism, Mesmerism, etc. —Miscellaneous Phenomena—Theories—Common Objections—Teachings —Spiritism, Pre-Existence, etc.—Psychometry—Cognate Facts and Phenomena. Cloth, 5s. 6d.

Proof Palpable of Immortality. Being an Account of the Materialisation Phenomena of Modern Spiritualism, with remarks on the relations of the Facts to Theology, Morals, and Religion. The work contains a wood-cut of the Materialised Spirit of Katie King, from a photograph taken in England. Cloth, 5s. 6d.

Does Matter do it all? A Reply to Professor Tyndall's latest attack on Spiritualism. Paper, 6d.

Harper's Cyclopœdia of British and American Poetry. Edited by Epes Sargent. This elegant volume of nearly 1000 pages is a wonderfully perfect work, combining rare judgment and knowledge of English literature; and, as the labour of the last years of Mr. Sargent's life, is fitly his crowning work. Cloth, illuminated cover, 20s.

WORKS OF P. B. RANDOLPH.

Curious Life of P. B. Randolph. Paper, 2s. 6d.

Pre-Adamite Man: Demonstrating the Existence of the Human Race upon this Earth 100,000 Years ago. 6s. 6d.

Woman's Book: A Life's issues of Love in all its Phases. 8s. 6d.

Seership !—The Magnetic Mirror. A Practical Guide to those who aspire to Clairvoyance Absolute. Original, and selected from various European and Asiatic adepts. Price 8s. 6d.

The Wonderful Story of Ravalette; also, Tom Clark and his Wife, and the curious things that befel them: being the Rosicrucian's Story. Two volumes in one. 8s. 6d.

The New Mola. The Secret of Mediumship. 2s. 6d.

"The Ghostly Land." The "Medium's Secret." Supplement to the "New Mola." 2s. 6d.

WORKS OF ROBERT DALE OWEN.

Threading my Way; or, Twenty-Seven Years of Autobiography. A most interesting volume; a narrative of the first twenty-seven years of the author's life; its adventures, errors, experiences; together with reminiscences of noted personages whom he met forty or fifty years since, etc. The *Boston Post* pronounces it "a fascinating autobiography." A handsome 12mo volume, beautifully printed and bound in cloth. Price 7s. 6d.; postage free.

Beyond the Breakers: a Story of the Present Day. 7s. 6d.

WORKS OF J. M. PEEBLES, M.D.

Immortality and our Employments Hereafter. 6s. 6d.

Travels Around the World; or, What I Saw in Polynesia, China, India, Arabia, Egypt, and other "Heathen" (?) Countries. This thrillingly interesting volume—describing the Manners, Customs, Laws, Religions, and Spiritual Manifestations of the Orientals—is the Author's masterpiece. Large 8vo, bevelled boards, gilt side and back, 8s. 6d.

Seers of the Ages; Ancient, Mediæval, and Modern Spiritualism. This volume, of nearly 400 pages, octavo, traces the Phenomena of Spiritualism through India, Egypt, Phœnicia, Syria, Persia, Greece, Rome, down to Christ's time, treating of the Mythic Jesus, the Churchal Jesus, the Natural Jesus. Bound in bevelled boards, 8s. 6d.

Spiritual Harmonies. Containing nearly one hundred popular Hymns and Songs, with the belief of Spiritualists, and Readings appropriate for funeral occasions. Without music. Designed to supply a want long felt in the ranks of Spiritualism. Cloth, illuminated cover, 6s.

Spiritual Harp. A fine collection of vocal music for the choir, congregation, and social circle; is especially adapted for use at Grove Meetings, Pic-Nics, etc. Edited by J. M. Peebles and J. O. Barrett. E. H. Bailey, Musical Editor. Cloth, 8s. 6d.

Jesus; Myth, Man, or God? Did Jesus Christ exist?—What are the proofs? —Was he man, begotten like other men?—What Julian and Celsus said of him—The moral influence of Christianity and Heathenism compared. These and other subjects are critically discussed. Cloth, 3s. 6d.

Parker Memorial Hall Lectures on Salvation; Prayer; the Methods of Spirit Influences, and the Nature of Death. The aim of these Lectures is to present the advance thought of Spiritualism in its religious aspects. Paper, 1s. 6d.

Christ the Corner-Stone of Spiritualism. Jewish Evidence of Jesus' Existence—Who was Jesus? and what the New Testament says of Him—What candid Freethinkers and Men generally think of Jesus—The Estimate that some of the leading and more cultured American Spiritualists put upon Jesus—Was Jesus, of the Gospels, the Christ?—The Commands, the Divine Gifts, and the Spiritual Teachings of Jesus Christ—The Belief of Spiritualists—The Baptized of Christ—The Church of the Future. Price 3d.

WORKS OF MRS. MARIA M. KING.

Principles of Nature, as Discovered in the Development and Structure of the Universe. 3 vols. Vol. I. Given inspirationally. One of the most important contributions to spiritual and physical science ever made by any seer or seeress. Cloth. Vol. II. continues the history of the development of Earth, commencing with the evolution of planetary conditions, giving a brief history of the planets' progress through successive eras to the present, with the Law of Evolution of Life, Species, and Man; stating principles to illustrate facts, and facts or events to illustrate principles. The Law of Life and Force is brought prominently to view—what it is, how it operates, the relations of Spirit and Matter, of God and Nature. etc. 268 pages, 8vo, cloth. Vol. III. discusses Magnetic Force and Spiritual Nature; treating specially of the practical questions of Modern Spiritual Manifestations and Mediumship, Life in Spirit, Spiritual Spheres. 261 pages, 8vo, cloth, 21s.

Real Life in Spirit-Land: Being Life-Experiences, Scenes, Incidents, and Conditions Illustrative of Spirit-Life and the principles of the Spiritual Philosophy. Of practical value to any who are anxious to study the theories of Spiritualists and mediums, etc. Given inspirationally. Cloth, 5s. 6d.

WORKS OF PROFESSOR WILLIAM DENTON.

Soul of Things; or, Psychometric Researches and Discoveries. 3 vols. By Wm. and Elizabeth M. F. Denton. A marvellous work. 6s. 6d. each.

What Was He; or, Jesus in the Light of the Nineteenth Century. 5s. 6d.

Radical Rhymes. Cloth. 5s. 6d.

Is Spiritualism True? This able and logical discourse, bringing in massive array the irrefutable evidences upon which the sublime truths of Spiritualism rest, ought to be read by the millions. 6d.

Christianity no Finality; or, Spiritualism Superior to Christianity. 6d.

Common Sense Thoughts on the Bible. 6d.

Is Darwin Right? or, The Origin of Man. This is a well-bound volume of two hundred pages, 12mo, handsomely illustrated. It shows that man is not of miraculous but of natural origin; yet that Darwin's theory is radically defective, because it leaves out the spiritual causes, which have been the most potent concerned in his production. Cloth. 4s. 6d.

Radical Discourses on Religious Subjects. 6s. 6d.

Man's True Saviours. A Lecture. 6d.

The Deluge in the Light of Modern Science. 6d.

Sermon from Shakespeare's Text. 6d.

What is Right? 6d.

Who are Christians? 6d.

The Irreconcilable Records; or, Genesis and Geology. 1s.

Orthodoxy False, since Spiritualism is True. 6d.

WORKS BY VARIOUS AUTHORS.

Question Settled: A Careful Comparison of Biblical and Modern Spiritualism. By MOSES HULL. The author's aim, faithfully to compare the Bible with modern phenomena and philosophy, has been ably accomplished. The Adaptation of Spiritualism to the Wants of Humanity; its moral Tendency; the Bible Doctrine of Angel Ministry; the Spiritual Nature of Man, and the Objections offered to Spiritualism, are all considered in the light of nature, history, reason, and common sense, and expressed clearly and forcibly. Cloth, bevelled boards, 6s. 6d.

Biography of Mrs. J. H. Conant, the World's Medium of the 19th Century. This book contains a history of the Mediumship of Mrs. Conant from childhood to the present time: together with extracts from the diary of her physician; selections from letters received verifying spirit communications given through her organism at the *Banner of Light* Free Circles; and spirit messages, essays, and invocations from various intelligences in the other life. The whole being prefaced with opening remarks from the pen of Allen Putnam, Esq. 6s. 6d.

Bible of Bibles; or, Twenty-Seven "Divine Revelations," containing a description of Twenty-Seven Bibles, and an Exposition of Two Thousand Biblical Errors in Science, History, Morals, Religion, and General Events. 8s. 6d.

Biography of Satan; or, A Historical Exposition of the Devil and his Fiery Dominions. Disclosing the Oriental origin of the belief in a Devil and future endless punishment. By K. GRAVES. Paper. 2s.

Blossoms of Our Spring. By HUDSON and EMMA TUTTLE. "America: a National Poem," 80 pages; "A Vision of Death"; forty short pieces, and "Life's Passion Story." 136 pages form the substance of this work. 4s. 6d.

Chapters from the Bible of the Ages. Fourteen Chapters. Selected from Hindoo Vedas, Buddha, Confucius, Mencius, Zoroaster, Egyptian Divine Pymander, Talmuds, Bible, Philo Judaeus, Orpheus, Plato, Pythagoras, Marcus Aurelius, Epictetus, Al Koran, Scandinavian Eddas, Swedenborg, Luther, Novalis, Renan, Talieson, Milton, Penn, Barclay, Mary Fletcher, Newman, Tyndall, Max Müller, Woolman, Elias Hicks, Channing, Garrison, H. C. Wright, Lucretia Mott, Higginson, Bushnell, Parker, A. J. Davis, Mary F. Davis, Emma Hardinge, Beecher, Tuttle, Abbot, Denton, and others. Gospels and Inspirations from Many Centuries and Peoples. Edited and Compiled by G. B. Stebbins, Detroit, Michigan. 400 pages, cloth, 6s. 6d.

Clock Struck One, and Christian Spiritualist. Revised and corrected. Being a Synopsis of the Investigations of Spirit Intercourse by an Episcopal Bishop, three ministers, five Doctors and others, at Memphis, Tenn., in 1855. Also, the Opinion of many eminent Divines, living and dead, on the subject, Communications received from a number of persons recently. By the Rev. Samuel Watson, of the Methodist Episcopal Church. Cloth, 4s. 6d.

Clock Struck Three; Being a Review of "Clock Struck One," and reply to it; and Part Second, showing the Harmony between Christianity, Science, and Spiritualism. By Rev. Samuel Watson. Cloth, 6s. 6d.

Ghost-Land; or, Researches into the Mysteries of Occult Spiritism. Illustrated in a series of auto-biographical papers, with extracts from records of magical seances, etc. Translated by Mrs. Emma H. Britten. Cloth, 9s.

Davenport Brothers, the World-renowned Spiritual Mediums. Their Biography and Adventures in Europe and America. Illustrated with numerous engravings, representing various phases of spiritual phenomena. These well authenticated and startling facts, concisely narrated, are valuable for reference, and interesting as any novel. Cloth, 6s. 6d.

Death, in the Light of the Harmonial Philosophy. By Mary F. Davis. A whole volume of philosophical truth is condensed into this little pamphlet. "The truth about Death," says the author, "never breaks upon us until the light of the Spiritual Universe shines into the deep darkness of the doubting mind. Until this higher revelation is given to the understanding, the outward fact of Death strikes one with the awful force of Fate." The following subjects are treated: Universal Unity of Things; Nature Without and Within Man; The Absolute Certainty of Death; The Soul's Supremacy to Death; Degrading Teachings of Theology, etc. 1s. 6d.

Danger Signals; An Address on the Uses and Abuses of Modern Spiritualism. Paper, 1s.

Future Life: As Described and Portrayed by Spirits. Through Mrs. Elizabeth Sweet. With an introduction by Judge Edmonds. Scenes and events in spirit-life are here narrated in a very pleasant manner, and the reader will be both instructed and harmonised by the perusal of this volume. Cloth, 6s. 6d.

Gazelle: A Tale of the Great Rebellion. The Great Lyrical Epic of the War. By EMMA TUTTLE. This poem has met with great favour, and is the most extended of the author's writings. Cloth, 6s. 6d.

Gadarene; or, Spirits in Prison. By J. O. BARRETT and J. M. PEEBLES. The motto of this critical work indicates its general drift—Try the Spirits! 6s. 6d.

How to Magnetize; or, Magnetism and Clairvoyance. A Practical Treatise on the Choice, Management, and Capabilities of Subjects, with Instructions on the Method of Procedure, etc. By JAMES VICTOR WILSON. Paper, 1s.

Health Manual. Devoted to Healing by means of Nature's Higher Forces; including the Health Guide, revised and improved. By EDWIN D. BABBITT, D.M., author of "Principles of Light and Colour," "Wonders of Light and Colour." Muslin, 4s. 6d.

Intuition. By Mrs. FRANCES KINGMAN. 6s. 6d.

Identity of Primitive Christianity and Modern Spiritualism. Two Vols. By EUGENE CROWELL, M.D. 8s. 6d. each.

The Spiritual Magazine, Bound Volumes of. By SAMUEL WATSON. (Formerly published in Memphis, Tenn.) Volume I.—Bound in cloth, 8vo, pp. 552, and containing a steel-plate engraving of Samuel Watson. Price $1,50, postage 15 cents. Volume II.—Bound in cloth, 4to, pp. 376. Price $1,50, postage 15 cents. Volume III.—Bound in cloth, 4to, pp. 384. Price $1,50, postage 15 cents. For sale by Colby & Rich.

Startling Facts in Modern Spiritualism, being a Graphic Account of Witches, Wizards, and Witchcraft; Table Tipping, Spirit Rapping, Spirit Speaking, Spirit Telegraphing, and *Spirit Materialisations* of Spirit Hands, Spirit Heads, Spirit Faces, Spirit Flowers, and every other Spirit Phenomenon that has occurred in Europe and America since the Advent of Modern Spiritualism, March 31st, 1848, to the present time. By Dr. N. B. WOLFE, Cincinnati, Ohio. Revised, enlarged, and appropriately illustrated. The "Startling Facts" recorded in this book are:—I. That there is a Spirit World as real and substantial as the Earth we live on, to which all go who die.—II. That Spiritual Science has discovered elementary laws which enable the inhabitants of the Spirit World to return to earth, to visit, and talk audibly to their friends, and tell of their experiences while dying—their awakening and realisations in the "Summer Land."—III. That the Inhabitants of the Spirit World build cities and beautify residences to please their varied tastes, and, in their spirit homes, remember and speak of their friends still on the earth just as we think and speak of them as they were known to us before they passed away.—IV. That families separated by death are reunited in the Spirit World, where the different members of the "Home Circle," still united in love, manifest the same diversified tastes and inclinations they exhibited while on earth.—V. That earthly riches do not necessarily add to the happiness of the individual in Spirit Life, but on the contrary, if not wisely employed, retard the development of the spirit, and mar its happiness more than abject poverty.—VI. That we are surrounded by spirit friends who are always striving to impress and guide us, that we may avert impending danger and premature death.—VII. That death is an ordinance written in the constitution of all things, and that instead of it being the "King of Terrors" it is an Angel of Peace, an event that ushers us into a higher and happier life than this. With these avowals of its teachings the book stands before the world, asking no favour but a reading—no consideration but the fair judgment of enlightened men and women. As Death is a heritage common alike to King, Pope, Priest, and People, all should be interested in knowing what it portends—of what becomes of us after we die. Those who have tasted death, our spirit friends, answer this great problem in this book of 600 pages. Cloth, 12mo, tinted paper, pp. 488, price 6s. 6d., postage free. In fine cloth, gold back and sides, 10s. 6d. per copy. The Psychological Press Association have been appointed the sole agents for the English sale. A parcel is now on the way, and orders will be executed in the order in which they are filed.

Life of Prof. William Denton, the Geologist and Radical. By J. H. POWELL. This biographical sketch of one of the ablest lecturers in the field of reform, is published in a neat pamphlet, comprising thirty-six pages. Those who would know more of this erudite scholar, bold thinker and radical reformer, should peruse its contents. Paper, 1s.

Modern American Spiritualism: A Twenty Years' Record of the Communion between Earth and the World of Spirits. By EMMA HARDINGE. This work is a Digest of Spiritualism in itself, and will repay the reader for every hour devoted to its pages. 15s.

Mystery of Edwin Drood, Completed by the Spirit-pen of Charles Dickens. The press declare this work to be written in "Dickens's happiest vein!" To show the demand there is for it, it may be well to state that the first edition of 10,000 copies was sold in advance of the press. There are forty-three chapters in the complete work, which embrace that portion of it written prior to the decease of the great author, making one complete volume of about 500 pages, in handsome cloth binding. Cloth, 6s.

Man and His Relations: Illustrating the Influence of the Mind on the Body; the relations of the faculties and affections to the elements, objects, and phenomena of the external world. By Prof. S. B. BRITTAN. 6s. 6d.

My Experience; or, Footprints of a Presbyterian to Spiritualism. By FRANCIS H. SMITH. An interesting account of sittings with various mediums, by a Baltimore gentleman. 3s. 6d.

Mediumship: Its Laws and Conditions. With brief instructions for the formation of spirit-circles. By J. H. POWELL. 1s. 6d.

Men, Women, and Ghosts. By ELIZABETH STUART PHELPS, author of "Gates Ajar," etc. 6s. 6d.

New Pilgrim's Progress. Purporting to be given by JOHN BUNYAN, through an Impressional Writing Medium. Cloth, 6s. 6d.

Occultism, Spiritism, Materialism: Demonstrated by the Logic of Facts; Showing Disembodied Man and Spirit Phases; also, The Immediate Condition Affecting Man After Death. Things of the most interest for man to know. By ALMIRA KIDD. Cloth, 4s. 6d.

Psychology: Re-Incarnation: Soul, and its Relations; or, the Laws of Being. By ALMIRA KIDD. 4s. 6d.

Works Published by the Psychological Press Association.

Modern Bethesda; or, The Gift of Healing Restored. Being some account of the Life and Labours of Dr. J. R. Newton, Healer, with observations on the Nature and Source of the Healing Power, and the conditions of its existence. Edited by A. E. NEWTON. Cloth, illustrated, 8s. 6d.

History of the Doctrine of a Future Life. By W. R. ALGER. A most comprehensive work of 800 pages, comprising Introduction; Barbarian, Druidic, Scandinavian, Etruscan, Egyptian, Brahmanic, Buddhistic, Persian, Hebrew, Rabbinical, Greek, Roman, and Mohammedan Doctrines of Future Life; New Testament Teachings, Doctrine of Peter, Paul, Christ, and John; Patristic, Mediæval, and Modern Doctrine, etc., etc. Tenth edition, with six new chapters, and a complete Bibliography of the subject, comprising 4977 books relating to Origin, Nature, and Destiny of the Soul. Compiled by Professor Ezra Abbot. 10s. 6d.

Personal Experiences of WILLIAM H. MUMLER in Spirit Photography. Written by himself. The demand for this work has induced the publishers to issue it in a cheap pamphlet form, and it will be found to be just the thing to hand to sceptics, as it contains a mass of reliable evidence of the truth of Spirit-Photography, such as no one can gainsay, and places the medium, MR. MUMLER, as the Pioneer Spirit-Photographer of the world. Paper, 1s.

Poems from the Inner Life. By LIZZIE DOTEN. Thirteenth edition. This handsome volume opens with the wonderful experiences of the author, who is peculiarly gifted as a trance medium and public speaker. 6s. 6d.

Poems. By the well-known medium, ACHSA W. SPRAGUE. A brief sketch of the gifted author precedes these poems. 4s. 6d.

Poems of Progress. By MISS LIZZIE DOTEN, author of "Poems from the Inner Life." Illustrated with a fine steel engraving of the inspired author. Gilt, 6s. 6d.

Poems of the Life Beyond and Within. Voices from many Lands and Centuries, saying, "Man, thou shalt never die." Edited and compiled by GILES B. STEBBINS. 6s. 6d.

Psalms of Life. A Compilation of Psalms, Hymns, Chants, Anthems, etc., embodying the Spiritual, Progressive, and Reformatory sentiment of the present age. By JOHN S. ADAMS. Fifth edition. 4s. 6d.

People from the Other World. Containing full and illustrative descriptions of the wonderful séances held by Colonel Olcott with the Eddys, Holmeses, and Mrs. Compton. The author confines himself almost exclusively to the phenomenal side of Spiritualism; to those facts which must elevate it sooner or later to the position of an established science. The work is highly illustrated. 12s. 6d.

Psycho-Physiological Sciences and their Assailants. Being a Response by Alfred R. Wallace, Prof. J. R. Buchanan, Darius Lyman, and Epes Sargent, to the Attacks of Prof. W. B. Carpenter, of England, and others. Paper, pp. 216. 2s. 6d.

Religion of Spiritualism; Its Phenomena and Philosophy. By SAMUEL WATSON, author of "The Clock Struck One, Two, and Three." Thirty-six years a Methodist minister. Cloth, 6s. 6d.

Religion of Spiritualism. By EUGENE CROWELL, M.D. Among the prime points of consideration in this work, may be mentioned, What is Religion? Spiritualism is a Religion; The Religion of Spiritualism identical with the Religion of Jesus. Paper, 1s.

Sixteen Saviours or None; or, The Explosion of a Great Theological Gun. By KERSEY GRAVES. Cloth, 4s. 6d.

Spirit - World; its Inhabitants, Nature, and Philosophy. By EUGENE CROWELL, M.D. Cloth, 6s. 6d.

Milton Keynes UK
Ingram Content Group UK Ltd.
UKHW022312170823
427072UK00005B/102